Tropical Forests - Ecology, Diversity and Conservation Status

Edited by Eusebio Cano Carmona, Carmelo Maria Musarella and Ana Cano Ortiz

Published in London, United Kingdom

Tropical Forests - Ecology, Diversity and Conservation Status
http://dx.doi.org/10.5772/intechopen.104009
Edited by Eusebio Cano Carmona, Carmelo Maria Musarella and Ana Cano Ortiz

Contributors
Suzanne L. Nelson, Nicola Justice, Kaitlynn M. Apple, Aidan H. Liddiard, Madeline R. Elias, Jon D. Reuter, George Muhamba Tryphone, Luseko Amos Chilagane, Michael Benedict Kazyoba, Frenk M. Reuben, Mugatsia H. Tsingalia, Ederson Augusto Zanetti, Allan Sbardelotto, Joesio Siqueira, Frederick N. Numbisi, Vithal Ramakrishna Karoshi, Roberto Rochadelli, Alain Levy Boussamba, Gabriel Costa, Cláudio Silva, Keila R. Mendes, Bergson Guedes Bezerra, Thiago R. Rodrigues, Jonh B. Silva, Higo J. Dalmagro, Hildo Nunes, Ana Gomes, Glauce Silva, Sarah Batalha, Larice Marinho, Gabriele Almeida, Losany Viana, Tiago Mandu, Demethrius P. L. de Oliveira, Mohanarangan Ashokkumar, Rajaratinavelu Nagarajan, Richa Kumar, Moharangan Ashokkumar, Alfayo Koskei, Gul Zareen Ghafoor, Memuna Ghafoor Shahid, Maryam Ali, Naghmana Ghafoor, Philippe Waldhoff, Saulo Eduardo Xavier Franco de Souza, Alfonso Avellaneda-Cusaría, Omar Cabrera, Pablo Ramón, Bernd Stimm, Sven Gunter, Reinhard Mosandl, Ruth Arias-Gutierrez, Andrés Tapia, Juan Pablo Reyes Puig, Kenneth Peprah, Raymond Aabeyir, Bismark Yeboah Boasu, Luiz Augusto Gomes de Souza, Rita de Cássia Paiva, Douglas Rodrigues Pinheiro da Costa, Eusebio Cano Carmona, Ana Cano Ortiz, Carmelo Maria Musarella

First published in London, United Kingdom, 2023 by IntechOpen
IntechOpen is the global imprint of INTECHOPEN LIMITED, registered in England and Wales, registration number: 11086078, 5 Princes Gate Court, London, SW7 2QJ, United Kingdom

British Library Cataloguing-in-Publication Data
A catalogue record for this book is available from the British Library

Additional hard and PDF copies can be obtained from orders@intechopen.com

Tropical Forests - Ecology, Diversity and Conservation Status
Edited by Eusebio Cano Carmona, Carmelo Maria Musarella and Ana Cano Ortiz
p. cm.
Print ISBN 978-1-83768-574-5
Online ISBN 978-1-83768-575-2
eBook (PDF) ISBN 978-1-83768-576-9

For EU product safety concerns:
IN TECH d.o.o., Prolaz Marije Krucifikse Kozulić 3, 51000 Rijeka, Croatia,
info@intechopen.com or visit our website at intechopen.com.

Meet the editors

Eusebio Cano Carmona obtained a Ph.D. in Science from the University of Granada, Spain. He is a Professor of Botany at the University of Jaén, Spain. His fundamental line of research is flora and vegetation and he has conducted investigations in Spain, Italy, Portugal, Palestine, the Caribbean islands, and Mexico. Dr. Carmona has directed 14 doctoral theses and published 255 articles, books, and book chapters. He participates in national and international congresses and has presented around 200 papers and communications. He has held several academic positions, including founder and director of the International Seminar on Management and Conservation of Biodiversity. Dr. Carmona is a member of the Spanish, Portuguese, and Italian Geobotany societies as well as a counselor of the Institute of Giennenses Studies. He teaches graduate courses in geobotany and biological sciences.

Ana Cano Ortiz obtained a Ph.D. in Botany from the University of Jaén, Spain. She has experience in both business and academia. She is a co-director of three doctoral theses. She has worked in Spain, Italy, Portugal, and Central America. Dr. Ortiz has more than 100 publications to her credit. She is currently a professor in the Department of Didactics of Experimental, Social and Mathematical Sciences (Complutense University of Madrid, Spain). She is also a journal co-editor and reviewer. Her research focuses on edaphic bioindicators, bioclimatology, agriculture, and science education.

Carmelo Maria Musarella, Ph.D., is a biologist, specializing in plant biology. He has studied and worked at several European universities in Italy, Spain, and Portugal. He is currently an Adjunct Professor of Plant Biology at the "Mediterranea" University of Reggio Calabria, Italy. His research interests include flora, vegetation, habitats, biogeography, taxonomy, ethnobotany, endemism, and biodiversity conservation. Dr. Musarella is the author of many research articles published in indexed journals and books. He is a guest editor for *Plant Biosystems* and a referee for many other journals. He is a member of the scientific committee of the International Seminar on Biodiversity Conservation and Management. He has participated in several international and national congresses, seminars, and workshops and presented oral communications and posters.

Contents

Preface

This book includes 17 chapters on the tropical biomes of Africa, Asia, and Central America. The studies address tropical diversity, the danger of species extinction, and the conservation and development of tropical forests. Some authors call for the management of tropical forests to be more in line with sustainable development, which is why greater efficiency in the management of ecosystem services is advocated. Due to the low per capita income of populations located in tropical areas, the research presented is dedicated to studies on the ecology, flora, and vegetation of tropical ecosystems. Chapters also highlight the relevance of ecosystem services in combining and accomplishing conservation and development.

The editors wish to thank the staff at IntechOpen, especially Author Service Manager Mrs. Maja Bozicevic, for the opportunity to edit this important book. We are also grateful to the authors for their excellent contributions.

Dr. Eusebio Cano Carmona
Professor of Botany,
Department of Animal and Plant Biology and Ecology,
Section of Botany,
University of Jaén Las Lagunillas,
Jaén, Spain

Dr. Ana Cano Ortiz
Professor of Experimental Sciences,
Department of Didactics of Experimental Social Sciences and Mathematics,
Section of Didactics of Experimental Sciences,
Complutense University of Madrid,
Faculty of Education,
Madrid, Spain

Dr. Carmelo Maria Musarella
Professor of Botany,
Department of AGRARIA,
"Mediterranea" University of Reggio Calabria,
Reggio Calabria, Italy

Section 1

Introduction

Chapter 1

Introductory Chapter: On the Need to Teach, Conserve and Develop the Diversity of Tropical Forests

Ana Cano Ortiz, Carmelo Maria Musarella and Eusebio Cano Carmona

1. Introduction

Tropical forests are located in one of the great biomes of the earth, in tropical and subtropical rainforests the average annual temperature ranges between 26 and 27°C, with thermal oscillations below 2°C. The precipitation, in general, is high, between 2000 and 4000 mm; however, there are shaded areas in which the precipitation is very low, being able to reach 200–300 mm. From the bioclimatic point of view, there is a dominance of the infra, thermo and mesotropical thermotypes. However, the rainy macrobioclimate of a pluvial nature becomes pluviseasonal due to the drop in temperature with altitude. For this reason, tropical environments are rainy in their basal zones and rainy seasons in the mountains. All these conditions, on the one hand, the broadleaf forest or rain forest and the semi-deciduous dry forest, as occurs on the Pacific coast of Mexico and in areas of Caribbean islands [1–5]. The high variability of ecological factors is the cause of the presence of interesting plant formations that present a high rate of endemism. Being areas of interest for conservation because they are hot spots on the planet, which must be studied, known and defended, both with regard to conservation and exploitation [6]. Thus in the tropical and subtropical areas, the pluvisilva forests are surrounded by laurisilva, and towards the central zones of the continents and shady zones [7] the savannahs in which grasses dominate are the centers of origin of this botanical family (Gramineae).

Tropical areas are of great interest worldwide, since they are CO_2 sinks, so in the face of global climate change, large areas of vegetation are needed to act as CO_2 sinks [8–13]. In addition, tropical forests have a high floristic diversity and a high rate of endemic species, being hot spots on the planet, which is why researchers present related works on tropical biomes.

2. Results and discussion

Due to the low purchasing power of the populations residing in the tropics, an increase in their per capita income is necessary. In a comparative analysis of per capita income between developed and poor countries according to the World Bank [14] (**Table 1**), high discrimination is observed with values ranging between USD 460 and 6000 for poor countries in Africa, Central America and Asian areas. However, the

Country	GDP per capita (USD)	Country	GDP per capita (USD)
Afghanistan	368,8	Isle of man	87.157,50
Haiti	1.829,60	Saint Martin Island	21.920,80
Albanian	6.492,90	Iceland	68.727,60
Germany	51.203,60	Cayman Islands	86.568,80
Andorra	42.137,30	Faroe Islands	69.010,30
Angola	1.953,50	Marshall Islands	6.172,10
Antigua and Barbuda	15.781,40	Solomon Islands	2.304,80
Saudi Arabia	23.185,90	Turks and Caicos Islands	20.908,60
Algeria	3.690,60	Virgin Islands (US)	39.552,20
Argentina	10.636,10	British Virgin Islands	75.152,60
Armenia	4.966,50	Israel	52.170,70
Aruba	29.342,10	Italy	35.657,50
Australia	60.443,10	Jamaica	5.183,60
Austria	53.637,70	Japan	39.312,70
Azerbaijan	5.388,00	Jordan	4.103,30
Bahamas	27.478,40	kazakhstan	10.373,80
Bahrain	26.563,00	kenya	2.081,80
Bangladesh	2.457,90	Kyrgyzstan	1.276,70
Barbados	17.225,50	Kiribati	1.606,50
Belarus	7.302,30	Kosovo	5.269,80
Belgium	51.247,00	Kuwait	24.300,30
Belize	6.228,30	Lesotho	1.094,10
Benin	1.319,20	Latvia	21.148,20
Bermuda	114.090,30	Lebanon	4.136,10
Bhutan	3.266,40	Liberia	675,7
Bolivian	3.345,20	Libya	6.357,20
Bosnia and Herzegovina	7.143,30	Liechtenstein	157.755,00
Botswana	6.805,20	Lithuania	23.723,30
Brazil	7.507,20	Luxembourg	133.590,10
Brunei Darussalam	31.449,10	North Macedonia	6.694,60
Bulgaria	12.221,50	Madagascar	500,5
Burkina Faso	893,1	Malaysia	11.109,30
Burundi	221,5	Malawian	634,8
Cape Verde	3.293,20	Maldives	10.366,30
Cambodia	1.625,20	Mali	873,8
Cameroon	1.666,90	Malt	33.486,70
Canada	51.987,90	Mariana	23.707,30

Country	GDP per capita (USD)	Country	GDP per capita (USD)
Chad	685,7	Morocco	3.795,40
Chili	16.265,10	Mauricio	9.106,20
China	12.556,30	Mauritania	2.166,00
Cyprus	31.551,80	Mexico	10.045,70
Colombia	6.104,10	Micronesia	3.571,30
Comoros	1.577,50	Monaco	234.315,50
Congo, Republic	577,2	Mongolia	4.566,10
Congo, Republic	2.290,40	Montenegro	9.465,70
Korea, Republic	34.997,80	Mozambique	491,8
Costa Rica	12.472,40	Myanmar	1.209,90
Côte d'Ivoire	2.549,00	Namibia	4.865,60
Croatia	17.685,30	Nauru	10.648,10
Cuba	9.499,60	Nepal	1.208,20
Curacao	17.717,60	Nicaragua	2.045,50
Denmark	68.007,80	Niger	590,6
Djiboutian	3.150,40	Nigeria	2.065,70
Dominica	7.653,20	Norway	89.154,30
Ecuador	5.965,10	New Caledonia	37.159,50
Egypt	3.698,80	New Zealand	48.781,00
The Savior	4.551,20	Oman	19.509,50
Arab Emirates	44.315,60	Netherlands	57.767,90
Eritrea	643,8	Pakistan	1.505,00
Slovenia	29.291,40	Palau	12.083,90
Spain	30.103,50	Panama	14.617,60
USA	70.248,60	Papua New Guinea	2.672,90
Estonia	27.943,70	Paraguayan	5.891,50
Swatini	3.978,40	Peru	6.621,60
Ethiopia	925,1	French Polynesia	19.914,60
Russia	12.194,80	Poland	17.999,90
Fiji	4.646,60	Portugal	24.567,50
Philippines	3.460,50	Puerto Rico	32.640,70
Finland	53.654,80	Qatar	66.838,40
France	43.659,00	Macau, China	43.873,60
Gabon	8.635,30	United Kingdom	46.510,30
Gambia	772,2	Syrian Arab Republic	533,4
Georgia	5.023,30	Central African Republic	461,1
Ghana	2.363,30	Czech Republic	26.821,20
Grenade	9.010,60	Lao republic	2.535,60
Greece	20.192,60	Dominican Republic	8.476,80

Country	GDP per capita (USD)	Country	GDP per capita (USD)
Greenland	54.571,20	Slovak Republic	21.391,90
Guam	35.904,90	Republic of Moldova	5.230,70
Guatemala	5.025,50	West Bank and Gaza	3.664,00
Guinea	1.189,20	Romania	14.858,20
Equatorial Guinea	7.506,70	Rwanda	822,3
Guinea-Bissau	795,1	Saint Kitts and Nevis	18.082,60
Guyana	9.998,50	Samoa	3.857,30
Haiti	1.829,60	American Samoa	15.743,30
Honduras	2.771,70	San Marino	45.320,20
Hong Kong	49.800,50	St. Vincent and the Grenadines	8.666,40
Hungary	18.728,10	St. Lucia	9.414,20
India	2.256,60	Sao Tome and Principe	2.360,50
Indonesia	4.332,70	Senegal	1.636,90
Iran	4.091,20	Serbian	9.230,20
Iraq	4.775,40	Seychelles	14.653,30
Ireland	100.172,10	Sierra Leone	480
		Singapore	72.794,00

Table 1.
Data on per capita income in dollars was obtained from World Bank.

developed countries that cause climate change have per capita incomes above USD 20,000, even reaching values above USD 200,000.

This wealth of developed countries based on a powerful industrial sector, due to the use of polluting energy (petroleum) and the cause of climate change, is a reason for all countries to abandon the current production model, and bet on a sustainable balanced development, combining productivity and conservation, so it is necessary to bet on ecosystem services [15–18].

Because in the tropics there are countries with low per capita income and a deficient educational and research system, it is essential to promote research, through which balanced development is achieved, conserving and exploiting at the same time. Exploitation and development are not mutually exclusive, it is only necessary to know the natural resources, their ecology, function, structure and state of conservation, to advocate said sustainable development. To be successful in the sustainable development model, education and research must be promoted, for which various teaching methodologies can be used [19–26].

The high diversity of plant species in tropical and subtropical areas, with a high rate of endemic plants, is reason enough for governments to establish conservation measures while advocating for sustainable development. For this, research must be based on knowledge of the flora and vegetation, as well as on the socioeconomic development model, especially with regard to the type of agriculture, since this, together with deforestation and fires, tend to be the causes of species loss.

Author details

Ana Cano Ortiz[1], Carmelo Maria Musarella[2] and Eusebio Cano Carmona[3]*

1 Department of Didactics of Experimental, Social and Mathematical Sciences, UCM, Madrid, Spain

2 Department of AGRARIA, "Mediterranea" University of Reggio Calabria, Reggio Calabria, Italy

3 Department of Animal and Plant Biology and Ecology, Section of Botany, University of Jaén, Jaén, Spain

*Address all correspondence to: ecano@ujaen.es

References

[1] Cano-Ortiz A, Musarella CM, Piñar Fuentes JC, Pinto Gomes CJ, Cano E. Analysis of the conservation of central American mangroves using the phytosociological method. In: Mangrove Ecosystem Ecology and Function. Croatia: INTECH PLUBISHER. University Campus; 2018. pp. 189-206. https://www. intechopen.com/chapters/62214

[2] Cano-Ortiz A, Musarella CM, Piñar Fuentes JC, Bartolomé Esteban C, Quinto Canas R, Pinto Gomes CJ et al. Advances in knowledge of vegetation of the Spanish island (Caribe-cento America). In: Vegetation. Croatia: INTECH PLUBISHER. University Campus; 2017. .pp. 83-99. https://www.intechopen.com/booksprocess/action/chapter/125184

[3] Cano E, Veloz Ramírez A, Cano-Ortiz A, Esteban FJ. Distribution of central American *Malastomataceae*: Biogeographical analysis of the Caribbean Islands. Acta Botanica Gallica. 2009;**156**(4):527-557

[4] Cano E, Veloz Ramírez A, Cano-Ortiz A, Alatorre J, Otero R. Comparative analysis of the mangrove swamps of the Dominican Republic and those of the state of de Guerrero. Mexico. Plant Biosystems146, suplemento. 2012;**1**:112-130. DOI: 10.1080/11263504.2012.704885

[5] Cano E, Veloz A, Cano-Ortiz A. Rain forests in subtropical mountains of Dominican Republic. American Journal of Plant Science. 2014;**5**:1459-1466

[6] Cano E, Veloz A, Cano-Ortiz A. The habitats of *Leptochloopsis virgata* in the Dominican Republic. Acta Botanica Gallica. 2010;**157**(4):645-658

[7] Cano E, Veloz A, Cano-Ortiz A. Phytosociological study of the *Pinus occidentalis* woods in the Dominican Republic. Plant Biosystems. 2011;**145**(2): 286-297. DOI: 10.1080/11263504.2010. 547685

[8] Spampinato G, Massimo DE, Musarella CM, De Paola P, Malerba A, Musolino M. Carbon sequestration by cork oak forests and raw material to built up post carbon city. In: Calabrò F, Della SL, Bevilacqua C, editors. New Metropolitan Perspectives. ISHT 2018. Smart Innovation, Systems and Technologies. Vol. 101. Cham: Springer; 2019. DOI: 10.1007/978-3-319-92102-0_72

[9] Cano-Ortiz A, Fuentes JCP, Gea FL, Ighbareyeh JMH, Quinto Canas RJ, Meireles CIR, et al. Climatology, bioclimatology and vegetation cover: Tools to mitigate climate change in olive groves. Agronomy. 2022;**12**:2707. DOI: 10.3390/agronomy12112707

[10] Dormindo L, Garrido I, Hotellerie-Fallois L & Santillán J. El cambio climático y la sostenibilidad del crecimiento: Iniciativas Internacionales y Políticas Europeas. Documentos Ocasionales N° 2213. 2022. pp. 1-42

[11] Lovelli S, Perniola M, Scalcione E, Troccoli A, Ziska LH. Future climate change in the Mediterranean area: Implications for water use and weed management. Italian Journal of Agronomy. 2012;**7**:44-49

[12] Macías F. Sumideros de carbono para el forzamiento climático Antropoceno. Una visión de alternativas de actuación desde la ciencia del suelo. Edafología. 2004;**11**:7-25

[13] Mota C, Alcaraz-López C, Iglesias M, Martínez-Ballesta MC, Carvajal M. Investigación Sobre la Absorción de CO_2 por los Cultivos Más Representativos de la Región de Murcia, Ed. CSIC: Madrid, Spain; 2011. pp. 1-41

[14] Spampinato G, Malerba A, Calabrò F, Bernardo C, Musarella CM. Cork oak Forest spatial valuation toward post Carbon City by CO2 sequestration. In: Bevilacqua C, Calabrò F, Della SL, editors. New Metropolitan Perspectives. NMP 2020. Smart Innovation, Systems and Technologies. Vol. 178. Cham, Switzerland: Springer; 2021. DOI: 10.1007/978-3-030-48279-4_123 https://datos.bancomundial.org/indicador/NY.GDP.PCAP.CD?locations=HT-AF. (Datos Obtenidos del Informe Banco Munidal 2021)

[15] Guerrero EM, Suarez MR. Integración de valores económicos y sociales de los servicios ecosistémicos del parque Migel Lillo (Necochea, Argentina). Revista Latinoamericana de Estudios Socioambientales. 2019;**26**:69-86. DOI: 10.17141/letrasverdes.26.2019.3945

[16] Iwan A, Guerrero EM, Romanielli A, Bocanegra E. Valoración económica de los servicios ecosistémicosde una laguna del Sudeste Bonarense (Argentina). Investigaciones Geográficas, diciembre. 2017;**2017**:1-18. DOI: 10.10.14198/INGEO2017.68.10

[17] Comisión Europea. Comunicación de la Comisión al Parlamento Europeo, al Consejo, al Comité Económico y Social Europeo y al Comité de las Regiones: Forjar una Europa resiliente al cambio climático-La nueva estrategia de adaptación al cambio climático de la UE. 2021. pp. 1-17

[18] Tenmgkeit T, Wikes A. Las Finanzas del Carbono de los Pastizales. World Agroforestry Center. 2008;**2008**:1-38

[19] Cano-Ortiz A, Piñar Fuentes JC, Cano E. Didactics of natural sciences in higher Scondary education. IJHSSE. 2021;**8**(10):6-10. DOI: 10.20431/2349-0381.0810002

[20] Cano-Ortiz A, Piñar Fuentes JC, Ighbareyeh JMH, Quinto Canas R, Cano E. Aspectos Didácticos en la Enseñanza de Conceptos Geobotánicos. IJHSSE. 2021;**8**(4):271-276. DOI: 10.20431/2349-0381.0804022

[21] Cano-Ortiz A, Piñar Fuentes JC, Cano E. Proposals for the learning of plant diversity. In: 5 th International Virtual Conference on Educational Research and Innovation- CIVINEDU. Conferencie Proceeding CIVINEDU 2021. Redine: Publisher Aldaya Press. Editor; 2021. pp. 325-327

[22] Caruso G. Calabrian native project: Botanical education applied to conservation and valorization of autochthonous Woody plants. Research Journal of Ecology and Environmental Sciences. 2022;**2**(2):47-59. DOI: 10.31586/rjees.2022.387

[23] Jiménez YM, Castro Acevedo G, Cebey Sánchez JA. The Cuabal of Callejón de Los Patos, in Santa Clara: Need for its sustainability. Research Journal of Ecology and Environmental Sciences. 2022;**2**(2):23-46. DOI: 10.31586/rjees.2022.236

[24] Montaner-Villalba S, Santiago R, Bergmann J. Aprender al revés. Flipped Learning 3.0 y metodologías activas en el aula. Barcelona: Faidós Educación pp. 240. Revista Interuniversitaria de Investigación en Tecnología Educativa (RIITE). 2018;7:98-99. DOI: 10.6018/riite.343561

[25] Martínez-Olvera W, Esquivel-Gámez I. Using the flipped learning model in a public high school. Revista de Educación

a Distancia RED. 2018;**58**:Artic. 11. DOI: 10.6018/red/58/11

[26] Domínguez Rodríguez FJ, Palomares Ruiz A. El "aula invertida" como metodología activa para fomentar la centralidad en el estudiante como protagonista de su aprendizaje. Contextos Educativos. 2020;**26**:261-279. DOI: 10.18172/con.4727

Section 2

Plant Diversity, Conservation and Climate Change

Chapter 2

The Relevance of Maintaining Standing Forests for Global Climate Balance: A Case Study in Brazilian Forests

Gabriel Costa, Cláudio Silva, Keila R. Mendes, Bergson Bezerra, Thiago R. Rodrigues, Jonh B. Silva, Higo J. Dalmagro, Hildo Nunes, Ana Gomes, Glauce Silva, Sarah Batalha, Larice Marinho, Gabriele Almeida, Losany Viana, Tiago Mandu and Demethrius P.L. de Oliveira

Abstract

Estimate Net Ecosystem Exchange (NEE) is important to better understand carbon exchanges between terrestrial ecosystems and the atmosphere. Comprehend these dynamics is essential to better understand the responses of environments to ongoing climatic changes. This study aims to analyze, with AMERIFLUX and LBA network measurements, the variability of NEE and climate variables in four different tropical coverages: Pantanal, Amazonia, Caatinga and Cerrado (savanna). Furthermore, was estimate the Gross Primary Productivity (GPP). We found a distinct seasonality of meteorological variables and CO_2 fluxes in each site. Despite acting mostly as a CO_2 sink, some environments already show worrying source data in certain periods, pointed out as a direct effect of the reduction of photosynthesis caused by land use changes. The preserved forest plays an important role in maintaining rainfall at a regional and global level, and its maintenance makes it possible, by the way, an important tool in combating global warming via carbon sequestration by trees, which requires commitment and public policies of environmental preservation and recovery of degraded areas.

Keywords: net ecosystem exchange, carbon balance, climate change, deforestation, micrometeorology

1. Introduction

In 1998, the World Meteorological Organization (WMO), a United Nations agency, and the United Nations Environment Programme (UNEP) started actions of

Intergovernmental Panel on Climate Change (IPCC), which is composed of a large group of researchers from various nationalities with the aim of developing scientific knowledge about climate change and its impacts on society. It is agreed that the increase in the concentration of greenhouse gases (GHG) is mainly caused by the burning of fossil fuels and changes in land use, such as deforestation. These and other human activities are responsible for climate change [1]. The alert for these changes has promoted several lines of research within climate science to debate and create understanding about the GHG cycle and how man would be influencing this cycle, which still has great uncertainties. To decrease these uncertainties, in situ measurements are necessary to better understand the particularities of each environment [2], as these data can be used for the evaluation of soil-vegetation-atmosphere interaction models, [3, 4], as well as in the analyses of satellite products that estimate components of the water and/or CO_2 balance [5–7], which can provide reliable information to monitor CO_2 exchange in tropical forests that have little coverage of towers equipped with Eddy Covariance (EC) system, for example Brazil.

One practical way to mitigate the effects of climate change would be to increase the vegetation cover in dryland and devastated areas, both through the replanting of large forest covers in order to capture atmospheric CO_2, and through the effect of forests on the hydrological cycle. With the emerging trade in carbon credits in the global market, there are several enterprises that benefit from the reforestation of large areas of land and the banks of hydroelectric power generation dams. Although carbon credit trading is accused of being an ineffective way to mitigate CO_2 emissions, because some people pay to continue polluting, this would be an advantageous way of transitioning to an economy with a clean energy matrix, especially in a country like Brazil, with extensive areas of tropical forest.

Therefore, it would be more economically advantageous to preserve untouched forests than to devastate them for logging or to start immediately profitable agricultural and cattle ranching enterprises. However, the prognoses for Brazilian forested areas related to global warming [8, 9], especially the Amazon, are not very good: Global warming is expected to increase temperatures in this region, which may make the climate drier, causing savannization of this forest, i.e., parts of the forests should be altered, changing their structure and approaching the Cerrado physiognomy.

River levels may present a great reduction, and the air should become drier during dry periods, which increases the risk of fires. In addition, the advance of the agricultural frontier, if maintained at current levels, is expected to reduce forest cover to 53% of the original by 2050. The number of studies on the response of Amazonian and Cerrado flora and fauna species to climate change is still very small, but they indicate that with an increase of 02–03°C in average temperature, up to 25% of the trees in the Cerrado and about 40% of the trees in the Amazon could disappear by the end of this century [10]. For the planet's temperature not to exceed an increase of 1.5°C and thus avoid drastic changes in the climate, "rapid, vast and unprecedented changes" will be necessary at the global level, warned a report released on 08/10/2018 by the UN IPCC.

Micrometeorological studies to obtain data that help understand the role of Brazilian tropical forests in the planet's climate balance exist since the 1980s until the LBA (Large Scale Biosphere-Atmosphere Experiment in Amazon) [11]. Some works pointed in the Amazon forest a potential sink of atmospheric CO_2 [12–14] with micrometeorological measurements. Saleska et al. [15] compared micrometeorological estimates and biometric measurements in Santarém- PA and reported a CO_2 source to the atmosphere of 1.3 Mg C ha^{-1} $year^{-1}$, associated with the prevalence of emissions by necromass decomposition in preceding episodes of high tree mortality in the region. Miller et al. [16] reported slightly positive ecosystem fluxes of 0.4 Mg C ha-1 year-1, closer to neutral, corroborated by biometric measurements. Espírito Santo et al. [17] combined satellite estimates and field data to suggest that uptake by living trees exceeds emission by dead trees, reinforcing evidence of the upland forest acting as a carbon sink in aboveground biomass. The differences between the contributions of different Brazilian biomes have been consistently reported in the literature [2]. Evapotranspiration rates are highest and greatest at Cerrado and Pantanal sites in wetter months. Even in the month with the highest evapotranspiration rates (October), the values of the Caatinga sites do not reach the magnitude scale of the measurements of the other sites, showing the peculiarity of this site as to its BSh (Arid/Stepp/Hot) climate by Köppen's climate classification. Generally the measurements show that evapotranspiration in the dry season is higher than in the wet season and Rn is the main control of evapotranspiration in humid tropical rainforests (like the Amazon site), which does not apply for the more arid Caatinga region nor for the Cerrado and Pantanal sites, which differ from the Amazon by showing depletion of evapotranspiration throughout the dry season, culminating with lower values in drier months such as August and September, the same period where the Amazon site presents its maximum [2].

All these particularities directly influence the local and regional carbon balance, showing the need to create a better understanding of the biogeochemical cycles in these locations to try to show the importance of forest climate control on climate and greenhouse gas emissions, in order to prove that keeping the forest standing can be the most viable alternative for a public policy on climate change mitigation, results that will be discussed below.

2. Methodology

2.1 Data policy and use license

The Ameriflux platform integrates the data monitored in three biomes, with the identifications: BR-Sa1 (Amazon), BR-CST (Caatinga) and BR-Npw (Pantanal). The data from the Cerrado site (BR-BI, Bananal Island - Javaés) are available at https://daac.ornl.gov. The data made available by ORNL DAAC are shared freely, without restriction, in agreement with NASA's Earth Science Program. Ameriflux data is shared under a CC-BY-4.0 data usage license (Creative Commons by Attribution 4.0 International). The CC-BY-4.0 license specifies that data usage is free to share (copy and redistribute the material in any medium or format) and/or adapt (remix, transform and build upon the material) for any purpose. The citation of the data sites is: BR-Sa1 [18], BR-CST [19] and BR-Npw [20].

2.2 Description of study áreas

2.2.1 Cerrado site

Measurements were made at an experimental floodplain site in Cantão State Park, 260 km west of Palmas, Tocantins, Brazil, in the context of the Large-Scale Biosphere-Atmosphere Experiment in Amazonia (LBA), at a micrometeorological tower with measurements of turbulent energy fluxes and meteorological variables. The tower was located 2 km east of the Araguaia river (9° 49′ 27.9“ S, 50° 08’ 92.8” W, at 120 m altitude), about 1 km east of Javaézinho river, on the northern border of Ilha do Bananal and south of the park. The region of the Araguaia plain, where Ilha do Bananal is located, stands out for its exuberant landscape with aspects of the Cerrado and Amazon biome with three conservation units, the Araguaia National Park, the Cantão State Park (PEC) and the Cantão Environmental Protection Area. Bananal island covers an area of about 21,000 km2 (approximately 80 x 260 km) and is the largest river island in the world, covered mostly by savannas and grasslands, in seasonal floods usually from February to June [21]. The measurement fetch area covers three types of physiognomy: cerradão and semideciduous forests (trees with an average height of about 20 m), cerrado s.s., and areas of clean field and isolated lagoons. The regional climate is hot and seasonally humid, with average annual precipitation of about 1466 mm year-1, approximately 90% of the annual rainfall in the rainy season between October and April [21] and therefore typical of the Cerrado biome in central-western Brazil. The measurements comprise the period from January 2004 to December 2006.

2.2.2 Caatinga site

The measurements were taken at a site managed by the Chico Mendes Institute for Biodiversity Conservation and the micrometeorological tower is part of the monitoring network of the National Observatory of Water and Carbon Dynamics in the Caatinga Biome (NOWCDCB) project. The period of measurements comprises January 1, 2014 to July 31, 2015. This project is located in a preserved Caatinga area (BR-CST), situated in the Pajeú river basin in Serra Talhada (7° 58′ 05.20“ S and 38° 23’ 02.62” W, 430 m), state of Pernambuco, northeastern Brazil. The climate is classified according to Köppen as BSwh (semi-arid), being characterized as hot and semi-arid, with summer rainfall concentrated between December and May (85%) according to Alcântara et al., [22]. The average annual precipitation is approximately 640 mm, and the average monthly air temperature ranges between 23.1 and 26.7°C [22]. The native species of the site are composed of Mimosa hostilis, Mimosa verrucosa and Croton sonderianus, and it is possible to find Anadenanthera macrocarpa, *Spondias tuberosa*, Caesalpinia pyramidalis and Ziziphus joazeiro, with a height of about 8.0 m [22].

2.2.3 Pantanal site

The study was conducted at the Brazilian Northern Pantanal Wetland (BR-Npw) flux tower (**Figure 1**) located approximately 35 km SE of Pocone, Mato Grosso, Brazil (16° 29′53.71” S: 56° 24′45.91” W; 120 m altitude). The site is part of a research station managed by the Federal University of Mato Grosso (UFMT) within a national reserve managed by the Brazilian Social Service of Commerce (SESC Pantanal) [23, 24]. Our data were collected from 1 January 2015 to 31 December 2016. Micrometeorological variables were measured 20 m aboveground, close to the eddy covariance sensors. Air temperature (*Ta*, °C) and relative humidity (*RH*, %) were measured using a thermohygrometer (HMP45AC, Vaisala Inc., Woburn, MA, USA). Precipitation (*Ppt*, mm) was measured 2 m above the ground using a micrometeorological station (WXT520, Vaisala Inc., Helsinki, Finland) installed in an open area to avoid interception by the tower or tree canopy. The flood stage was determined by measuring water levels (*WL*) above the ground at the study site. These inundation

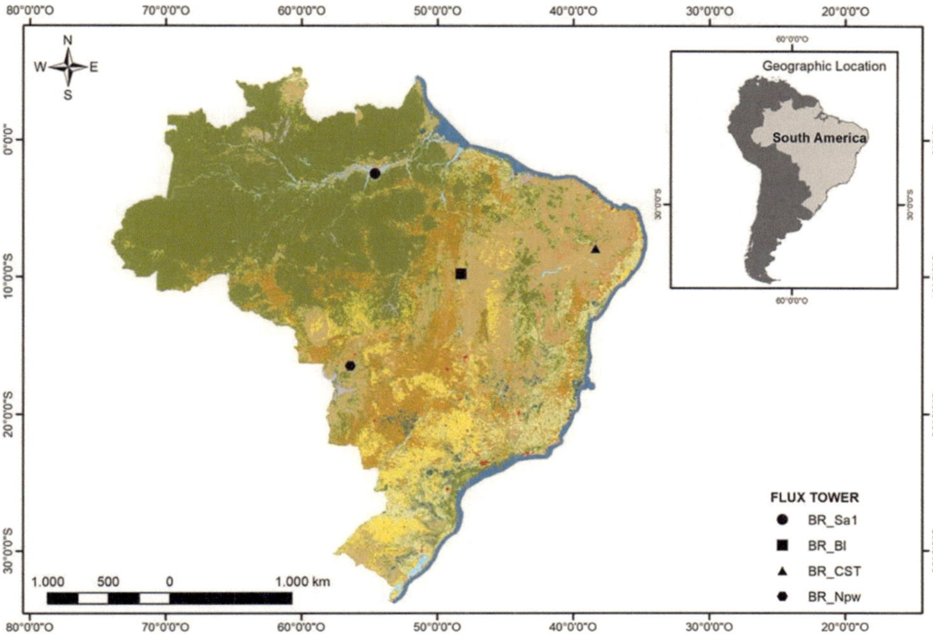

Figure 1.
Location of the study sites.

levels (± 1%) were measured along with water temperature (± 0.3°C using a CTD-10 [Decagon Devices Inc., Pullman, WA, USA], ± 0.05% full scale at 20°C) in 2015 and 2016. Due to instrument malfunction in 2014, the data for this year are not available. The start of each flood cycle began with the first reading of standing water at the site and ended when sensors indicated the absence of standing water. These flood cycles were then compared to the stage of the Cuiaba River collected by the RPPN-SESC Pantanal park rangers (pers. comm.) approximately 1 km away.

2.2.4 Amazonia site

Measurements were made at a site located in the Tapajós National Forest (FNT, 2° 51' S, 54° 58' W), which is situated near the Santarém-Cuiabá Highway (BR-163), at km 67. The NTF is bounded by the Tapajós River to the west and the BR-163 highway to the east, extending 50 km to 150 km south of the city of Santarém-PA. On the eastern side of the BR-163 highway the landscape is extensively developed for agriculture. The tower was installed approximately 6 km west of the BR-163 highway. The data analyzed are CO_2 and energy fluxes associated with meteorological measurements. Measurements range from January 2009 to December 2011, with daily and monthly averages of hourly data. CO_2 fluxes were measured at 58 m using a closed path analyzer (LICOR- 6262) and a Campbell CSAT3 Anemometer was used for the three-dimensional wind measurements. The 65 m micrometeorological tower is located in an area emerging from within a primary forest with a closed canopy of approximately 40 m in height, and can reach up to 55 m with some emergent trees [25]. **Figure 1** shows the location of the four study sites.

2.3 Instrumentation and data processing

Site instrumentations are given in previous publications [21, 22, 24, 26]. The gaps arising from the exclusion of spurious data during the rigorous screening process, were filled using the gap-filling algorithm of marginal distribution sampling (MDS) described by Reichstein et al. [27], which takes into account the covariation of fluxes with meteorological variables and also the temporal self-correlation of fluxes. In this algorithm, three conditions are identified with their respective procedures: when flux data are missing, but meteorological data are available (Rg, Ta and VPD): the missing data is replaced by the average value under similar meteorological conditions in a 7-day window. If similar conditions are not available, the window is increased to 14 days; (2) when only radiation values are available: the missing data is replaced by the average value under similar meteorological conditions within a 7-day window; (3) when no meteorological data is available: the missing value is replaced by the average value of the last hour, thus considering the diurnal variability of each variable. If, after these steps, the data are not filled, the procedure is repeated with larger window sizes until the value can be filled. For filling the gaps, an automated online tool made available by the Max Plank Institute (Max Planck Institute for Biogeochemistry - http://www.bgc-jena.mpg.de/~MDIwork/eddyproc/) was used.

2.4 Fluxes partitioning

Gross primary productivity (GPP) and ecosystem respiration (Reco) were partitioned from the CO_2 flux data (NEE). For the Cerrado and Pantanal sites, NEE is given as a proxy measure of turbulent flux (Fc). At the other sites NEE is composed of turbulent flux + storage. We used a nighttime-based flow partitioning method [27]. Since GPP =0 in night time, NEE corresponds to:

$$NEE = R_{eco}, \text{for night hours} \tag{1}$$

$$NEE = R_{eco} - GPP, \text{for daytime hours} \tag{2}$$

Reco (μmol m^{-2} s^{-2}) being the sum of autotrophic and heterotrophic respiration. Reco and GPP were calculated using the online tool provided by the Max Plank Institute (Max Planck Institute for Biogeochemistry - http://www.bgc-jena.mpg.de/~MDIwork/eddyproc/).

Diurnal corrections for missing NEE data were modeled based on diurnal data using the common rectangular hyperbolic light response curve model [28, 29]:

$$NEE = \frac{\alpha \cdot \beta \cdot R_g}{\alpha \cdot R_g + \beta} + \gamma \tag{3}$$

where α (μmol C J^{-1}) is the light utilization efficiency and represents the initial slope of the light response curve, β (μmol C m^{-2} s^{-1}) is the maximum canopy CO_2 uptake rate at light saturation, γ (μmol C m^{-2} s^{-1}) is the ecosystem respiration and Rg (W m^{-2}) is the global radiation. GPP was calculated as:

$$\text{GPP} = NEE + \text{R}_{\text{eco}} \tag{4}$$

3. Results

3.1 Climate variables

Air temperature and precipitation are strong controls on ecosystem productivity [2], and all study sites showed effective seasonality in terms of rainfall and air temperature (**Figures 2–5**), with precipitation events (rainy days) more frequent in the Amazon and Cerrado sites. The distribution of average air temperature throughout the year is linked to the intensity of precipitation at the sites, with great variability throughout the year. The Caatinga and Pantanal sites are the warmest throughout the day and in the first months of the year (**Figure 6**), a pattern that is inverted from

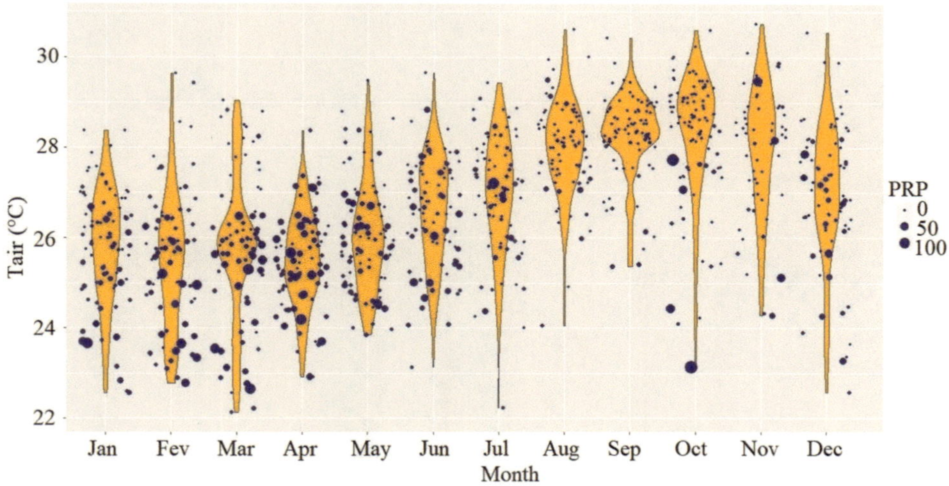

Figure 2.
Distribution of monthly air temperature data density by precipitation intensity at the Amazon site.

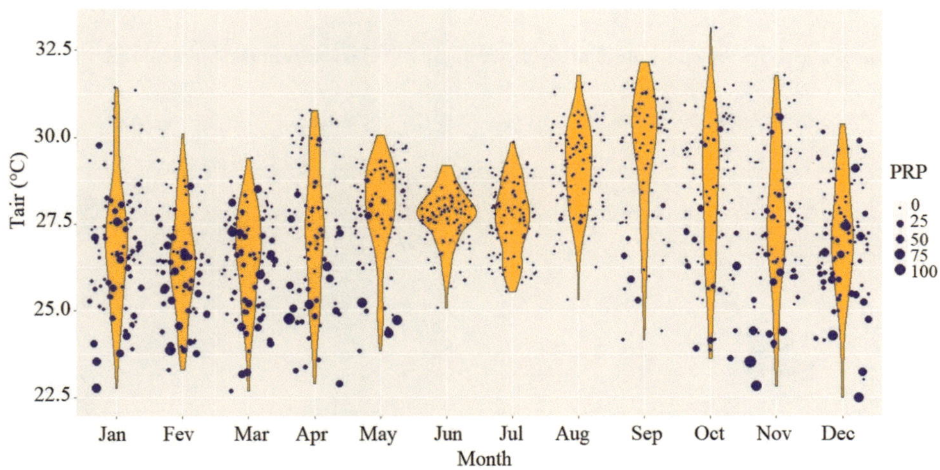

Figure 3.
Distribution of monthly air temperature data density by precipitation intensity at the cerrado site.

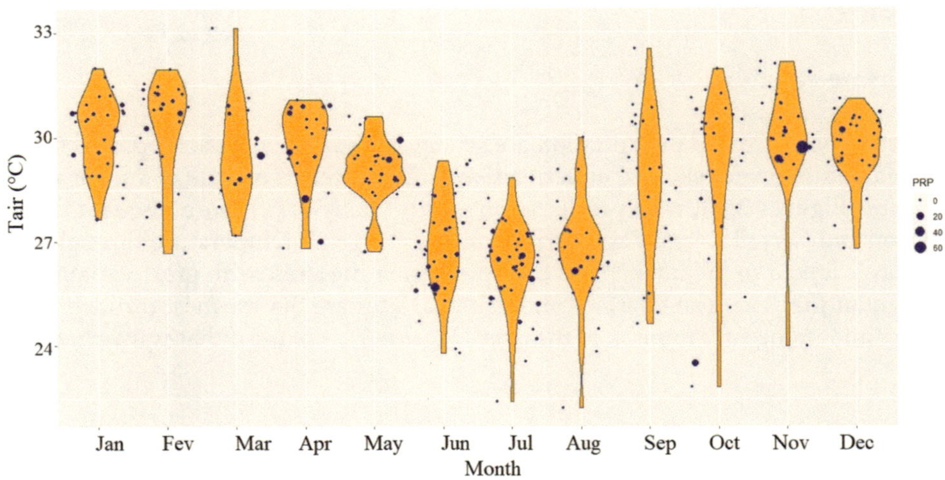

Figure 4.
Distribution of monthly air temperature data density by precipitation intensity at the Caatinga site.

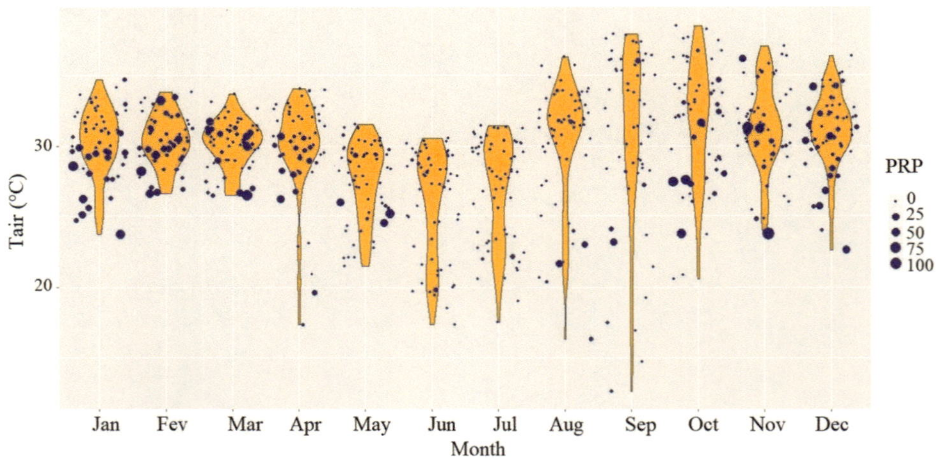

Figure 5.
Distribution of monthly air temperature data density by precipitation intensity at the Pantanal site.

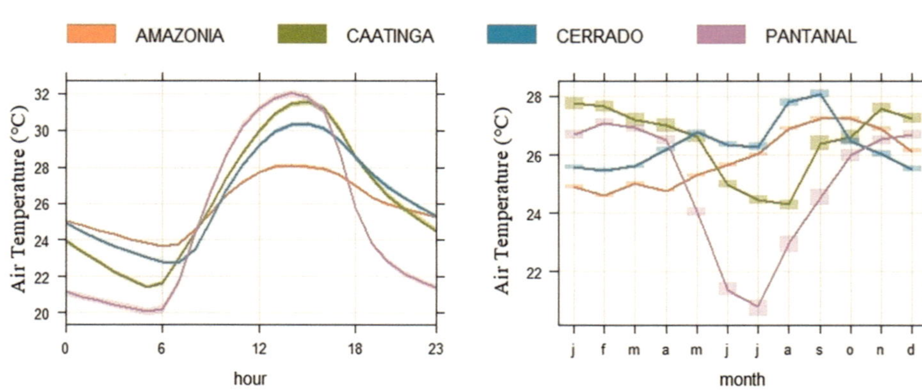

Figure 6.
Hourly (left) and monthly (right) variation of air temperature at the analyzed sites.

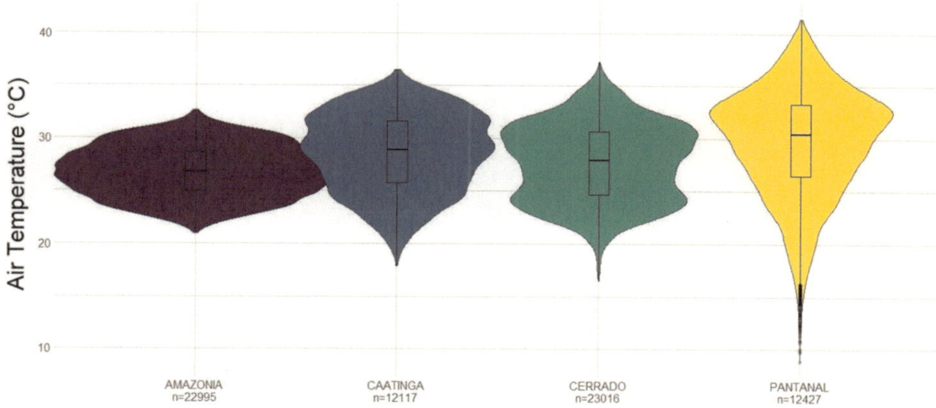

Figure 7.
Violin plot of air temperature at the analyzed sites.

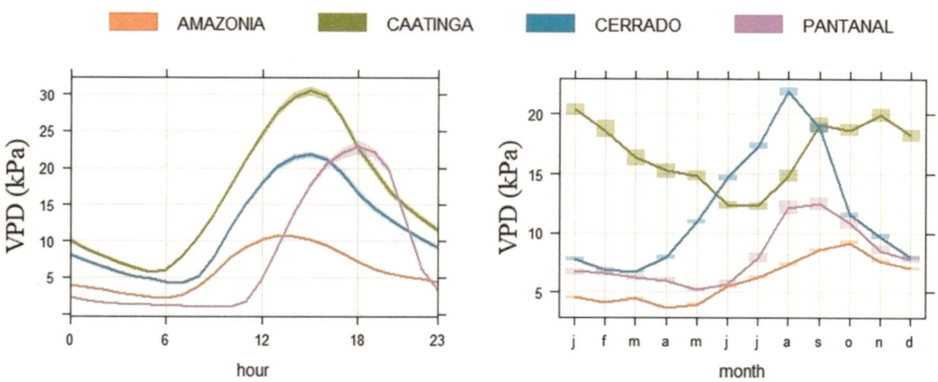

Figure 8.
Hourly (left) and monthly (right) variation of vapor pressure deficit at the analyzed sites.

June until October, when the Amazon and Cerrado sites are warmer than the others. The site with the highest average daily temperature is the Pantanal (**Figure 7**), with 31.6°C in October, while the Amazon site has the lowest temperatures (25.4°C) in April.

Despite being warmer during the day, the Pantanal site is colder at night, and due to its geographical location being susceptible to cold front events, it presents the lowest temperatures in the months of June to August. The Caatinga and Cerrado sites present values up to 3 times higher in VPD (**Figure 8**) than the others, while the Pantanal is close to the Amazon site in this variable. The VPD in the Amazon site practically doubles in value until October, concomitant with the lower precipitation rates and higher air temperature in the year.

3.2 CO_2 fluxes

Monthly averages of NEE and GPP are presented in **Figure 9**, where seasonal changes in NEE are shown to be more intense in the Caatinga and Pantanal sites than in the Amazon and Cerrado sites. The maximum monthly mean GPP values exceed

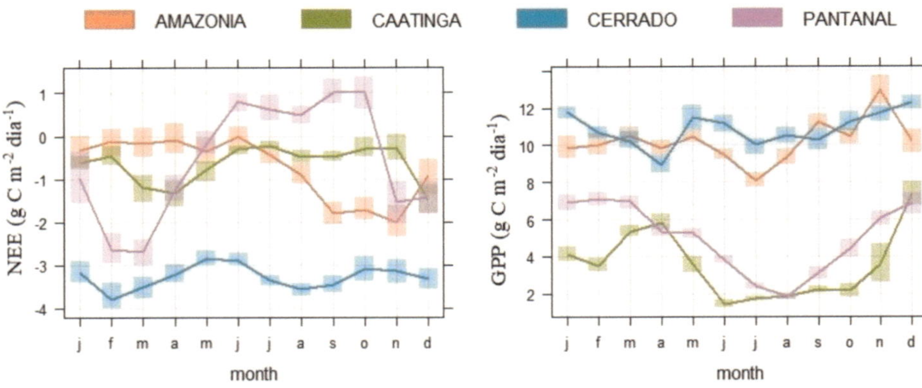

Figure 9.
Monthly variation of NEE (left) and GPP (right) at the analyzed sites.

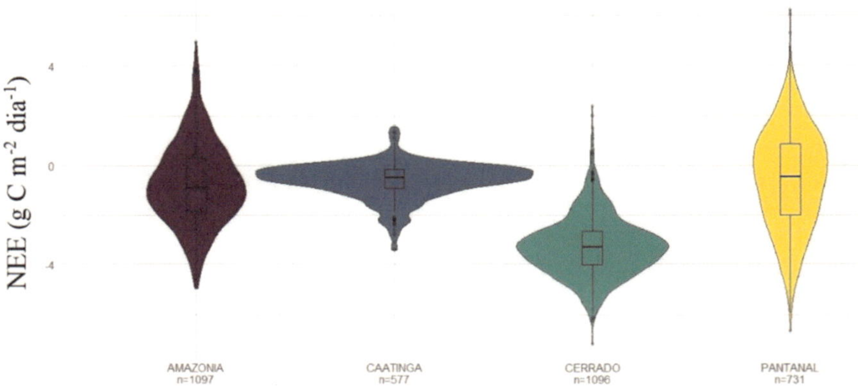

Figure 10.
Violin plot of air temperature at the analyzed sites.

14.0 gC m^{-2} d^{-1} in the Cerrado and Amazon sites, while the lowest GPP values reach about 1.0 gC m^{-2} d^{-1} in the Caatinga site. The seasonality and magnitude of the CO_2 exchange values found in the Amazon and Cerrado resemble each other, while the Caatinga and Pantanal patterns show very close values and variability. The maximum assimilation of the sites occurs in the month of February, when Pantanal and Cerrado show assimilations close to 4.0 gC m^{-2} d^{-1}, and the minimum is very close to 0.5 gC m^{-2} d^{-1} in the month of August at the Pantanal site, after the coldest and least rainy month of the site. Sites susceptible to seasonal flooding [21, 30] in the early months of the year (Cerrado and Pantanal) show increases in assimilation (more negative NEE) between January and February, while Amazon and Caatinga sites have marked increases in assimilation in the warmer months of the year. The density of NEE data (**Figure 10**) shows that most sites act as CO_2 sinks, with higher variability at the Pantanal site and lower variability at the Caatinga site. The medians of GPP (**Figure 11**) are higher in the Amazon, Cerrado and Pantanal sites compared to Caatinga. However, the same does not occur for NEE, where the medians of the Caatinga site have higher values compared to the other sites. Despite this, the large density of data below 0 gC m^{-2} d^{-1} indicates that the Caatinga biome is much more

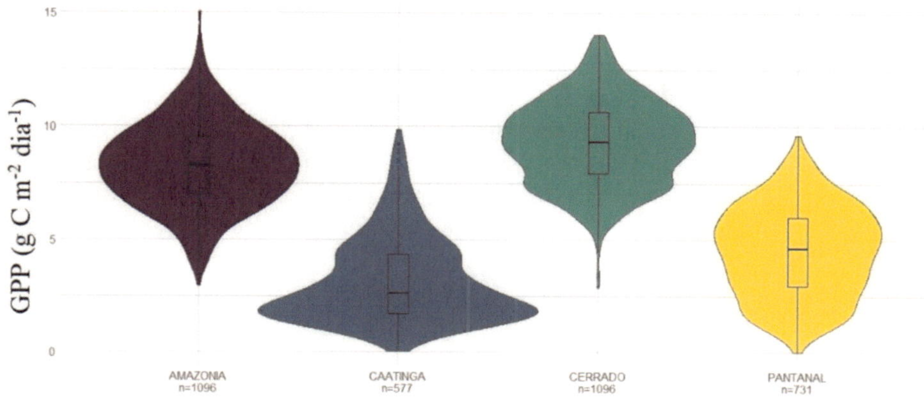

Figure 11.
Violin plot of air temperature at the analyzed sites.

firmly established as a CO_2 sink, despite its smaller vegetative size, than the Amazon site, for example, which have a bimodal NEE data density distribution pattern, showing a secondary maximum above 0 gC m^{-2} d^{-1}, indicating that the biome can act as a source in some seasons, probably controlled by CO_2 emissions higher than assimilation due to the intensification of wildfires in the region, included within the arc of deforestation of the Amazon.

4. Discussion

The results show that most of the sites act as CO_2 sink, corroborating works in the literature [12, 23, 31, 32]. There are data of higher CO_2 emissions (positive NEE) mainly at the Amazon site, where other work has shown the possibility of the environment acting as a moderate source [26, 33]. The overall magnitude of the carbon source/sink is generally highly sensitive to the choice of filter u∗ (to measure turbulence intensity), and even with all the corrections recommended by the literature, the pattern was reinforced by biometric measurements [16], which makes us look for alternative explanations for this unusual pattern, since the vast majority of forest sites in different areas of the same biome, point to a CO_2 sink. Heyek et al. [33] suggested that continuous integrated responses to changes in meteorology, with increased humidity and decreased sunlight, rather than a temporary disturbance, were responsible for the high carbon source at the site. The authors also suggested that reduced photosynthesis, rather than increased respiration, contributed to the high NEE source in specific years at the site. This suggests that partial drought-induced damage to still-living trees can adversely affect ecosystem-wide photosynthesis for several years, which is consistent with results from forest biometric studies at regional and global scales [34, 35]. Tian [36], analyzing series from 1980 to 1994, notes that the carbon balance of the Amazon forest can have great variability, sometimes positive and sometimes negative, depending on variables such as sunlight incidence, CO_2 concentration in the atmosphere and rainfall volume. Gatti et al. [37] point out that regions of the Amazon forest, such as the site region of this study, are affected by environmental degradation and are leading the Amazon as a whole to emit more carbon than it can absorb. The authors point out that a secondary effect of deforestation has been created: the indirect carbon emission caused by the impact of reduced rainfall on photosynthesis, corroborating what was pointed out [33]. Indirect emission happens because deforested regions have a greater loss of rainfall, especially in the dry season (August to October). With the drop in rainfall volume, the temperature rose 2°C in the northeast of the forest and 2.5°C in the southeast, and this "stress" affected photosynthesis, causing the trees to emit more CO_2 than in normal situations to compensate for the imbalance, showing the importance of maintaining the forest, including for the maintenance of rainfall important for agriculture and cattle ranching in the region.

5. Conclusions

The study showed an important meteorological control on the carbon cycle in the biomes studied, which infers that changes in surface cover will directly affect these variables and, consequently, the local carbon balance. Despite acting mostly as a CO_2 sink, some environments already show worrying source data in certain periods, pointed out as a direct effect of the reduction of photosynthesis caused by land use changes. The preserved forest plays an important role in maintaining rainfall at a regional and global level, and its maintenance makes it possible, by the way, an important tool in combating global warming via carbon sequestration by trees, which requires commitment and public policies of environmental preservation and recovery of degraded areas.

Acknowledgements

The authors are also thankful to the Coordination for the Improvement of Higher Education Personnel (CAPES) for the postdoctoral funding granted to KRM and to the National Council for Scientific and Technological Development (CNPq) for the research productivity grant of C.M.S.S (Process n° 303802/2017-0), the financial support of CNPq, through undergraduate research project (PIBIC-UFOPA for L. B. V and G. V. A) and the project NOWCDCB: National Observatory of Water and Carbon Dynamics in the Caatinga Biome (INCT-MCTI/CNPq/CAPES/FAPs 16/2014, grant: 465764/2014-2) and (MCTI/CNPq N° 28/2018, grant 420854/2018-5).

Conflict of interest

The authors declare no conflict of interest and the funders had no role in the design of the study; in the collection, analyses, or interpretation of data; in the writing of the manuscript, or in the decision to publish the results.

Author details

Gabriel Costa[1,2*], Cláudio Silva[2], Keila R. Mendes[2], Bergson Bezerra[2], Thiago R. Rodrigues[3], Jonh B. Silva[4], Higo J. Dalmagro[5], Hildo Nunes[6], Ana Gomes[7], Glauce Silva[7], Sarah Batalha[7], Larice Marinho[7], Gabriele Almeida[7], Losany Viana[7], Tiago Mandu[8] and Demethrius P.L. de Oliveira[9]

1 Biosciences Post-Graduate Program (PPG-BIO), Federal University of Western Pará (UFOPA), Santarém, Brazil

2 Climate Sciences Post-Graduate Program (PPGCC), Federal University of Rio Grande do Norte (UFRN), Natal, Brazil

3 Atmospheric Science Lab, Federal University of Mato Grosso Do Sul, Campo Grande, MS, Brazil

4 Post-Graduate Program in Environmental Physics, Federal University of Mato Grosso, Cuiabá, MT, Brazil

5 Post-Graduate Program in Environmental Sciences, University of Cuiabá, Cuiabá, Brazil

6 Socio-Environmental and Water Resources, Federal Rural University of the Amazon, Belém, Brazil

7 Federal University of Western Pará (UFOPA), Santarém, Brazil

8 National Institute for Space Research, São José dos Campos, Brazil

9 Secretary of Science, Technology and Higher, Professional and Technological Education, Belém, Pará, Brazil

*Address all correspondence to: gabriel.costa@ufopa.edu.br

References

[1] Intergovernmental Panel On Climate Change - IPCC. Climate Change 2007: The Physical Science Basis. Geneva: Intergovernmental Panel on Climate Change; 2007

[2] Costa GB, Santos e Silva CM, Mendes KR, dos Santos JGM, Neves TTAT, Silva AS, et al. WUE and CO_2 estimations by Eddy covariance and remote sensing in different tropical biomes. Remote Sensing. 2022;**14**:3241. DOI: 10.3390/rs14143241

[3] Bai Y, Li X, Zhou S, et al. Quantifying plant transpiration and canopy conductance using eddy flux data: An underlying water use efficiency method. Agricultural and Forest Meteorology. 2019;**271**:375-384

[4] Mendes KR, Campos S, Mutti PR, Ferreira RR, Ramos TM, Marques TV, et al. Assessment of SITE for CO_2 and energy fluxes simulations in a seasonally dry tropical Forest (Caatinga ecosystem). Forests. 2021;**12**:86. DOI: 10.3390/f12010086

[5] Fonseca LDM, Dalagnol R, Malhi Y, Rifai SW, Costa GB, Silva TSF, et al. Phenology and seasonal ecosystem productivity in an Amazonian flood-plain Forest. Remote Sensing. 2019; **11**:1530

[6] Moreira AA, Ruhoff AL, Roberti DR, de Arruda Souza V, Da Rocha HR, Paiva RCD. Assessment of terrestrial water balance using remote sensing data in South America. Journal of Hydrology. 2019;**575**:131-147

[7] Laipelt L, Ruhoff AL, Fleischmann AS, Kayser RHB, Kich EM, da Rocha HR, et al. Assessment of an automated calibration of the SEBAL algorithm to estimate dry-season surface-energy partitioning in a Forest–savanna transition in Brazil. Remote Sensing. 2020;**12**:1108

[8] Marengo Jose A, Souza Carlos M, Kirsten T, Chantelle B, Kate H, Betts Richard A, et al. Changes in climate and land use over the Amazon region: Current and future variability and trends. Frontiers in Earth Science. 2018;**6**:1

[9] Marengo JA, Drought F. Climate change, and Forest loss in the Amazon region: A present and future danger? Frontiers for Young Minds. 2020;**7**:147

[10] Nobre C, Sampaio GS. Mudanças climáticas. In: Mudanças climáticas e mudanças socioambientais globais: reflexões sobre alternativas de futuro. Brasília: UNESCO, IBECC; 2008. pp. 15-31

[11] Nobre CA, Marengo JA Artaxo P Understanding the climate of Amazonia: Progress from LBA. In: Keller M, Bustamante M, Gash J, Silva Dias P. (Orgs.). Amazonia and Global Change. Geophysical Monograph Ser. Vol. 186. Washington, D.C.: American Geophysical Union Books; 2010. pp. 145-147

[12] Araújo AC, Nobre AD, Kruijt B, Elbers JA, Dallarosa R, Stefani P, et al. Comparative measurements of carbon dioxide fluxes from two nearby towers in a central Amazonian rainforest: The Manaus LBA site. Journal of Geophysical Research. 2002;**107**:8090

[13] Grace J. Carbon dioxide uptake by an undisturbed tropical rain forest in Southwest Amazonia. Science. 1995: 778-780

[14] Malhi Y. Carbon dioxide transfer over a central Amazonian rain forest.

Journal of Geophysical Research-Atmospheres. 1998:593-612

[15] Saleska SRSR, Miller SD, Matross DM, Goulden ML, Wofsy SC, Rocha HR, et al. Carbon in Amazon forests: Unexpected seasonal fluxes and disturbance-induced losses. Science. 2003:1554-1557

[16] Miller SD, Goulden ML, Menton MC, Da Rocha HR, De Freitas HC, Figueira AMS, et al. Biometric and micrometeorological measurements of tropical forest carbon balance. Ecological Applications. 2004:S114-S126

[17] Espírito-Santo FDB, Gloor M, Keller M, Malhi Y, Saatchi S, et al. Size and frequency of natural forest disturbances and the Amazon forest carbon balance. Nature Communications. 2014;**5**(3434):1-6

[18] Saleska S. AmeriFlux BASE BR-Sa1 Santarem-Km67-Primary Forest, Ver. 5-5, AmeriFlux AMP, (Dataset). 2019. DOI: 10.17190/AMF/1245994

[19] Antonino ADCD. AmeriFlux BASE BR-CST Caatinga Serra Talhada, Ver. 1-5, AmeriFlux AMP, (Dataset). 2019. DOI: 10.17190/AMF/1562386

[20] Vourlitis G, Dalmagro H, Nogueira JS, Johnson M, Arruda P. AmeriFlux BASE BR-Npw Northern Pantanal Wetland, Ver. 1-5, AmeriFlux AMP, (Dataset). 2019. DOI: 10.17190/AMF/1579716.

[21] Borma LS et al. Atmosphere and hydrological controls of the evapotranspiration over afloodplain forestin the Bananal Island region, Amazonia. Journal of Geophysical Research: Biogeosciences. 2009; **114**:G01003. DOI: 10.1029/2007jg000641

[22] Alcântara LRP, Coutinho AP, dos Santos Neto SM, de Gusmão C, da Cunha Rabelo AE, Antonino ACD. Modeling of the hydrological processes in Caatinga and pasture areas in the Brazilian semi-arid. Water. 2021;**13**:1877

[23] Dalmagro HJ, Zanella de Arruda PH, Vourlitis GL, Lathuilliere MJ, de Nogueira SJ, Couto EG, et al. Radiative forcing of methane fluxes offsets net carbon dioxide uptake for a tropical flooded forest. Global Change Biology. 2019;**25**:1967-1981

[24] Da Silva JB, Valle Junior LCG, Faria TO, Marques JB, Dalmagro HJ, Nogueira JS, et al. Temporal variability in evapotranspiration and energy partitioning over a seasonally flooded scrub Forest of the Brazilian Pantanal. Agricultural and Forest Meteorology. 2021;**308–309**:108559

[25] Stark S, Leitold V, Wu J, Hunter M, Castilho C, Costa F, et al. Amazon forest carbon dynamics predicted by profiles of canopy leaf area and light environment. Ecology Letters. 2012;**15**:1406-1414

[26] Hutyra LR, Munger JW, Saleska SR, Gottlieb E, Daube BC, Dunn AL, et al. Seasonal controls on the exchange of carbon and water in an Amazonian rain forest. Journal of Geophysical Research. 2007;**112**

[27] Reichstein M, Falge E, Baldocchi D, Papale D, Aubinet M, Berbigier P, et al. On the separation of net ecosystem exchange into assimilation and ecosystem respiration: Review and improved algorithm. Global Change Biology. 2005;**11**:1424-1439. DOI: 10.1111/j.1365-2486.2005.001002.x

[28] Falge E, Baldocchi D, Olson R, Anthoni P, Aubinet M, Bernhofer C, et al. Gap filling strategies for defensible annual sums of net ecosystem exchange.

Agricultural and Forest Meteorology. 2001;**107**:43-69

[29] Lasslop G, Reichstein M, Papale D, Richardson AD, Arneth A, Barr A, et al. Separation of net ecosystem exchange into assimilation and respiration using a light response curve approach: Critical issues and global evaluation. Global Change Biology. 2010;**16**:187-208

[30] Rebellato L, Nunes Da Cunha C. Efeito do fluxo sazonal mínimo da inundação sobre a composição e estrutura de um campo inundável no Pantanal de Poconé, MT, Brasil. Acta Botanica Brasilica. 2005;**19**(4):789-799

[31] Mendes KR, Campos S, da Silva LL, et al. Seasonal variation in net ecosystem CO_2 exchange of a Brazilian seasonally dry tropical forest. Scientific Reports. 2020;**10**:9454

[32] Fu Z et al. The surface-atmosphere exchange of carbon dioxide in tropical rainforests: Sensitivity to environmental drivers and flux measurement methodology. Agricultural and Forest Meteorology. 2018;**263**:292-307

[33] Hayek MN, Longo M, Wu J, Smith MN, Restrepo-Coupe N, Tapajós R, et al. Carbon exchange in an Amazon forest: Fromhours to years. Journal of Geophysical Research: Biogeosciences. 2018;**15**:4833-4848. DOI: 10.5194/bg-15-4833-2018

[34] Anderegg WRL, Schwalm C, Biondi F, Camarero JJ, Koch G, Litvak M, et al. Pervasive drought legacies in forest ecosystems and their implications for carbon cycle models. Science. 2015;**349**: 528-532

[35] Brienen RJW, Phillips OL, Feldpausch TR, Gloor E, Baker TR, Lloyd J, et al. Long term decline of the Amazon carbon sink. Nature. 2015;**519**: 344-348. DOI: 10.1038/nature14283

[36] Tian H, Melillo JM, Kicklighter DW, McGuire AD, Helfrich JVK, Moore B, et al. Effect of interannual climate variability on carbon storage in Amazonian ecosystems. Nature. 1998; **396**:664-667

[37] Gatti LV, Basso LS, Miller JB, Gloor E, Domingues LG, Cassol HLG, et al. Amazonia as a carbon source linked to deforestation and climate change. Nature. 2021;**595**:388-393

Chapter 3

CDR and Tropical Forestry: Fighting Climate Change One Cubic Meter a Time

Ederson Augusto Zanetti, Frederick N. Numbisi, Vithal Karoshi, Roberto Rochadelli, Allan Sbardelotto, Joesio Siqueira and Alain Levy Boussamba

Abstract

In the coming decades, there will be a global increase in demand for biomass and in advocating GHG emission removal technology and practices. In the agriculture and forestry context, intensification of land use is the most promising solution—together with processing efficiency—in balancing consumption, rated as human appropriation of net primary production (HANPP), with Net Primary Production (NPP) from atmospheric CO_2 fertilization. Forest plantations, croplands, cultivated pastures, lianas, palms and other secondary vegetation have shown yield gains from CO_2 fertilization, while native forest (trees) experience short-lived increases in growth rates and are out-competed by fast-growing components—secondary vegetation. There is evident path of degradation in non-managed, native tropical forests fueled by atmospheric CO_2 fertilization. Following such BAU scenario, tropical forests would experience important dwindling in tree cover on a temporal scale. An alternative IFM scenario is proposed combining contemporary silviculture techniques, adapted land use intensification and HWP increase. This would contribute additional atmospheric CO_2 removals, certifiable as CDR goods able to generate carbon credits and financial incentive for cultivation of improved native tree species. These CDR credits can be included in tropical countries' NDC and presented at UNFCCC as an ITMO for fighting global climate change.

Keywords: tropical forest, HWP, IFM, CDR, NDC, ITMO

1. Introduction

Carbon Dioxide Removal (CDR) comprises anthropogenic activities that remove CO_2 from the atmosphere and store it in a durable way in geological, oceanic and terrestrial reservoirs or in products. Such activities include existing and potential anthropogenic enhancement of biological or geochemical sinks and direct air capture and storage, but they exclude natural absorption of CO_2 not directly caused by human activities. The latest report from Working Group III (WG III) of the

Intergovernmental Panel on Climate Change (IPCC) includes CDR to offset hard-to-abate residual emissions [1]. In the European Union (EU), the European Climate Law commits the Union to reach climate neutrality by 2050. Both greenhouse gas (GHG) emission reductions and CDR will be needed to achieve the objective of climate neutrality by 2050. The European Green Deal includes rules on certifying carbon removals to expand sustainable carbon removals and encourage the use of innovative solutions to capture, recycle and store CO_2 by farmers, foresters and industries [2]. Carbon removal certification is proposed as a potential preamble to establishing a carbon trading system for land sector removals, as from 2030 [3]. In the USA, Assembly Bill A8597NYS Enacts the carbon dioxide removal leadership act. § 76–0103, establishing a market for certified CDR of a minimum of 0.1 M tCO2e in 2025 and which can reach up to 60 M tCO2e/year, at a maximum price of US$ 350/ tCO2e [4]. Industrial tropical hardwood timber is a clear example of a high-quality CDR, coming from forestry, one of the best available Nature Based Solutions NBS.

In the next four decades, population is expected to grow by 40%, the world economy by a factor of 3, and agricultural production by 60–100%. The capacity of land to produce biomass is one critical limiting resource, and humans can influence through inputs and management, resulting on large Net Primary Production (NPP). Biomass provides humans with food, fiber, and fuel, and generates the terrestrial carbon sink that helps to mitigate climate change. The total Human Appropriation of Net Primary Production (HANPP) grew from 13% in 1910 to 25% in 2005. Global biomass harvest ($HANPP_{harv}$) and consumption of biomass products have risen in almost perfect correlation with global population growth.[1] Over the twentieth century, human induced land use productivity was the main anthropogenic action responsible for reducing global HANPP from 2.1 to 1.6 t. Thus, less production land may be needed to supply 1 ton of biomass for human consumption. By increasing yields over the last 50 years, farmers brought cropland closer to replicating the productivity of native vegetation.[2] In cultivated pastures, in 1961 there were 29 tons of grazed biomass (dry matter) used for the production of 1 ton of animal products; by 2005, this ratio was down to 17. Asia, Africa, and Latin America experienced[3] the expansion of agriculture and cropland yield gains[4], increased from 41–69%[5]([5] By contrast, with cropland and cultivated pastures, land use efficiency of natural pasture and native forest did not increase, with a $HANPP_{luc}$ = zero for woodland and for non-degraded natural grasslands. Although the importance of rising yields has been well known, HANPP

1 "HANPP" is the total carbon produced annually by plant growth. Total HANPP, measured in units of carbon, is the sum of two subcategories: HANPPluc and HANPPharv. HANPPharv is the quantity of carbon in biomass harvested or otherwise consumed by people, including crops, timber, harvested crop residues, forest slash, forages consumed by livestock, and biomass lost to human-induced fires. HANPPluc is the change in NPP, also measured as annual carbon flow, as a result of human-induced land use change. The calculation of HANPPluc requires the estimation of the NPP that would be generated by the potential natural vegetation if vegetation were left unaltered—NPPpot. From NPPpot, we can also calculate HANPP as a percentage of the potential productivity. Global HANPP measured in GtC/y grew by 116% and by 2005 reached 14.8 GtC/y. As a percentage of the potential plant growth of native vegetation (NPPpot).

2 Which meant that HANPPluc decreased.

3 Very high growth rates in HANPP; as a percentage, HANPP doubled or even tripled in these regions during the last century.

4 Measured in HANPPharv on cropland as a ratio of NPPpot.

5 That increase, spread-out over-all cropland in 2005, generated 2.5 GtC/y of crops, which met 49% of the total increase in human consumption from 1910.

provides a useful measure of these efficiency gains because it equates all crops based on their carbon content, relates it to the productivity in global land ecosystems and, hence, demonstrates the magnitude of human-induced changes to the global carbon cycle [5].

Human induced land use improves HANPP also in the tropics. At the tropics, there is large availability of hardwoods. Tropical hardwoods are more durable, rot and marine animals resistant, stronger and cheaper than overall global hardwoods [1], which makes them standout when competing on the global markets. They have characteristic longer lifespans among timber species making them very attractable to consumers wishing lumber with special qualities such as durable, colorful, fragrant wood at their homes, offices and industries. When it comes to forestry and the tropics, the role of timber in removing atmospheric CO_2 and their transformation into industrial and energy wood might offer opportunity in accounting emission removals in tropical countries. An acknowledgement of such roles might influence national policies and decision making towards including CDR as part of goals to reach carbon neutrality and in accounting NDC (National Determined Contributions).

The consumption of tropical timber products has potentials in reducing the overall carbon footprint of construction globally, and the supply of woods with long lifespan will enhance the stocks of carbon in society or urban settlements. Tropical forest productivity is directly linked to applied silvicultural practices. In managed forests, forest biological processes react to silvicultural treatments that determine the short- and long-term productivity and stock increase or decrease. Replacing natural regeneration by human induced silviculture practices increases standing stocks and the positive effects of contemporary silvicultural techniques is improvement in harvesting volumes. Globally, about ¾ of forest plantation are established using country's native species [6]. Increasing productivity is a way to remove atmospheric CO_2 and transform it into industrial and energy wood. Both processes can be certified as CDR. This tropical industrial and energy hardwoods certified as CDRs can contribute to reduce emissions at the consumers level. Tropical wood CDRs are goods which include potential carbon credits that can, therefore, be used by consumers to reduce their overall negative GHG balance due to consumption rates and value chains.

In this study, we address the uncertainties and challenges of GHG accounting and monitoring in the forestry sector by jointly reviewing different components that may contribute to effective GHG assessment in tropical forests context, especially with consideration of local needs and spatial dynamics of land use activities, vegetation and forest transitions fueled by climate change and increasing atmospheric CO_2 stock, and aligning sustainable forest management models to both ecosystem enhancement and economic opportunities of CDR in NDCs.

2. Methodology

This work is a review of current themes under the UNFCCC process that are relevant to forestry. This review, in collating studies across different dimensions that can potentially contribution to CDR assessment and accounting in tropical forestry, investigates current trend of primary and secondary forest transitions in the context of tropical forest management and HWP, the approaches of quantifying anthropogenic activities and their contribution to CO_2 emissions and removals, the integration of sustainable forest management models with livelihood opportunities and incentives for CDRs.

Making reference to different reviewed studies in literature, which only provide different single assessments of the components of forest GHG assessment, we conduct an multi-disciplinary and quantitative review of the potential of tropical forest as CO_2 sinks in relation to nature- and made-induced changes in land use, and as well other socioeconomic needs fueled by local and global market transitions. With a target towards tropical forests, we review current themes under the UNFCCC process that are relevant to forestry. Thus, some aspects of GHG estimation in forests were not within the scope of this review. For instance, we do not include components such as atmospheric CO_2 reduction by direct removal such as CO_2 Capture and Removal and CO_2 flux measurements.

Central to this review, among others, is the increasing CO_2 fertilization of secondary vegetation growth which out-competing old forest trees, and the need for assessing CDR contribution from anthropogenic actions, which have high uncertainty and variability between local contexts and across geographic scales. Build upon reported evidence of atmospheric CO_2 enrichment of vegetation growth and transition in tropical forests, we compare a Business As Usual (BAU) scenario of native forests without management interventions versus an alternative scenario of human interference with Sustainable Forest Management practices based on contemporary silviculture and HWP production and consumption. In the following sections, we present the different themes of our integrated review. Based on these multi-dimensional components, together with qualitative and quantitative insights from the reviewed studies, we highlight potential options for both sustainable forest managements and monitoring, and as well enhancement of GHG estimations for NDC and CDR incentives.

3. Results and discussion

3.1 Tropical forest biomass and GHG monitoring

Forests have been recognized and acknowledge in both IPCC reports and the Paris Agreement to contribute substantially in achieving climate change mitigation goals [7–9]. However, it is currently challenging to spatially quantify and temporally monitor the extent to which forests impact atmospheric greenhouse gas (GHG) concentrations. In terms of atmospheric CO_2, loss (CO_2 emission) and gain (CO_2 removal) can co-occur on pixels or areas undergoing forest management or other forms of disturbance and regrowth.[6] These actions and dynamic land-use patterns occur at spatial and temporal scales not often captured by global models and estimates of GHG flux [10]. Nonetheless, estimates of GHG emissions and sinks by most developing countries, translated into NDCs, are mainly based on default emission factors that do not necessarily reflect country specifics in terms of forest structure and status of forest transitions.[7]

Global models and maps of GHG fluxes are based of inventory database that do not reliably represent the contexts in tropical forests with consideration of the high local or regional variability in forest structure and anthropogenic changes [10, 11].

[6] If we are unable to quantify them, we would not reach the goal of reliable monitoring and sustainable management. Regarding GHG fluxes, opposing fluxes simultaneously occur at local and regional scales at magnitudes that depend on the location and time of disturbance or management actions.

[7] For instance, the nature of forest degradation, the composition of intact forests, and state type of secondary forests.

Based on two decades temporal series of observational spatial data, Harris et al. [10] introduced a global spatial framework for GHG fluxes in forests of any geography. However, existing forest GHG flux assessment frameworks and models are unable to discriminated the contributions from anthropogenic versus non-anthropogenic effects and likewise, between managed and unmanaged land. To achieve such distinctions, adaptable combinations of field inventory and spatial data are needed to unravel and aggregate local to regional estimates. In the context of tropical forests, the use of spatial data from radar or synthetic aperture radar (SAR) systems have remarkable potentials in providing weather- and daylight-independent information of land features.

In spatial data modeling, remote sensing procedures offer unprecedented advantages (resources, time, and cost) in large-scale biomass and GHG estimation in forests and other land uses.[8] However, by design, current satellite-based earth observation platforms have orbiting patterns and image capturing intervals over tropical forests that provide low temporal data, which do not guaranty the parallel monitoring of anthropogenic activities within tropical forests in a timely manner. Thus, there is large time-lapse between the on-going deforestation actions and potential remote sensing data[9] to support the monitoring of both GHG emission and CO_2 removal[10] [12].

In the Brazilian Amazon for instance, most data used today are still from old studies carried out by RADAMBRASIL surveys, from the late 1950s to the early 1970s using side-looking airborne radar imagery combined with 1-ha ground plots at approximately 3000 points, often reached by helicopter. Even with these limitations, the use of the RADAMBRASIL surveys[11] is still not easily compensated for by applying more sophisticated remote sensing interpretation to a small set of ground-based plots[12] [13].

Unlike spatial data captured from optical sensors, radar sensors have the characteristic advantage of penetrating cloud cover, which is predominant over most tropical

[8] In the context of tropical forests, the application of remote sensing procedure for biomass mapping and monitoring is receiving wide attention and progress. Several compounding factors may be accountable for low rate of remote sensing application and technology transfer to tropical forest monitoring. Among these factors, technical capacity is increasingly a lesser hurdle compared the situation a decade prior. There are growing freely accessible archives of satellite-based remote sensing images (data) such as data provided from NASA Landsat missions and the operational mission of the European Space Agency (ESA) Copernicus program, which have jointly reduced the hurdle of access to remote sensing data.

[9] The closest being a minimum of 12 days across the tropics for ESA's Space-borne Sentinel operational satellites.

[10] Thus, in current times, most of the challenges in remote sensing monitoring of tropical forest GHG flux may center around the nature of available data for applications in tropical forest contexts—different data are needed for different contexts and as well in addressing the wide uncertainties for tropical forests in global projections and maps of GHG flux.

[11] It has been daunting to many research groups: the reports are a vast labyrinth of over 50,000 pages, written in Portuguese and historically with limited availability at any single location. However, ignoring this enormous body of work represents a loss that

[12] Unlike spatial data captured from optical sensors, radar sensors have the characteristic advantage of penetrating cloud cover, which is predominant over most tropical forests during seasonal monsoons and vegetation proliferation. Though radar images can potentially capture vegetation information across seasons, radar or SAR image processing workflows have been largely unreported for the myriad potential applications in tropical forests. Using satellite-based radar data.

forests during seasonal monsoons and vegetation proliferation. Though radar images can potentially capture vegetation information across seasons, radar or SAR image processing workflows have been largely unreported for the myriad potential applications in tropical forests. Using satellite-based radar data, there are increasing efforts in delineating and quantifying sectoral anthropogenic actions and land use [14, 15] and above-ground biomass [16, 17] in tropical forests.[13] Notwithstanding these growing efforts, most estimates of forest biomass and GHG emission for project-based and national assessments still rely largely on extrapolations from often scant field inventories using either allometric models or application of remote sensing data at spatial scales that mask variability across geographies and local details—the scale of most anthropogenic activities. In tropical forest landscapes, the majority of anthropogenic actions and land use changes occur widely at smallholder scales that range in spatial extent between 0 and 2 hectares; this is undermining the increasing tendencies of large-scale plantation establishment beyond the aforementioned range. Context-dependent information on anthropogenic contributions to atmospheric CO_2 emissions and removals are, therefore, needed to reliably account and aggregate impacts at a global level. Thus, multi- and cross-sectoral efforts towards climate change adaptation and achieving the objectives of the Paris Agreement should consider a per-hectare assessment, m^3 production and monitoring frameworks to match the realities and needs in tropical forests, and offer a more inclusive incentive for communities to engage in and benefit from CDR activities.

3.2 Carbon fertilization and tropical forestry NPP

Net Primary Productivity (NPP) refers to the balance between carbon gain through photosynthesis (gross primary productivity, GPP) and losses through autotrophic respiration (Ra).[14] Practically, it is not possible to precisely measure forest NPP in terms of this difference. At the ecosystem scale, NPP is measured over a long period such as a year. As per Clark et al. [18], NPP comprises new biomass produced by plants, soluble organic compounds that diffuse or are secreted into the environment such as root or phytoplankton exudation[15], carbon transfers to microbes through symbiotic association with roots as found in mycorrhizae and nitrogen-fixing bacteria, and the volatile emissions that are lost from leaves to the atmosphere. However, most field measurements of NPP consider only 'new plant biomass produced' and, therefore, probably underestimate the true NPP by at least 30%. For our practical understanding, we can say 'NPP is the net carbon gain by plants'. NPP is an important

[13] Information on forests biomass and GHG emission for project-based and national assessments still rely on making extrapolations from scant field inventories through either allometric models or application of remote sensing data at spatial scales that mask variability across geographies and local details—the scale of most anthropogenic activities. In tropical forest landscapes, the majority of anthropogenic actions and land use changes occur at a smallholder scales range in spatial extent between 0 and 2 hectares, this is without undermining the increasing tendencies of large-scale plantation establishment above the aforementioned range.

[14] Practically, it is not possible to measure precise forest NPP in terms of this difference. At the ecosystem scale, NPP is measured over a long time-interval such as a year.

[15] Carbon transfers to microbes through symbiotic association with roots as found in mycorrhizae and nitrogen-fixing bacteria, and the volatile emissions that are lost from leaves to the atmosphere. However, most of the field measurements of NPP considers only 'new plant biomass produced' and therefore probably underestimate the true NPP by at least 30%.

parameter in many forestry models that are used to assess the future mitigation potential of the sector. A forest management project can exploit NPP for carbon sequestration in forests and biomass production for climate change mitigation [19].

The increasing in human-induced CO_2 emissions indirectly implies that forest worldwide will grow faster and reduce the amount of atmospheric CO_2 which stays airborne—an effect known carbon fertilization, which is high in the tropics. There has been increasing carbon sink on land since the 1980s; living woody plants were responsible for more than 80% of the sources and sinks on land[16,17] [20]. Globally, vegetation is locking away more carbon as atmospheric CO_2 levels rise. Plants are growing faster, fueled by a more CO_2-fertile atmosphere. Carbon stocks (i.e., standing plant production) are related to, but do not refer to the same thing as, NPP (i.e., rate of growth of plant production). Increased CO_2 concentrations reduce photorespiration, which translates into greater plant productivity—NPP.[18] Giving plants more CO_2 increased net primary productivity by 24% on average.[19] Factorial simulations with multiple global ecosystem models suggest that CO_2 fertilization effects explain 70% of the observed greening trend.[20] CO_2 fertilization effects explain most of the greening trends in the tropics. Results show a considerable increase in net primary production (NPP) over the last century, mainly due to the CO_2 fertilization effect.

Pastures uptake 5–50 tCO_2-eq/ha/year of atmospheric CO_2 and also hold Nitrogen, which is turned by each animal into something like 0,5 tCO_2-eq/year of carbon-based products—protein.[21] The associated methane emissions is a result of the balance between the atmospheric CO_2 removed by pastures and what gets process through animals' digestive system. The more animals are grazing, the more atmospheric CO_2 is turned into protein and other products—including fertilizers [21], resulting on a process of removing the gas and returning into Society as useful goods.

In identifying potential plant species with higher NPP capabilities for carbon sequestration, Vithal and Nadagoudar [19] found that Bamboo has the highest NPP(17.523).[22] Secondary vegetation like Lianas, palms, bamboo and other non-tree life forms have been omitted from a number of Amazonian biomass studies, which often fail to report what biomass components are included[23] [13]. For Brazil as a whole, an average aboveground carbon stock of 120 tCO2e/ha in savanna woodlands

[16] with soil, leaf litter, and decaying organic matter making up the rest. But they also saw that vegetation retained a far smaller fraction of the carbon than the scientists originally thought.

[17] Globally, vegetation is locking away more carbon as atmospheric CO_2 levels rise. Plants are growing faster, fueled by a more CO_2-fertile atmosphere. Carbon *stocks* (i.e., standing plant production) are not the same thing as NPP (i.e., rate of growth of plant production).

[18] although warmer temperatures counteract this effect by increasing photorespiration somewhat

[19] Terrestrial Ecosystem Model (TEM) predicts that doubled CO_2 will increase 16.3% of the global NPP. Under real conditions on the large scale where water and nutrient availability are also important factors influencing plant growth, experiments show increases under unstressed conditions.

[20] followed by nitrogen deposition (9%), climate change (8%) and land cover change (LCC) (4%).

[21] while returning fertilizers and gases to the environment.

[22] followed by rubber (15.970), oil-palm (14.500), *Samanea* and *Erythrina* (13.350), coconut (12.150), *cassia* (10.350), *eucalyptus* (10.009), alnus (10.000), *sesbania* (9.433), *prunus* (9.000), *leucaena* (8.739), *acacia* (9.000) and *casuarinas* (7.550).

[23] Standardization for non-tree components, together with trees <10 cm DBH, removes almost all of the difference between aboveground live biomass.

classified as "forestland"[24], and 45 tCO_2-eq/ha in those classified as "shrublands" (65.6% of the area), giving a weighted average of 75 tCO_2-eq/ha[25] [13].

Net primary productivity (NPP) of a closed-canopy[26] forest stand was assessed for three years in a free-air CO_2-enrichment (FACE) experiment. NPP increased 21% in stands exposed to elevated CO_2, and there was no loss of response over time. Wood increment rate cumulated significantly during the first year of exposure, but subsequently return to its initial value, reducing the potential of the forest stand to sequester additional C in response to atmospheric CO_2 enrichment[27] [22]. Currently, there is limited pool of knowledge regarding the long-term impacts of CO_2 enrichment in tropical rainforests [22]. Young trees and other small plants responded well to higher CO_2, but it remains undetermined how more mature trees would react. Brazilian Amazonian trees are dying faster than they are growing. On land, reports suggest a decline in the tropical forest CO_2 sink, increased plant mortality and decreased plant productivity. Under low Nitrogen conditions[28], plants will have difficulties to transform elevated CO_2 into production[29] [5].

Standing undisturbed tropical forest sites over the last 50 years lost total volume of trees to secondary invasive vegetation, making them naturally net emitters of CO_2. As regards above-ground live biomass and carbon flux, the world's remaining intact tropical forests have been reported to be largely out-of-equilibrium [23, 24]. Following the current increasing trend of forest degradation over deforestation [25, 26] and dwindling resilience to changing climates and rainfall patterns [27, 28], it is anticipated in the next 50 years that tropical forest sites are to yet another part of its volume stocks to the increasing competition from secondary vegetation. With this trend, large protected areas at isolated areas in the Amazon region that are retained in unmanaged conditions should hold less biomass volume yearly than their managed and plantation counterparts. The CO_2 fertilization up-take is much faster by secondary vegetation, and old forests trees are losing their competitiveness every year, without harvesting and silvicultural treatments.

Today[30], spatial biomass analyses[31] show major differences between all of the resulting biomass maps, including those with largely overlapping ground-based

---

[24] 34.4% of the total savanna woodland area.

[25] Conversion from the original text in Mgha-1 using 1:1 ratio for m, and 3,67 factors for C-CO_2.

[26] Liquidambar styraciflua (sweetgum).

[27] Most of the extra C was allocated to production of leaves and fine roots. These pools turn over more rapidly than wood.

[28] CO_2 may not much affect plant productivity because of lack of Nitrogen in the soil. Plant acclimatization and water availability.

[29] Moreover, in the long term, elevated CO_2 condition may cause the accumulation of carbohydrates in the plant tissues which may reduce the photosynthetic rates or decrease photosynthetic response to elevated CO_2.

[30] Usually, continuous Forest Inventory data, with a proportion of 0.1% (for the effective area) of sampling, is used to determined standing stocks volumes from which biomass estimates are made. A number of fixed size plots of 10 meters wide by 250 meters long, used for monitoring tree increment and mortality.

[31] Using space-borne LiDAR (Light Detection and Ranging) from the US National Aeronautics and Space Agency (NASA) Geoscience Laser Altimeter System (GLAS) on the Cloud and Land Elevation Satellite (ICESat), together with optical data from MODIS imagery and radar data from the Global Quick Scatter meter (OSCAT).

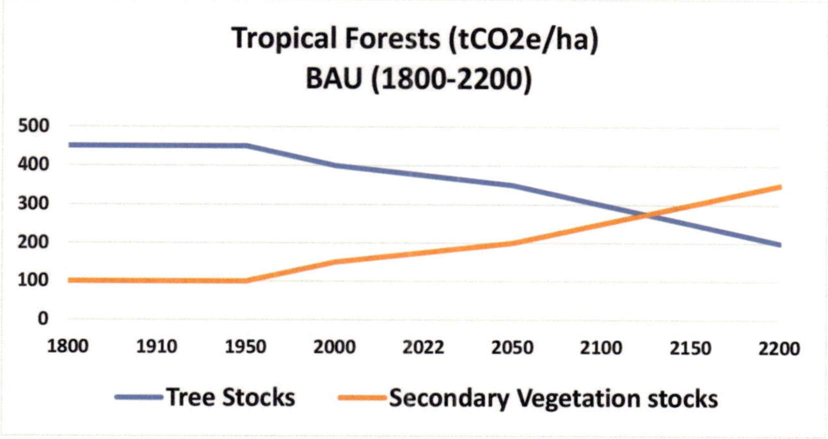

Figure 1.
Tropical forests standing stocks BAU scenario over 400 years (Illustrative proxy by authors).

datasets.[32] The way forward will require using remote sensing data together with ground-based measurements, with progress needed in both areas [13]. From RADAM Brasil studies, the highest dense tropical forest at the Amazon region used to hold average between 420 and 480 m^3/ha, toping 520–580 m^3/ha of tree biomass [29]. Nowadays, average biomass of standing stocks range from 248.92 ± 61.78 t/ha, passing by 293.19 ± 27.74 t/ha, and reaching up to 356 ± 47 t/ha [30, 31], based on measurements for trees ≥10 cm DBH (diameter at breast height: diameter at 1.3 m above the ground or above any buttresses) with a 12% correction for small trees [13]; roughly, making it 270 to 320 m^3/ha and topping 310 to 400 m^3/ha[33], circa of 25 to 35% less volume than 50 years earlier, as stated at RADAM Brasil early studies.

As illustrated in **Figure 1**, following the current BAU (Business As Usual) scenario and considering a 400 years' time frame, tropical forests are going to become less and less tree covered as the atmospheric CO_2 levels rises and if no management interventions are implemented (the data is illustrative, a proxy from the previous findings and this section highlighting increasing secondary vegetation in natural unmanaged tropical forests).

The graph in **Figure 1** illustrates that secondary vegetation gains competitiveness over trees as the atmospheric CO_2 becomes more and more available, reducing the overall carbon stock of forest stands. The process is ongoing and tends to speed up with the increase of CO_2 and reduction of tree cover, which favors even more secondary vegetation growth. The associated gains in productivity of secondary vegetation can be can be compared to those from Croplands. The Brazilian agricultural sector, for instance, has portrayed continuous productivity increase over the last 30–40 years [18], showcasing the positive effect of atmospheric CO_2 enrichment on plant NPP. Meanwhile, the ability of tropical forests trees to absorb massive amounts of carbon has waned [32].

SFM + HWP = CDR.

Degradation is translated as: "change between forest classes (i.e. from "close" to "open") which negatively affects the site and, in particular, reduces its productivity

---

32 Expanding the network of ground-based inventories is essential.

33 Assuming1:1 ratio from biomass to m,

capacity.[34] The Intergovernmental Panel on Climate Change—IPCC2006 guidelines for GHG inventories from different sectors includes accounting procedures for Dead Wood—DW and Harvested Wood Products—HWP.[35] Thus, wood used for project activities such as fencing of boundaries, furniture, construction, energy and others must be accounted as DW when determining the carbon sequestration and storage in forest areas, including from those without a formal Sustainable Forest Management Plan—SFMP. For areas holding SFMP, the rule is the same regarding DW, and besides this logs, timber, firewood and others imports and exports are also to be accounted for as HWP for the balance of forest carbon areas, carbon sequestration and carbon storage [33]. At harvesting, a large portion of aerial biomass carbon is transferred to HWP (Harvested Wood Products) and will be available at one of the forest product categories. Forest areas biomass volume is used as starting point for HWP carbon estimates, applying specific conversion factors for each log destination. Estimates related to wood products baseline are available under the format of volumes delivered to industrial plants or in terms of their outputs, comprising industrial logs or primary HWP (boards, planks, panels or paper). Carbon availability at those HWP over the years is then estimate allocating other parameters which indicate carbon amount 'in use' and destined to landfills. Thus, HWP Carbon estimates, including recycling, rely largely on data availability.[36]

3.3 Improved forest management in the tropics

Tropical forests are accountable for about 35% of global net primary productivity (NPP).[37] The CO_2 fertilization effect that increases CO_2 concentrations in leaves enhances plants' capacity in fixing carbon through photosynthesis has been considered as a primary mechanism that maintains and enhances tropical forest productivity [34].

The human appropriation of net primary production (HANPP) provides a useful measure of human intervention into the biosphere. The productive capacity of land is appropriated by harvesting or burning biomass and by converting natural ecosystems to managed lands. HANPP has still risen from 6.9 Gt of carbon per y in 1910 to 14.8 GtC/y in 2005, i.e., from 13 to 25% of the net primary production of

[34] Deforestation means: "changing on land use with reduction of tree crow cover below 10% by hectare" while resulting in land degradation afterwards according to IPCC.

[35] Within IPCC2006 Dead Wood (DW) is classified as all kinds of branches, leaves, roots, dead trees and other types of biomass not included as litter or soil. Harvested Wood Products (HWP) are all wood material leaving project activities boundaries—other materials remaining within boundaries are to be accounted as DW.

[36] Estimates of forest products contribution, in terms of carbon, use generic variables, including (i) domestic HWP and imports (tCO_2-eq/year); (ii) annual variation of HWP produced domestically, including annual variations on exported HWP (tCO_2e/year); (iii) annual imports of all kinds of wood and paper (tCO_2e/year); (iv) annual exports of all kinds of wood and paper (tCO_2e/year); and (v) annual HWP (tCO_2e/year). The level of lost on solid products and paper, in a given year, are specified towards the use of a lost constant (k), which by convenience is expressed in terms of half-life in services, in years. Half-life in service describes the number of year necessary for half of the material to change environment, which can be, for example, from a home to landfill, within that sector where it remains stored. Solid wood and paper production, imports and exports are converted from m or tons into tCO_2-eq. For annual estimates calculation the method uses yield data (Consumption = Domestic Production + Imports—Exports).

[37] And store about 72% of global forest biomass carbon (C).

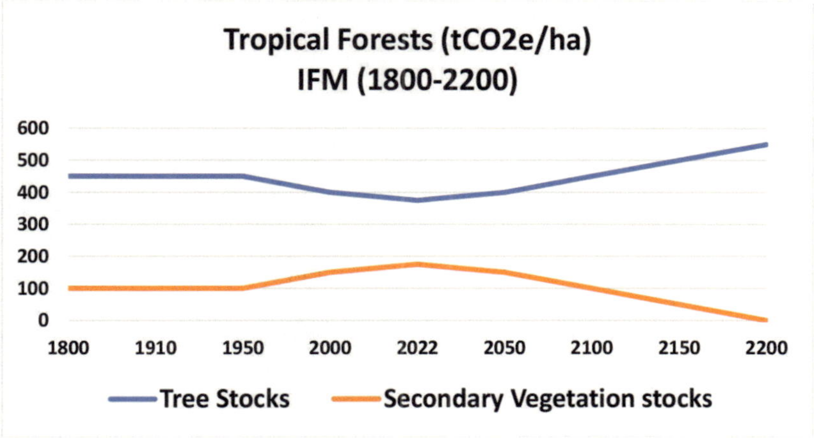

Figure 2.
Tropical forests standing stocks IFM scenario over 400 years (Illustrative proxy by authors).

potential vegetation. Biomass harvested per capita and year has slightly declined despite growth in consumption because of a higher conversion efficiency of primary biomass to products.[38] The rise in efficiency is overwhelmingly due to increased crop yields. HANPP might only grow to 27–29% by 2050, but providing large amounts of bioenergy could increase global HANPP to 44%. This result calls for strategies that foster continuous and increasing land-use efficiency.

Harvesting and consumption of tropical timber products stimulates management opportunities of increased productivity, reverting the degradation process due to increase of secondary invasive species volume, and generate profits. With a profitable forestry activity in place, there is incentive to practice forestry and reduce conversion to other land uses. Therefore, tropical timber is a value added CDR that can reduce forest degradation and conversion of forests to other land uses, while increasing CO_2 removals. Advanced silvicultural techniques can be applied to improve productivity [6], taking advantage of the CO_2 fertilization. As illustrated in **Figure 2**, following Improved Forest Management (IFM) contemporary silviculture techniques, the scenario considering a 400 years' time frame shows tropical forests recovering tree volume against the competing secondary vegetation.

Silvicultural practices—planning; individuals' selection; seed collection, genetic improvement; seedling development; fertilization; maintenance; weed, insects and diseases control; harvesting—are applied to reduce the presence of secondary vegetation and introduce CO_2 enriched environment with adapted trees` varieties. This will result in increasing yields and therefore reducing HANPP while supplying society with more industrial, energy wood and other Non-Timber Forest Products (NTFP). The positive effect of IFM techniques are widely known globally, and these have promoted the cultivation of native tree species all over the world [6].

Brazil holds the largest stock of hardwoods in the planet. Some of these tropical hardwoods have characteristics that make them therapeutic, comfortable, charming as well as immune to fungi and insect attacks. Brazilian tropical hardwoods, just as softwoods, possess a diversity of qualities that are hardly reachable by tree species in other parts of the world. These unique qualities are competitive advantages that can be used to enhance Amazon biodiversity cultivation and tropical timber consumption

[38] And decline in reliance on bioenergy.

role. With a growing and promoted timber consumption, rural landholders have markets available to justify necessary investments on Brazilian native tropical timber species cultivation, with the use of IFM[39] [33].

3.4 Quantifying anthropogenic contributions to GHG flux

Following the UNFCCC global estimates of anthropogenic net land-use emission, there is a discrepancy of about 4 Gt CO_2 per year between aggregated national GHG inventories and the global models in IPCC assessment reports. According to Grassi et al. [35] a great proportion of the discrepancy (3.2 Gt CO_2 pr^{-1}) is due to differences in concepts implemented in estimating anthropogenic forest sinks—a representation of environmental changes and managed areas. Such differences between inventories and models of GHG emissions and sinks need to be addressed [36] to enhance monitoring and achieve collective progress towards the goal of the Paris Agreement on global temperature.[40] Following global estimate of GHG fluxes from forest, uncertainties in global gross removals and net flux are mostly attributable to extremely high uncertainty in applying the removal factors from the IPCC Guidelines to old secondary forests[41,42] [10]. Global models and maps of GHG emissions and fluxes are based on spatially clustered inventories that are translated to make predictions across geographies and forest types [37]. Although they provide important insights on global trends, such models may support misleading applications such as using some default values in IPCC recommendations [38] management actions and policy as they do not capture variability and uncertainties across geographies and generalize assumptions of carbon flux to unknown local and regional spaces [12]. Globally, forests store approximately 8.4 billion tCO_2-eq and are capable of retaining some further billions; meanwhile, about 4.2 to 20 billion tCO_2-eq are estimated to be stored within HWP "in use".[43] The 3.4 billion m^3 of yearly global harvested wood is equivalent to just 20% of total yields (some 17 billion m^3/year) [33]. A lot from what is harvested is used for direct and inefficient burning as fuel wood. Increasing the sustainable removal of senescing biomass from forests and harvesting yields would have a profound positive effect to fight global warming. With the use of extra 2 billion m^3/year, industrial woods will be possible to reduce between 14 and 31% of all cement and steel GHG emissions and between 12 to 19% of all fossil fuel consumption by the use of residues from industrial wood production chains for clean energy appliances. With the intensification of sustainable forest management, more CO_2 is sequestered and stored

[39] Biodiversity banking regional strategies implementation and the use of contemporary industries (MDF, HDF etc.) value aggregation will increase social inclusion chances and, by that, project activity sustainability over time.

[40] By and large, inventory data is scarce or absent for tropical forests, and there is large variability in the methodology for and quality of existing data.

[41] To make estimates at large scales, inventory (activity) data vital in making extrapolation from information and models based on spatially continuous data collected from airborne or satellite-based remote sensing procedures.

[42] The absence of 'activity data' constitutes a key impediment to and source of error (over- or under-estimation) in estimating GHG emissions and CDR in tropical forests.

[43] World wood production includes more than 1.5 billion m/year of industrial logs, accounting for something like 1.1 billion tCO_2-eq/year, with 420 million m of sawed lumber and 220 million m on plywood and panels—representing some 20% of total in long life-spam forest products, which sequester and store close to 200 million tCO_2-eq each year.

avoiding emissions from alternative materials and still producing renewable energy from harvesting residues. Besides, harvested volumes are renewed. Brazil has by far the largest global stock and growth of "hardwoods", which have the longest life-span between tree species, making them relevant suppliers of HWP storing carbon for many years.

The International Wood Culture Society (IWCS) is a non-profit organization formed by wood enthusiasts, dedicated to research, education and promotion of wood culture. IWCS advocates for a harmonious living between people and nature, explores the value of wood use from a cultural perspective and supplies a platform for studying wood culture, encouraging its practice and promotion.[44] IWCS established March, 21st as Wood World Day, a data to disseminate the value wood aggregates to daily life [33]. Tropical forestry must be accompanied by similar public and private efforts towards trade and use of tropical hardwoods, creating the synergies that might help in removing huge amounts of atmospheric CO_2 and returning to society noble wood products. The current stocks of billions of m^3 of dying mature trees, ready to be harvested on Brazilian Amazon region alone, have the capacity to remove billions tCO_2-eq from the atmosphere, just by turning them into timber and having new trees planted. Thus, cutting down trees do not necessarily implicate on GHG emissions, and neither is the change of land use directly linked to atmospheric CO_2 generation. The use of wood could also generate millions of jobs and trillions of dollars in revenue over the next decades. Tropical forests hold capacity to regenerate after harvesting, and the magnitude and benefits that this capacity would mean is directly related to silvicultural practices, which will impact global GHG balance positively with broad use of tropical HWP.

3.5 Tropical forestry and the certification of HWP: CDR

As the world will face, in the next few decades, further increase in global population and economic output resulting on large new demands for food, fuel and fiber, this stresses the importance of developing improved practices for sustainable intensification of land use. Production of CDR from increasing forest and HWP atmospheric CO_2 removals, at the same, copes with reducing emissions targets, which makes it a highly competitive credit for global carbon markets. CDR production also represents a significant opportunity to private investors on engaging in Environmental and Social Governance (ESG) activities and into the international carbon markets. Registered carbon credits can supply an income source for landowners, support rural development and facilitate IFM implementation. Logs produced to supply industries with sustainable sources can receive payments directed to improve technology of silviculture, trade and finance towards inclusion of payments for carbon credits. When tropical timber used by society comes from sustainable origins, it increases forestlands atmospheric CO_2 removal capacity.

Production and consumption of tropical timbers need to be within the framework of accepted CDR for global carbon Market development within countries 'National

[44] "Tackle Climate Change: Use Wood" is a European Parliament program directed to strength societal use of wood as a way of fighting atmospheric CO2 accumulation. France has "de Bois-Construction-Environment", England the "Wood for Good", Netherlands "Centrum Hout", Denmark "Trae Information", Finland "Puuinfo", Belgium "Wood Forum", Spain's "Viver Con Madera", Australia "Wood Naturally Better" and Austria and Italy "Promo Legno" are few from national, binational and multilateral networks for the promotion of wood use as a form of global climate change mitigation.

Determined Contribution (NDC) to UNFCCC. Countries around the globe could include tropical timber products as CDRs and purchase these credits as part of the acceptable contributions—Internationally Tradeable Mitigation Opportunities (ITMO) to alter forest degradation and land use change. With Tropical HWP accepted as CDR, global carbon markets can promote increasing carbon stocks within society as a way to reduce global GHG emissions (from cement, iron etc.) while increasing removal of atmospheric CO_2 at the same time. The more tropical timber is sustainably consumed, the better the potentials for the climate. The same goes for all tropical agriculture and pastures products, which are carbon-based products resulting from up-take of atmospheric CO2 and its conversion into useful goods for humanity.

The Bio-economy of Brazilian Amazon ecosystems sustainable management rely on technological interventions. With investments directed to appropriate silvicultural technologies, national wood products from Brazilian native tropical timbers will be highly competitive at international Green Economy markets. Brazilian tropical timber species diversity, productivity and qualities being cultivated under contemporary silvicultural techniques are capable of placing native forest sector among world's greatest. Native forest species biodiversity cultivation ,contributed by the use of Brazilian woods, will be a direct result from consumption incentives. National regulations must incentivize the use and consumption of native timber from sustainable sources as a way of assuring the sustainability of forest biodiversity cultivation.

4. Conclusion

Over the next decades, there will be an increase in the global biomass for biomass and GHG emissions` reduction and removals, and intensification of land use is the most promising solution—together with processing efficiency—for balancing HANPP consumption with NPP from atmospheric CO_2 fertilization. Forest plantations, croplands, cultivated pastures, lianas, palms and other secondary vegetation have shown yield gains from CO_2 fertilization, while trees respond somehow at first, losing the capacity afterwards.

There is evidence showcasing a path of native tropical forest degradation given atmospheric CO_2 fertilization, which is mainly due to favoring secondary vegetation competitiveness against trees at un-managed standing stocks. Following the BAU scenario, tropical forest should become less and less covered with trees over the next century. An alternative IFM scenario is perceived, where IFM plus contemporary silvicultural techniques can reverse the process and produce HWP and NTFP as a result of land use intensification. This will generate additional atmospheric CO_2 removals, certifiable as CDR goods, which are able to generate carbon credits for financing the reduction of secondary vegetation and promote cultivation of improved native tree species. These CDR credits can be included in tropical countries` NDC and presented at UNFCCC as an ITMO for fighting global climate change.

Author details

Ederson Augusto Zanetti[1*], Frederick N. Numbisi[2], Vithal Karoshi[3], Roberto Rochadelli[4], Allan Sbardelotto[5], Joesio Siqueira[6] and Alain Levy Boussamba[7]

1 Green Farm CO2FREE, Curitiba, Parana, Brazil

2 Independent Researcher, Freiburg in Breisgau, Auggen, Germany

3 Lilongwe University of Agriculture and Natural Resources (LUANAR), Lilongwe, Malawi

4 Federal Paraná University (UFPR), Coordinator of "MRV-FUPEF Programm", Curitiba, Parana, Brazil

5 Independent Researcher, Rio Branco, Acre, Brazil

6 STCP Engenharia de Projetos, Curitiba, Parana, Brazil

7 ONG Humanitas, Libreville, Gabon

*Address all correspondence to: eder.zanetti@fulbrightmail.org

References

[1] Aggarwala RT. Tropical Hardwood Reduction Plan. Memorandum to Mayor Michael R. Bloomberg. New York: Long-Term Planning and Sustainability Office, NYC; 2008. p. 19. Available from: http://www.nyc.gov/html/om/pdf/tropical_hardwoods_report.pdf

[2] European Commission Certification of Carbon Removals—EU Rules. Bruxells, Belgium: European Commission; 2022. Available from: https://ec.europa.eu/info/law/better-regulation/have-your-say/initiatives/13172-Certification-of-carbon-removals-EU-rules_en

[3] EU Parliament (2022) Legislative Proposal on Carbon Removal Certification / before 2023-1. p. 2. Available from: https://www.europarl.europa.eu/legislative-train/carriage/carbon-removal-certification/report?sid=6301

[4] New York State (2022) Assembly Bill A8597. Enacts the Carbon Dioxide Removal Leadership Act. p. 10. Available from: https://www.nysenate.gov/legislation/bills/2021/A8597

[5] Krausman F, Erb K-H, Gingrich S, Haberl H, Bondeau A, Gaube V, et al. Global human appropriation of net primary production doubled in the 20th century. PNAS. 2013;**110**(25):10324-10329. In Tomasik, Brian. (2018) Effects of CO_2 and Climate Change on Terrestrial Net Primary Productivity. Available from: https://reducing-suffering.org/effects-climate-change-terrestrial-net-primary-productivity/

[6] Zanetti EA. Indicators for Sustainable Forest Management: Brazilian Amazon within Global Scenery—NEA. Saarbrücken, Germany: Edições OmniScriptum GmbH & Co. KG; 2015, 2015

[7] IPCC. In: Core Writing Team, Pachauri RK, Meyer LA, editors. IPCC Climate Change 2014: Synthesis Report. Bonn, GE: IPCC; 2014

[8] IPCC. IPCC Special Report on Climate Change, Desertification, Land Degradation, Sustainable Land Management, Food Security, and Greenhouse Gas Fluxes in Terrestrial Ecosystems. Bonn, GE: IPCC; 2019

[9] UNFCCC, 2015. Adoption of the Paris Agreement FCCC/CP/2015/10/add.1

[10] Harris NL, Gibbs DA, Baccini A, et al. Global maps of twenty-first century forest carbon fluxes. Nature Climate Change. 2021;**11**:234-240. DOI: 10.1038/s41558-020-00976-6

[11] Saatchi SS, Harris NL, Brown S, Lefsky M, Mitchard ET, Salas W, et al. Benchmark map of forest carbon stocks in tropical regions across three continents P. Proceedings of the National Academy of Sciences. 2011;**108**:9899-9904. DOI: 10.1073/pnas.1019576108

[12] Houghton RA, House JI, Pongratz J, van der Werf GR, DeFries RS, Hansen MC, et al. Carbon emissions from land use and land-cover change. Biogeosciences. 2012;**9**:5125-5142. DOI: 10.5194/bg-9-5125-2012

[13] Fearnside PM. Brazil's Amazonian forest carbon: The key to southern Amazonia's significance for global climate. Regional Environmental Change. 2018;**18**:47-61. DOI: 10.1007/s10113-016-1007-2

[14] Numbisi FN, Van Coillie FMB, De Wulf R. Delineation of cocoa agroforests using multiseason Sentinel-1 SAR images: A low Grey level range reduces

uncertainties in GLCM texture-based mapping. ISPRS International Journal of Geo-Information. 2019;**8**:179. DOI: 10.3390/ijgi8040179

[15] Kalischek N, Lang N, Renier C, Daudt RC, Addoah T, Thompson W, et al. Satellite-Based High-Resolution Maps of Cocoa Planted Area for Côte d'Ivoire and Ghana; Ithaca, NY: Cornell University, 2022. arXiv:2206.06119 [cs. CV]. DOI: 10.48550/arXiv.2206.06119.

[16] Macave OA, Ribeiro NS, Ribeiro AI, Chaúque A, Bandeira R, Branquinho C, et al. Modelling aboveground biomass of Miombo woodlands in Niassa special reserve, northern Mozambique. Forests. 2022;**13**:311. DOI: 10.3390/f13020311

[17] Migolet P, Goïta K, Pambo AFK, Mambimba AN. Estimation of the total dry aboveground biomass in the tropical forests of Congo Basin using optical, LiDAR, and radar data. GIScience & Remote Sensing. 2022;**59**(1):431-460. DOI: 10.1080/15481603.2022.2026636

[18] Clark Deborah A, Brown S, Kicklighter DW, Chambers JQ, Thomlinson JR, Ni J. Measuring net primary production in forests: Concepts and field methods. Ecological Applications. 2001;**11**(2):356-370

[19] Karoshi VR, Nadagoudar BS. Forest plantations for climate change mitigation—Reviewing estimates of net primary productivity in Forest plantations. Indian Journal of Agricultural Economics. 2012;**67**(1):157-162

[20] Reiny S. Carbon Dioxide Fertilization Greening Earth, Study Finds. 2019. Available from: https://www.nasa.gov/feature/goddard/2016/carbon-dioxide-fertilization-greening-earth

[21] Grise MM. PARTIÇÃO DA BIOMASSA E DE NUTRIENTES NA PASTAGEM DE BAHIAGRASS cv. PENSACOLA EM DIFERENTES SISTEMAS DE MANEJO COM NOVILHOS. Thesis presented to the Postgraduate Course in Agronomy, Area of Concentration in Plant Production, Department of Plant Science and Phytosanitary, Sector of Agrarian Sciences, Federal University of Paraná, as part of the requirements for obtaining the title of Doctor in Sciences. Advisor: Prof. Dr. Adelino Pelissari. 2005, 150

[22] Norby RJ, Hanson PJ, O'Neill EG, Tschaplinski TJ, Weltzin JF, Hansen RA, et al. Net primary productivity of a Co2-enriched deciduous Forest and the implications for carbon storage. Ecological Applications. 2002, 2002;**12**(5):1261-1266 q 2002 by the Ecological Society of America. Available from: https://people.ucsc.edu/~wxcheng/2002%20Norby%20et%20al.%20EA.pdf

[23] Qie L, Lewis SL, Sullivan MJP, et al. Long-term carbon sink in Borneo's forests halted by drought and vulnerable to edge effects. Nature Communications. 2017;**8**:1966. DOI: 10.1038/s41467-017-01997-0

[24] Boulton CA, Lenton TM, Boers N. Pronounced loss of Amazon rainforest resilience since the early 2000s. Nature Climate Change. 2022;**12**:271-278. DOI: 10.1038/s41558-022-01287-8

[25] Qin Y, Xiao X, Wigneron JP, et al. Carbon loss from forest degradation exceeds that from deforestation in the Brazilian Amazon. Nature Climate Change. 2021;**11**:442-448. DOI: 10.1038/s41558-021-01026-5

[26] Feng Y, Zeng Z, Searchinger TD, et al. Doubling of annual forest carbon loss over the tropics during the early twenty-first century. Nature Sustainability. 2022;**5**:444-451. DOI: 10.1038/s41893-022-00854-3

[27] Ciemer C, Boers N, Hirota M, et al. Higher resilience to climatic disturbances in tropical vegetation exposed to more variable rainfall. Nature Geoscience. 2019;**12**:174-179. DOI: 10.1038/s41561-019-0312-z

[28] Smith T, Traxl D, Boers N. Empirical evidence for recent global shifts in vegetation resilience. Nature Climate Change. 2022;**12**:477-484. DOI: 10.1038/s41558-022-01352-2

[29] Saatchi SS, Houghton RA, Dos Santos Alval RC, Soares ÁJV, Yu Y. Distribution of aboveground live biomass in the Amazon basin. Global Change Biology. 2007;**13**(4):816-837

[30] Santos FG, Camargo PB, Oliveira RCJ, Santos DB, Oliveira DR. ESTOQUE E DINÂMICA DE BIOMASSA ARBÓREA EM FLORESTA OMBRÓFILA DENSA NA FLONA TAPAJÓS: AMAZÔNIA ORIENTAL. Congresso Técnico Científico da Engenharia e da Agronomia CONTECC'2018 Maceió -AL 21 a 24 de agosto de 20182018. p. 5. Available from: https://www.confea.org.br/sites/default/files/antigos/contecc2018/agronomia/121_eeddbaefodnftao.pdf

[31] Santos FG, Camargo PB, Oliveira RCJ. ESTOQUE E DINÂMICA DE BIOMASSA ARBÓREA EM FLORESTA OMBRÓFILA DENSA NA FLONA TAPAJÓS: AMAZÔNIA ORIENTAL. Ciência Florestal, Santa Maria. 2018;**28**(3):1049-1059. DOI: 10.5902/1980509833388

[32] NASA (2021) NASA Study Finds Tropical Forests' Ability to Absorb Carbon Dioxide Is Waning. 2021-151. JPL/CIT. Available from: https://www.jpl.nasa.gov/news/nasa-study-finds-tropical-forests-ability-to-absorb-carbon-dioxide-is-waning

[33] Zanetti EA. Wood is good for REDD+. In: Pandey K, Ramakantha V, Chauhan S, Arun Kumar A, editors. Wood Is Good. Singapore: Springer; 2017. DOI: 10.1007/978-981-10-3115-1_41. Print ISBN978-981-10-3113-7. Online ISBN978-981-10-3115-1.

[34] Wang Z, Tian H, Pan S, Shi H, Yang J, Liang N, et al. Phosphorus Limit to the CO2 Fertilization Effect in Tropical Forests as Informed from a Coupled Biogeochemical Model. Poster: Auburn University; 2021. Available from: https://cce-datasharing.gsfc.nasa.gov/files/conference_presentations/Poster_Wang_0_159_21.pdf

[35] Grassi G, House J, Kurz WA. et al. Reconciling global-model estimates and country reporting of anthropogenic forest CO_2 sinks. Nature Climate Change. 2018;**8**:914-920. DOI: 10.1038/s41558-018-0283-x

[36] Hansis E, Davis SJ, Pongratz J. Relevance of methodological choices for accounting of land use change carbon fluxes. Global Biogeochemical Cycles. 2015;**29**:1230-1246

[37] Zhu K, Zhang J, Niu S, et al. Limits to growth of forest biomass carbon sink under climate change. Nature Communications. 2018;**9**:2709. DOI: 10.1038/s41467-018-05132-5

[38] Cook-Patton SC, Leavitt SM, Gibbs D, et al. Mapping carbon accumulation potential from global natural forest regrowth. Nature. 2020;**585**:545-550. DOI: 10.1038/s41586-020-2686-x

Chapter 4

Damaged Trees Caused by Selective Thinning in Two Tropical Mountain Rain Forest Types in Southern Ecuador

Omar Cabrera, Pablo Ramón, Bernd Stimm, Sven Gunter and Reinhard Mosandl

Abstract

The proportion of damaged trees and the type caused by the selective thinning can have serious impacts on the ecological and economic sustainability of forests. So far, the damage caused by thinning operations in montane tropical forests is unknown. In this study, we try to quantify the damages caused by selective thinning in two types of humid montane forests in southern Ecuador. For this, we installed 52 permanent plots of 50 m x 50 m in two force types. In the valley forest (VF), we extracted an average basal area of 2.75 m^2 ha^{-1}, in the ridge forest (RF) an average basal area of 0.8 m^2 ha^{-1} was extracted. For each felled tree, we counted and categorized the damage separated by large (>20cm DBH) and small (<20cm DBH) residual trees. Using a generalized linear model (GLM) we could determine that the diameter of the felled tree significantly influences the number of large as well as small trees temporarily or permanently damaged. Basal area thinned in the VF significantly influenced the proportion of trees < 20 cm DBH affected by Permanente Damage (PD) and Temporal Damage (TD). In the RF, the crown area of felled trees influences the proportion of (PD) trees with DBH > 20 cm. The extracted basal area significantly influences the proportion of (TD) trees with DBH < 20 cm. In general, the proportion of temporarily damaged (TD) trees is greater than the proportion of permanently damaged (PD) trees in both types of forest. Considering only the heavily damaged trees we can conclude that the damage in total is acceptable.

Keywords: tropical mountain rain forest, sustainable forest management, selective thinning, damage type and probability, damaged trees

1. Introduction

Currently, sustainable forest management (SFM) practices in some tropical forests are poorly implemented for several reasons, including high cost, lack of government incentives to promote sustainable practices, conversion of exploited (unmanaged) forests into pastures, and land for agriculture between others [1, 2].

Some SFM practices include activities, new methods, and technology development for the sustainable use of forest resources and to minimize any negative impact on the environment [3]. Within these activities and as an important component to achieve ecological sustainability, thinning and harvesting operations must be planned to avoid the negative impact of forestry operations on the soil, on regeneration, and on the residual mass [3, 4].

Activities such as directed felling and optimized skidding methods have shown that the percentage to remaining forest has decreased [5]. We should also mention that the size of felled trees and the size of residual trees affect the amount and types of damage to residual trees, i.e., larger felled trees cause more damage to residual trees [6, 7].

To make SFM compatible with biodiversity conservation, the impact of these activities on the remaining forest must be evaluated. The damage caused by felling and thinning on residual trees should be considered one of the fundamental factors to assess the sustainability of forest management. Nevertheless, no study has predicted different degrees of logging damage with consideration of the size of both residual and felled trees simultaneously for tropical forests [8].

General damage (severe or light) caused to residual trees could jeopardize the goal of selective thinning and show a higher probability of mortality for severely damaged trees [9]. In some cases, thinning logging damage to residual trees can result in their death, and the frequency of wounded trees and severity of wounds during logging operations may have detrimental impacts on stand growth and forest sustainability [10].

In order to know the magnitude in which the mechanical damage caused by selective thinning affects the remaining forest, it is necessary to determine the levels of damage that could be determined as indicators of the intensity of thinning and exploitation [11, 12] in large forest concessions as well as on private farms to minimize the impacts of thinning and ensure that it is included as a good SFM practices [13].

Another consequence of implementing selective thinning is the opening canopy; in forests, without intervention the fall of old trees naturally produces clearings and this process allows the renewal of tree populations [14, 15]. On the other hand, the intensity and frequency of the intervention are factors that influence the structure and composition of the species that colonize the gaps.

Authors like [16] determined that the intensity of intervention and the different silvicultural treatments have a significant impact on the density and growth of regeneration as well as saplings and seedlings of timber species. The loss of diversity [17] and number of individuals decrease due to post-disturbance mortality and its limited recruitment, and Dickinson et al. [18] state that the gaps produced naturally and those produced by the effect of the intervention have different disturbance conditions, with natural gaps favorable for the colonization of shade-tolerant species. Under the previous premises, we consider that the impact of selective thinning on natural regeneration must be monitored [19], with natural regeneration being a factor that has always been a bottleneck for foresters in tropical forests.

In several neotropical countries, forestry activities have been implemented to reduce the negative effects of harvesting and treatments to improve forest conditions. Reduced impact logging (RIL) includes pre- and post-activities that minimize unnecessary damage to remnant forest and soil [1]. Under this framework in Ecuador, the norms to regulate the use and management of forests are relatively new (approximately since 2001). They briefly mention the activities that promote sustainable forest management and use. All efforts to develop sustainable forest management practices focus on issues related to the felling and use of forests, ecosystem services, and forest plantations, but silvicultural experiments have not been carried out to evaluate the effectiveness, for example, selective thinning for the production of wood in natural forests and the impact or damage to the remaining forest that they produce, considering that this process is a fundamental element to assess the sustainability of both thinning and logging of the forest [8].

The purpose of this study is to evaluate the type and extent of mechanical damage caused by the fall of thinned trees over the remaining forest. The hypothesis that we propose to develop the work are as follows: (i) the damage type of residual trees is correlated with the intensity of thinning; (ii) the number and size of affected trees is related to the DBH of felled trees and the total basal area extracted; and (iii) the structure and spatial distribution of the residual forest determine the types of damage and number of trees affected by the felling operation.

2. Materials and methods

2.1 Study area and biophysical conditions

The study was carried out in the San Francisco Biologic Reserve (ECSF), which is located at 03°58′ S, 79°04′ W, 1850 m a.s.l. [20], i.e., north of Podocarpus National Park (PNP) in Southern Ecuador. ECSF is situated within the eastern cordillera of the Andes [21]. A comprehensive and detailed description of the San Francisco Reserve and its surroundings can be found in [22], which also detail a series of the research activities in this tropical mountain forest and present the different experiments conceived with the purpose of knowing the functionality and management of the forest. Refs. [23, 24] found a description of the scope and achievements of the forestry project installed in the ECSF.

2.2 Plot installation

Fifty-two plots of 2500 m^2 were installed in ECSF and located in three sites (quebradas) named Q2-Q3-Q5 at different altitudes. Authors like [25, 26] determined several forest types covering the ECSF. Installed plots were grouped and floristically ordered in two forest types. The first group called valley forest (VF) is characterized by the presence of *Tabebuia chrysantha* (Jacq.) G. Nicholson, *Cedrela montana* Moritz ex Turcz., *Inga acreana* Harms., and *Ficus citrifolia* Mill. These species are involved in the application of selective thinning, and ridge forest (RF) that is characterized by the presence of *Podocarpus oleifolius* D. Don ex Lamb., *Hyeronima moritziana* Mull. Arg, *Clusia ducuoides* Engl. which are released species. In Q2 (control block), we installed 20 plots (6 plots in RF and 14 plots in VF), in Q3 (only RF) we installed 16 plots, and in Q5 (only VF) we installed 16 plots (**Figure 1**). To implement the selective thinning (n = 30), we used only the plots installed in Q3 and Q5.

2.3 Thinning intensity

In VF plots, an average of 24.6 ± 0.66 stems ha^{-1} was extracted. The lowest value was 8 trees ha^{-1} and the highest value was 56 trees ha^{-1}, the basal area extracted is = $2.75\ m^2 \pm 1.1\ ha^{-1}$, and the highest extraction intensity is 4.8 $m^2\ ha^{-1}$. In RF, an average of $= 9.67 \pm 0.44$ trees ha^{-1} was extracted. The lowest value was 4 trees ha^{-1}, the highest value was 24 trees ha^{-1}, the average basal area extracted was $0.89\ m^2 \pm 0.1\ ha^{-1}$, and the highest extraction intensity was 1.1–2 $m^2\ ha^{-1}$, and the remaining plots are control without extraction.

In VF, the diameter of the felled trees ranged from 14.8 to 61.7 cm DBH (n = 97). In RF, the diameter of the felled trees ranged from 10.2 to 60 cm DBH (n = 60); in classes V and VI, there were no cut trees. The next table shows the number of felled trees in each forest type. The entire thinning operation was carried out under the condition of directional felling in order to minimize damage to residual trees (**Table 1**).

2.4 Assessing damages to residual trees caused by selective thinning

Immediately after the implementation of selective thinning in both forest types (16 plots in VF and 14 plots in RF), the number of damaged trees and the damage type caused by felled trees on residual large and small trees were monitored. We did not discriminate between trees affected by the trunk or by the crown of the felled tree; we only counted the trees affected by both sections. A separation of the damages caused by the felling of the trees and those caused by dragging the trunks out of the forest was not necessary, because in our case, we did not remove the fallen trunks and they were left inside the forest.

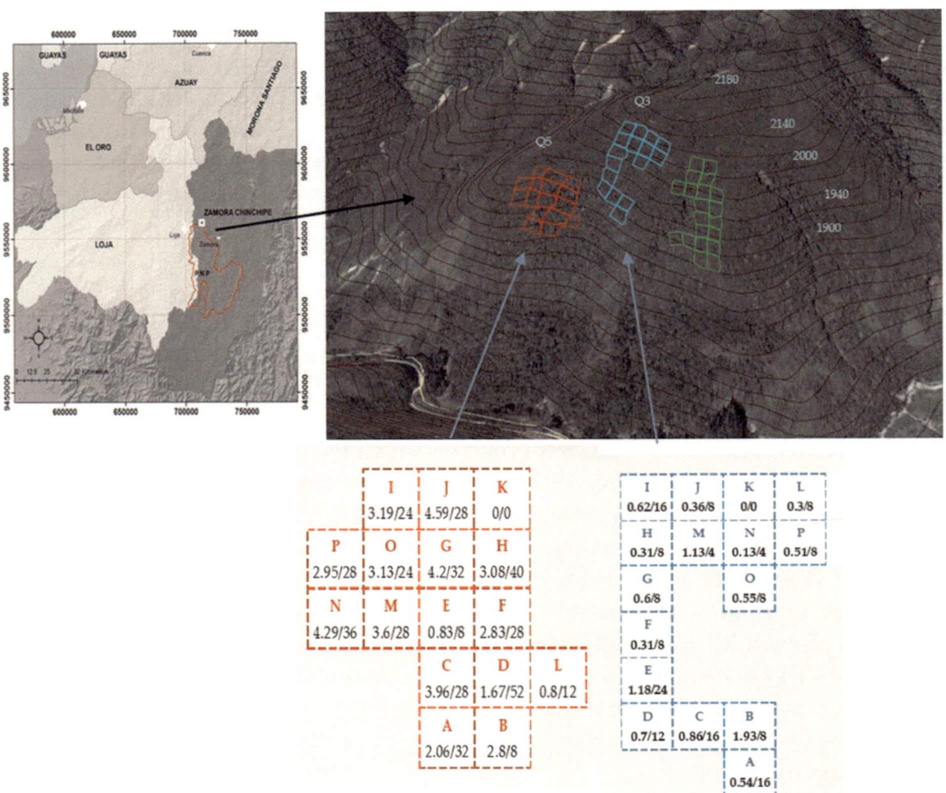

Figure 1.
Location of study area. The diagram shows the plots of each block and the internal numbers show the extracted basal area and the number of trees cut/ha^{-1}.

Damaged trees were classified according to their size in small trees (up to 20 cm DBH) and big trees (>20 cm DBH) trees. The damage was categorized as temporary (TD) when the trees were moved from their original position but were not uprooted, when branches were broken or the bark of the trunk was removed by the falling tree. The damage was described as permanent damage (PD) when trees were uprooted or their trunk was broken.

2.5 Assessing damages to residual trees caused by selective thinning

After counting the damaged large and small trees and determining the type of damage, we proceeded to calculate the proportion of damaged trees in each of the different diameter classes (large and small trees; permanent damage and temporal damage). The calculation of the proportion allows us to include in the evaluation model other parameters that are related to the structure of the residual forest. For each felled tree, the proportion of damaged residual trees was calculated using the following formula:

Proportion of trees temporarily damaged:

$$P_{Temp} = \frac{X_{ic}}{N_c} \qquad (1)$$

X_{ic}: affected trees by the felled tree in classc

N_c: total trees belonging to diametric classc (>20 cm, <20 cm DBH).

This proportion was computed in both forest types (ridge forest and valley forest). The same was done for permanent damages too.

To evaluate the effect of the variables on the variability of the proportion of damage on the residual trees, we fit generalized linear models (GLMs) of the binomial family. Such models are appropriate for binary or proportional data. Significant effects were elucidated at the alpha = 0.05 level. All calculations were performed in R version 3.6.3 (R 2020 core development team). **Table 2** summarizes the predictor and response variables involved in the model.

3. Results

A total of 157 trees were thinned, 97 in VF and 60 in RF. In the plots that belong to the VF, we found $\bar{X}$ = 285.4 ± 46.7 SD trees > 20 cm DBH and $\bar{X}$ = 1084.6 ± 390.4 SD trees < 20 cm DBH. In the RF plots, we found $\bar{X}$ = 157.6 ± 40.6 SD trees > 20 cm DBH and $\bar{X}$ = 1717.1 ± 390.6 SD trees < 20 cm DBH. These results are important because

	Diametric classes (cm)						
Forest type	**I (10.1–20)**	**II (20.1–30)**	**III (30.1–40)**	**IV (40.1–50)**	**V (50.1–60)**	**VI (>60)**	**Total**
Valley forest	7	45	27	10	7	1	97
Ridge forest	24	26	9	1			60
							157

Table 1.
Number of thinned trees for each diametric class in both types of forest.

Variables	Description	Type	Factor scales
DBH	DBH of felled tree	Predictive	cm
Tree density	Trees/ha in each 2500 m^2 plots (50 m × 50 m) where selective thinning was implemented	Predictive	Trees/ha
Basal area	Basal area extracted/ha	Predictive	m^2/ha
Crown area	Crown area of each of the felled trees.	Predictive	m^2-m^2/ha
Proportion of damaged trees < 20 cm	Proportion of trees temporarily or permanently damaged by the implementation of selective thinning with respect to total trees of sample plot	Response	Rate of damaged trees
Proportion of damaged trees > 20 cm	Proportion of trees temporarily or permanently damaged by the implementation of selective thinning with respect to total trees of sample plot	Response	Rate of damaged trees

Table 2.
Variables that were used to model the GLM.

the numbers of large and small trees making up the structure of each forest are key variables to estimate the amount of damage of residual trees.

A total of 1061 affected trees were monitored, 710 trees in VF and 351 in RF. In the VF, 457 trees were damaged temporarily (>5 cm DBH: n = 252; <5 cm DBH: n = 205) and 253 trees had a permanent damage (>5 cm DBH: n = 92 and <5 cm DBH: n = 161). In the RF, 239 trees with temporary damage were observed (> 5 cm DBH: n = 139; < 5 cm DBH: n = 100) and with 112 trees had a permanent damage (> 5 cm DBH: n = 53; < 5 cm DBH: n = 59).

3.1 Assessing the damages

In both forest types, TD were higher than PD as well in large as in small trees. In both forest types, permanent damages are higher in large trees than in small trees, the difference being more pronounced in the RF. In both forest types, the mechanical damage pattern can be explained. When directional felling was applied, it was paid attention that medium and large trees (in this study trees > 20 cm DBH) were not affected. These trees were better visible than the smaller ones (< 20 cm DBH) (**Figure 2**).

Regarding the damage rates related to the size of the thinned tree in trees >20 cm DBH and trees <20 cm DBH, the TD rate is higher than the PD rate. Trees felled within class III are those that cause the highest TD and PD damage rates. The felled trees of the larger diameter classes are less frequent in the thinning, since they generally reach upper canopy strata and do not affect the trees of the middle strata where the released trees are located (**Figure 3a**). As for small trees, the pattern is the same; that is, thinned trees belonging to diameter class III are those that present the highest rate of TD inflicted on residual trees (**Figure 3b**).

In VF, felled trees of diameter classes IV–V are the ones that produce the highest rates of TD and PD in large (**Figure 4a**) and small (**Figure 4b**) trees. As in the previous forest, the felled trees of the smaller classes produce a lower damage rate, while the larger trees cause TD and PD at a higher rate. The implementation of selective thinning implies the release of trees mainly from classes II–IV, and the competitors are pigeonholed in the same diameter classes and are the most visible at the time of planning the directed felling.

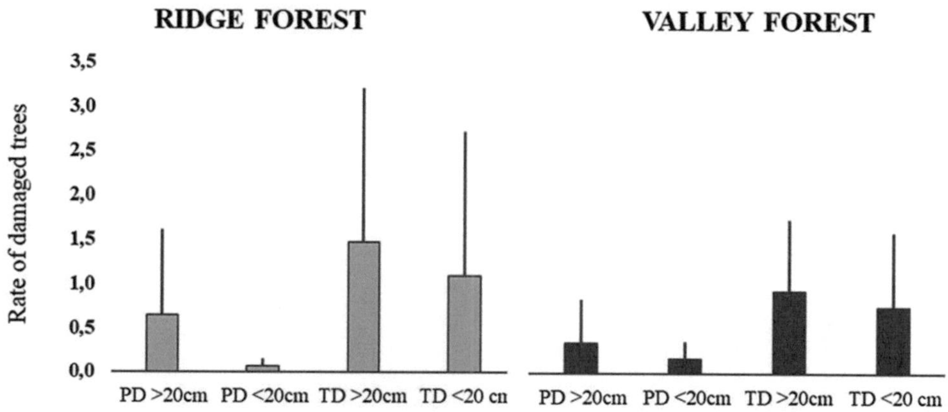

Figure 2.
Average rate of TD and PD in large and small trees in the two forest types (RF and VF). The lines above the bars represent the standard deviation.

3.2 Tree damage proportion

The proportion of total damage is significantly related to the diameter of the felled tree in the VF (***p*** = <0.0001, **Figure 5a**) and in RF (*p* = <0.0001, **Figure 5b**). In the VF, the proportion of damaged trees is significantly related to the diameter of the extracted tree and to the total basal area extracted, which also significantly influences the proportion of trees permanently damaged < 20 cm DBH (*p* = 0.0059) and affected with temporary damage < 20 cm DBH (*p* = 0.001) (**Table 3**).

In the RF, the proportion of damaged trees is significantly related to the diameter of the harvested tree and the total basal area harvested, which also significantly influences the proportion of temporarily damaged trees < 20 cm DBH (*p* = 0.005), and the proportion of trees with permanent damage >20 cm DBH is also influenced by the crown area of the felled tree (*p* = 0.006) (**Table 4**).

4. Discussion

The implementation of activities to improve the performance of natural forests and the effects of harvesting wood also directly affect the residual stand. Knowing the number of trees affected and the type of affectation suffered by the residual stand are two important parameters that will allow better management of natural forests [5].

The relationship that exists between the volume of wood harvested, the size (mainly DBH expressed in basal area), the number and type of damage to residual trees plus the harvesting system (traditional vs. reduced impact logging) have been monitored for some time and in some types of forest [27–29]; however, the impact produced, for example, by selective thinning as an activity to improve forest yield is less known and documented, especially in neotropical countries where sustainable forest management practices, although regulated and mandatory, have not been monitored and evaluated, which has resulted in small- and medium-sized areas (< 50 ha) where the forest has been exploited and degraded.

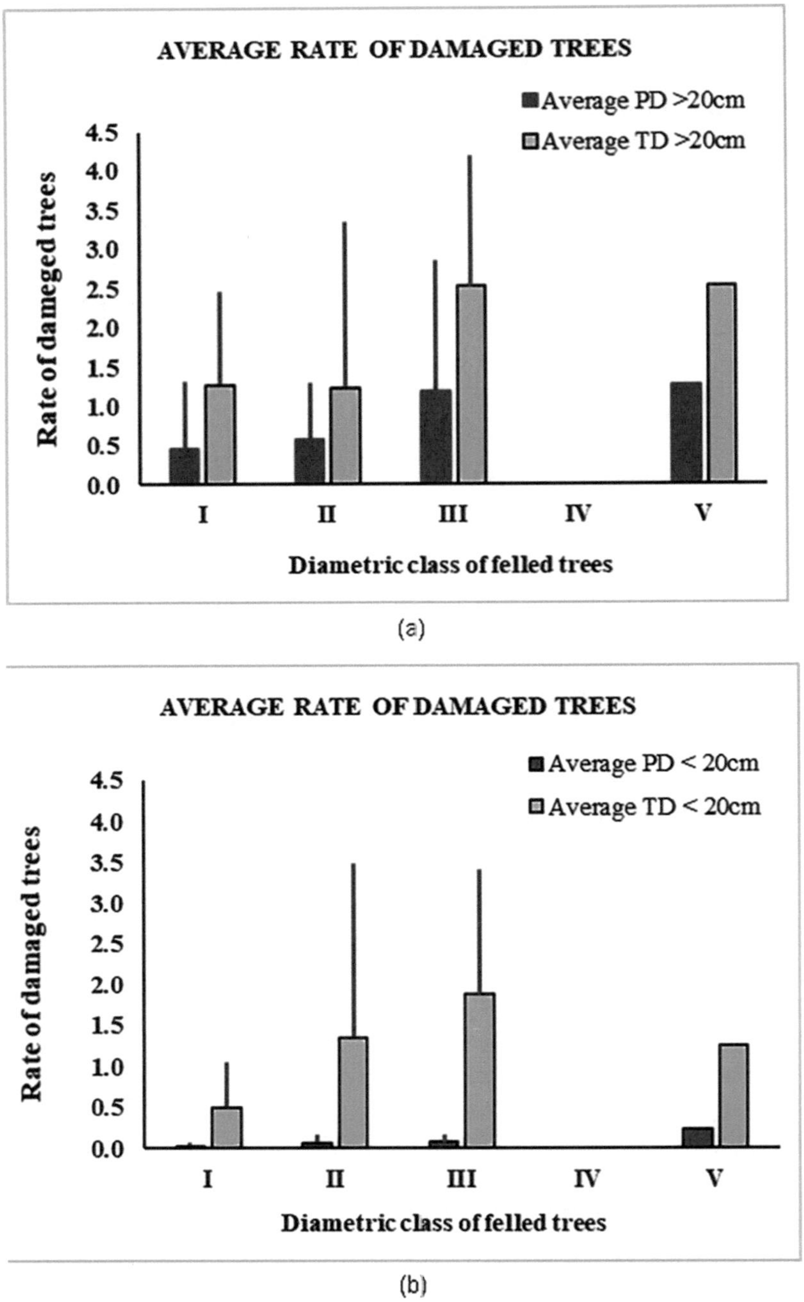

Figure 3.
TD and PD rates of big and small residuals trees in RF.

4.1 Mechanical damage

In Ref. [30], an average of 6.3 trees ha^{-1} (3.4 m^2 ha^{-1}) was extracted; however, the variation was much higher (up to 9.6 m^2 ha^{-1} in some plots) and the values of affected trees were greater than those reported in our study (15.6 ± 8.3 vs. 31). However, the aforementioned authors correlate in a positive and significant way the number of trees

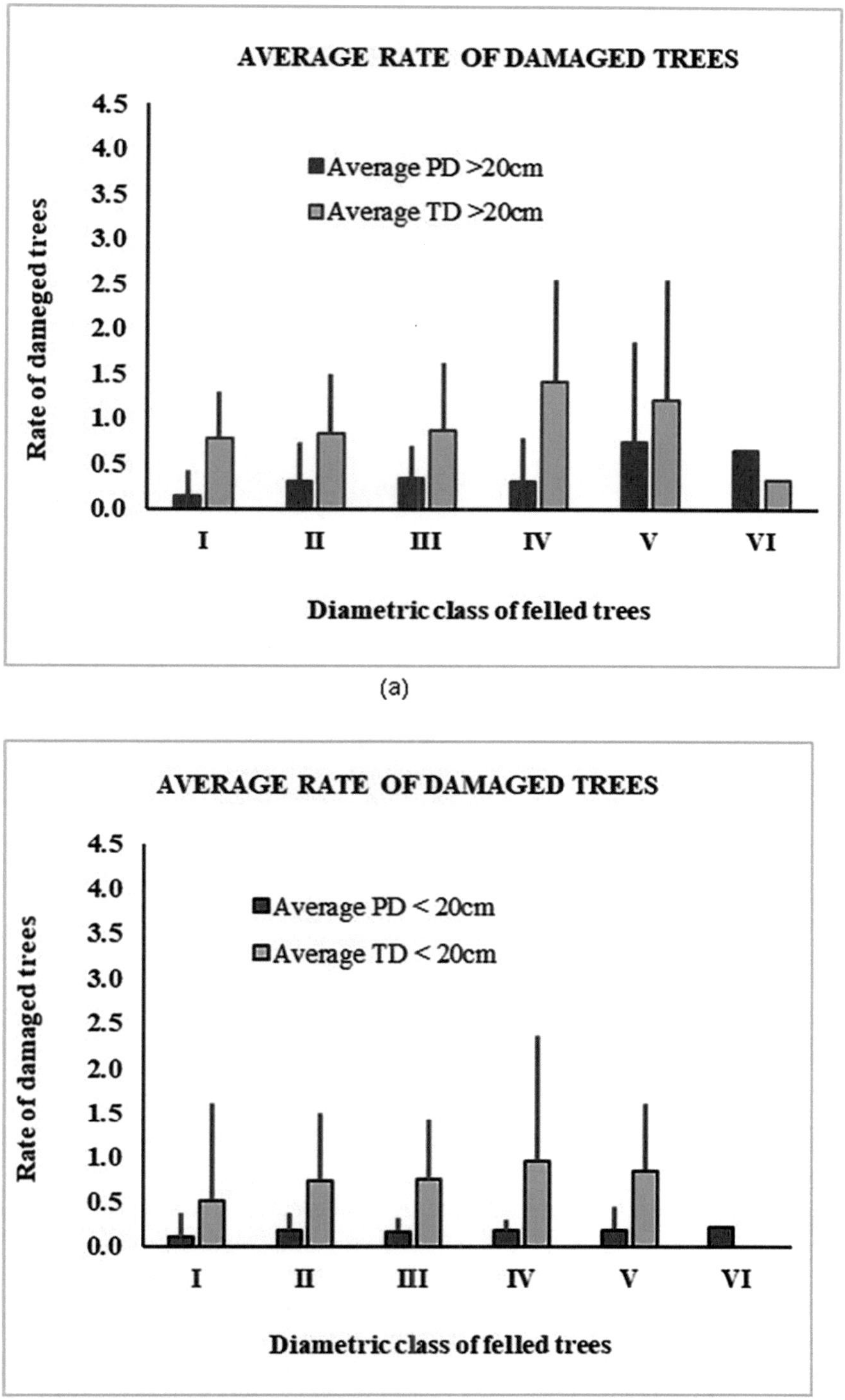

Figure 4.
TD and PD rates in big and small residual trees of VF.

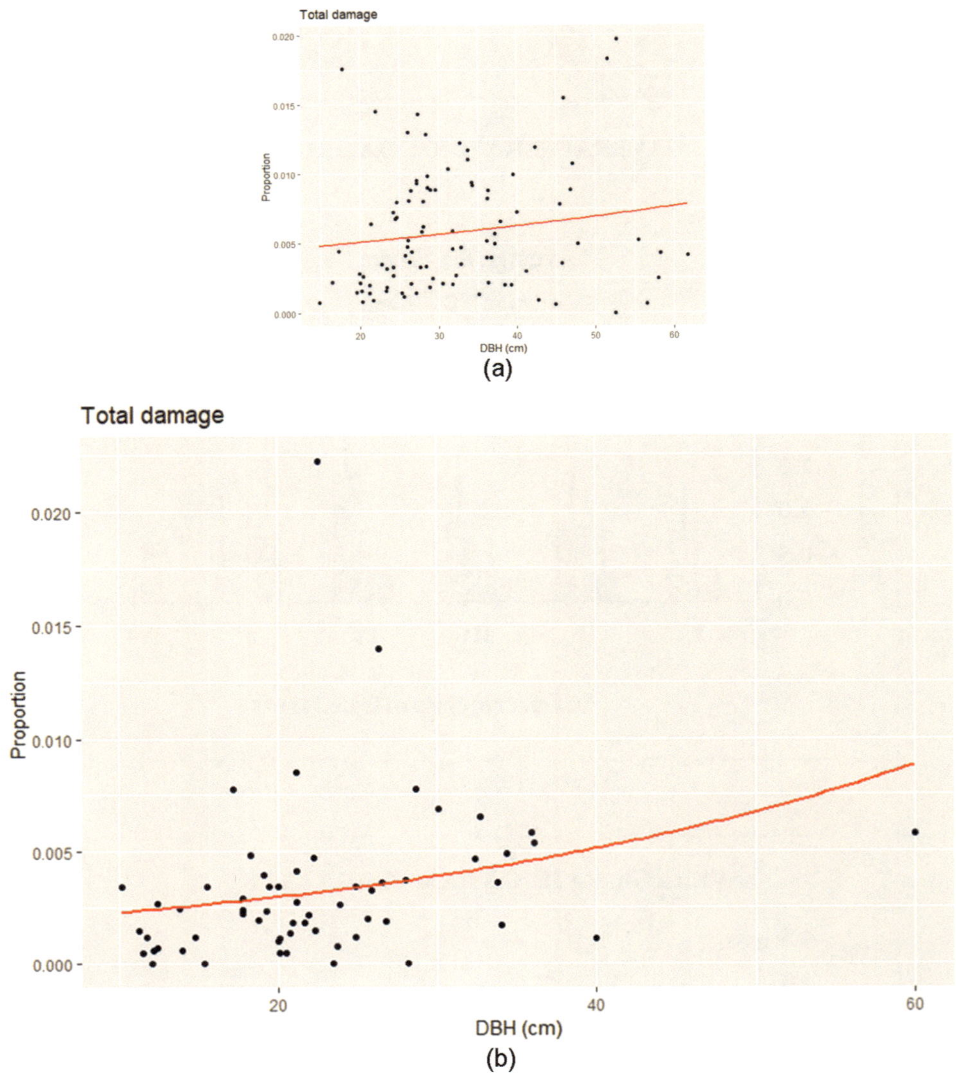

Figure 5.
Damaged trees by DBH of felled trees. The lines show a significant correlation (p = 0.05).

harvested with the number of trees destroyed and the number of damaged trees; in this case, the damaged trees correspond to the TD type and the trees destroyed correspond to the PD type.

Regarding the type of damage suffered by trees in the remaining forest [27] determined six damage types, of which four classes are evaluated in this work, and which can be included within the categories of damage that we determine here (e.g., damage in the bark and damage in the crown of the trees can be equated in the category of temporary and permanent damage, respectively). The most frequent type of damage reported by Ref. [27] is the damage to tree crowns and also uprooted trees both on the trawl roads and in the gaps within the forest, while in our study the damage more frequently in both intensities of thinning is broken trunk, uprooted trees in our evaluation represent between 10 and 20% of the total trees affected in both intensities of thinning; in the same study, the percentage represents between 15

Source of variation	Estimate	Standard error	P	
	Permanent damage (DBH > 20 cm)			
DBH	0.024	0.0089	0.00729	**
	Permanent damage (DBH < 20 cm)			
DBH	0.154	0.0523	0.00314	**
Basal area extracted	−24.644	89.517	0.00590	**
	Temporal damage (DBH > 20 cm)			
DBH	0.014	0.0056	0.00932	**
	Temporal damage (DBH < 20 cm)			
DBH	0.113	0.0449	0.0121	*
Basal area extracted	−18.352	77.241	0.0175	*
	Total damage			
DBH	0.112	0.0237	< 0.0001	***
Basal area extracted	−16.849	40.131	< 0.0001	***

DBH: diameter at breast height, BA: basal area.
0.05.0.05.***<0.001.*

Table 3.
Summary table of the binomial GLM, for permanent, temporal, and total damage to trees of different diameter class in VF.

Source of variation	Estimate	SE	P	
	Permanent damage (DBH > 20 cm)			
DBH	0.0353	0.0124	0.0044	**
Crown area	0.0154	0.0057	0.0067	**
	Permanent damage (DBH < 20 cm)			
DBH	0.0403	0.0105	0.0001	***
	Temporal damage (DBH > 20 cm)			
DBH	0.0185	0.0082	0.0247	*
	Temporal damage (DBH < 20 cm)			
DBH	0.1754	0.0499	0.0004	***
Basal area extracted	−297.217	106.027	0.0050	**
	Total damage			
DBH	0.09136	0.0215	< 0.0001	***
Basal area extracted	−122.738	41.269	0.0029	**

DBH: diameter at breast height, BA: basal area, CA: crown area.
0.05.0.05.***<0.001.*

Table 4.
Summary table of the binomial GLM, for permanent, temporal, and total damage to trees of different diameter class in RF.

and 30% in all the scenarios evaluated. Regarding the number of individuals affected by type of damage, the average values in VF range from 1 to 9 uprooted trees and in RF, it ranges between 0.99 and 2 uprooted trees, values much lower than those reported in the mentioned study.

Despite the fact that there is evidence to suggest that the ecological impact of felled, or lying trees, have the so-called competitors, the actual values show a decrease with respect to those obtained by Veríssimo [31], who determined that in a forest in the Brazilian Amazon for every tree harvested, and there were 23 trees over 10 cm DBH, which were severely damaged. However, this includes openings in the road for hauling. Likewise, Johns [5] evaluated an unplanned method of harvesting trees. Their study found that severely damaged conditions led to significant differences between the planned and unplanned operation (7.4 vs. 4.5 trees / tree lying).

On the other hand, Abebe and Holm [32] evaluated the effect of selective harvesting on residual stand by classifying the type of damage in three categories, which according to [33] show a slight modification: damage type I or slight damage, damage type II (defined as injury in the trunk surface or more than 2 m long, or affecting 20% of the circumference, or a broken tree crown of over two-thirds), and finally damage type III, which is when the tree is broken.

It was determined that an average of 8 trees > 5 cm DBH per lying tree was damaged during the fall. Sixty-two percent of all damaged trees were broken. And, compared with our study, this result can be equated with PD, which does not equate the percentage or number of trees and the way they died—this being a parameter with which no information could be found for evaluation and comparison.

It was observed that in the two sites, damage percentages have values in the three forms of mortality assessed. In RF plots, the highest value corresponds to broken trees, while the values of the other two forms have similar values (9.8 and 8.9 dead standing and uprooting). In RF plots, the values of uprooting and standing dead have the same values (20.1 and 21.5). In VF plots, the values of uprooting and dead standing show a double percentage of occurrences in RF plots, which is calculated by determining the difference of large trees that exist between both sites.

Ref. [11] indicates that the damage produced by the effect of harvesting and selective felling depends on many factors including some that we have analyzed in this first part of the discussion; among them (those studied here) are the size of the tree lying down, the implementation or not of directed fall, the intensity in this case of selective thinning, even though the values and percentages of damage reported in the literature are very variable and recommend to indicate the type of damage as an important factor in the evaluation of the sustainability of forest operations.

Comparing with the results obtained by Ref. [34], the percentages of trees with permanent damage due to the harvest reach 10.1%, a value that is higher than the average percentage determined in our work (3.5%); however, when comparing the traditional methods with reduced impact methods, there is a lower percentage of damage in the second method, which also corroborates the importance of RIL methods to ensure the sustainability of forest management. Ref. [10] correlates the logging method with the percentage of damaged trees, the lowest percentage being the one that uses reduced impact methods with 4.1% of total damage, which confirms that the RIL damages residual trees to a lesser extent. Another coincident answer in the evaluation of damages on residual trees is the correlation between the diameter of the thinned tree with the affected trees, and Ref [4] reports percentages similar to those obtained in our work (average total damage 6.7%) and linearity between the basal area extracted and the percentage of damage of residual trees.

5. Conclusions

Finally, we can conclude that there is a correlation between the intensity of thinning and the temporary damage rates of the residual trees. The damage type of the residual trees is correlated with the intensity of thinning, with the DBH of the thinned tree and with the structure of the forest, since the damage rate will depend exclusively on the number of large and small trees that make up the total number of stand trees. The damage rates obtained in our study are within the percentages achieved in other tropical areas using mainly RIL methods, which is undoubtedly an ecological framework that allows planning harvest activities in Ecuador.

Acknowledgements

The first author thanks the German Research Foundation (DFG) for funding field work and logistical support, the Universidad Técnica Particular de Loja for support in finishing the manuscript. The authors thank MSc Galo Guamán for helping in GIS analyses. The authors thank Dr. Diana Szekely for the revision of English language.

Conflict of interest

The authors declare no conflict of interest.

Author details

Omar Cabrera[1,2]*, Pablo Ramón[2], Bernd Stimm[2], Sven Gunter[2,3]
and Reinhard Mosandl[1]

1 Department of Life Science Systems, Institute of Silviculture, Technical University of Munich, Freising, Germany

2 Department of Biological and Agricultural Sciences, Private Technical University of Loja, Loja, Ecuador

3 Institut für Internationale Waldwirtschaft und Forstökonomie, Hamburg, Germany

*Address all correspondence to: hocabrera@utpl.edu.ec

References

[1] Putz FE, Dykstra DP, Heinrich R. Why poor logging practices persist in the tropics. Conservation Biology. 2000;**14**: 951-956

[2] Petrokofsky G, Sist P, Blanc L, Doucet JL, Finegan B, Gourlet-Fleury S, et al. Comparative effectiveness of silvicultural interventions for increasing timber production and sustaining conservation values in natural tropical production forests. A systematic review protocols. Environmental Evidence. 2015;**4**:8

[3] Picchio R, Mederski PS, Tavankar F. How and how much, do harvesting activities affect forest soil, regeneration and stands? Current Forestry Reports. 2020;**6**(2):115-128

[4] Khai TC, Mizoue N, Ota T. Harvesting intensity and disturbance to residual trees and ground under Myanmar selection system; comparison of four sites. Global Ecology and Conservation. 2020;**24**: e01214

[5] Johns JS, Barreto P, Uhl C. Logging damage during planned and unplanned logging operations in the eastern Amazon. Forest Ecology and Management. 1996;**89**:59-77

[6] Jackson SM, Fredericksen TS, Malcolm JR. Area disturbed and residual stand damage following logging in a Bolivian tropical forest. Forest Ecology and Management. 2002;**166**(1–3): 271-283

[7] Medjibe VP, Putz FE, Starkey MP, Ndouna AA, Memiaghe HR. Impacts of selective logging on above-ground forest biomass in the Monts de Cristal in Gabon. Forest Ecology and Management. 2011;**262**(9):1799-1806

[8] Chheng K, Mizoue N, Khorn S, Kao D, Sasaki N. Tree-based approach to evaluate size dependence of residual tree damage caused by selective logging: Case study in tropical semi-evergreen forests of Cambodia. Forest Ecology and Management. 2015;**356**:285-292

[9] Tavankar F, Bonyad AE, Nikooy M, Picchio R, Venanzi R, Calienno L. Damages to soil and tree species by cable-skidding in Caspian forests of Iran. Forest Systems. 2017;**26**(1):1-9. DOI: 10.5424/ fs/2017261-09100

[10] Bodaghi AI, Nikooy M, Naghdi R, Tavankar F. Logging damage to residual trees during sustainable harvesting of uneven-age stands in the Hyrcanian forests of Iran. New Zealand Journal of Forestry Science. 2020;**50**

[11] Picard N, Gourlet-Fleury S, Forni É. Estimating damage from selective logging and implications for tropical forest management. Canadian Journal of Forest Research. 2012;**42**: 605-613

[12] Roopsind A, Sist P, Peña-Claros M, Thomas R, Putz FE. Beyond equitable data sharing to improve tropical forest management. International Forestry Review. 2014;**16**:497-503

[13] Guitet S, Pithon S, Brunaux O, Jubelin G, Gond V. Impacts of logging on the canopy and the consequences for forest management in French Guiana. Forest Ecology and Management. 2012; **277**:124-131

[14] De Graaf NR, Poels RLH, Van Rompaey RSAR. Effect of silvicultural treatment on growth and mortality of rainforest in Surinam over long periods. Forest Ecology and Management. 1999; **124**:123-135

[15] Rivett SL, Bicknell JE, Davies ZG. Effect of reduced-impact logging on seedling recruitment in a neotropical forest. Forest Ecology and Management. 2016;**367**:71-79

[16] Peña-Claros M, Peters EM, Justiniano MJ, Bongers F, Blate GM, Fredericksen TS, et al. Regeneration of commercial tree species following silvicultural treatments in a moist tropical forest. Forest Ecology and Management. 2008;**255**:1283-1293

[17] Kariuki M, Kooyman RM. Floristic changes and regeneration patterns for a 12-year period during the 3rd and 4th decades following selection logging in a subtropical rainforest. Austral Ecology. 2005;**30**:844-855

[18] Dickinson MB, Whigham DF, Hermann SM. Tree regeneration in felling and natural treefall disturbances in a semideciduous tropical forest in Mexico. Forest Ecology and Management. 2000;**134**(1–3):137-151

[19] Dolanc CR, Gorchov DL, Cornejo F. The effects of silvicultural thinning on trees regenerating in strip clear-cuts in the Peruvian Amazon. Forest Ecology and Management. 2003;**182**:103-116

[20] Ohl C, Bussmann R. Recolonization of natural landslides in tropical mountain forests of Southern Ecuador. Feddes Repertorium: Zeitschrift für botanische Taxonomie und Geobotanik. 2004;**115**(3–4):248-264

[21] Brehm G, Homeier J, Fiedler K. Beta diversity of geometrid moths (Lepidoptera: Geometridae) in an Andean montane rainforest. Diversity and Distributions. 2003;**9**:351-366

[22] Beck E, Bendix J, Kottke I, Makeschin F, Mosandl R, editors. Gradients in a Tropical Mountain Ecosystem of Ecuador. Ecological Studies 198. Berlin, Germany: Springer-Verlag; 2008. pp. 347-359

[23] Günter S, Cabrera O, Weber M, Stimm B, Zimmermann M, Fiedler K, et al. Chapter 26: Natural forest management in neotropical mountain rain forests - An ecological experiment. In: Beck E, Bendix J, Kottke I, Makeschin F, Mosandl R, editors. Gradients in a Tropical Mountain Ecosystem of Ecuador. Ecological Studies 198. Berlin, Germany: Springer-Verlag; 2008. pp. 347-359

[24] Cabrera O, Fries A, Hildebrandt P, Günter S, Mosandl R. Early growth response of nine timber species to release in a tropical mountain forest of Southern Ecuador. Forests. 2019;**10**:254

[25] Paulsch A. Development and Application of a Classification System for Undisturbed and Disturbed Tropical Montane Forests Based on Vegetation Structure. Germany: Fakultat fur Biologie, Chemie und Geowissenschaften der Universitat Bayreuth; 2002

[26] Homeier J, Breckle SW, Günter S, Rollenbeck RT, Leuschner C. Tree diversity, forest structure and productivity along altitudinal and topographical gradients in a species-rich Ecuadorian montane rain forest. Biotropica. 2010;**42**:140-148

[27] Iskandar H, Snook LK, Toma T, MacDicken KG, Kanninen M. A comparison of damage due to logging under different forms of resource access in East Kalimantan, Indonesia. Forest Ecology and Management. 2006;**237**: 83-93

[28] Khai TC, Mizoue N, Kajisa T, Ota T, Yoshida S. Using a tree-based approach to evaluate logging damage in a tropical mixed deciduous forest of Myanmar:

Comparison with cases in Cambodia. Journal of Forest Research. 2017;**22**(3): 185-190

[29] Matangaran JR, Putra EI, Diatin I, Mujahid M, Adlan Q. Residual stand damage from selective logging of tropical forests: A comparative case study in central Kalimantan and West Sumatra. Indonesia. Global Ecology and Conservation. 2019;**19**:e00688

[30] Sist P, Ferreira FN. Sustainability of reduced-impact logging in the Eastern Amazon. Forest Ecology and Management. 2007;**243**:199-209

[31] Veríssimo A, Barreto P, Mattos M, Tarifa R, Uhl C. Logging impacts and prospects for sustainable forest management in an old Amazonian frontier: The case of Paragominas. Forest Ecology and Management. 1992;**55**: 169-199

[32] Abebe T, Holm S. The effect of commercial selective logging on a residual stand in tropical rain forest of south-western Ethiopia. Journal of Tropical Forest Science. 2003;**2003**: 387-398

[33] Nicholson DI. The Development of Silviculture in North Queensland Rainforests. Australia: The Australian National University, Canberra, Australia; 1985. pp. 203-223

[34] Bertault JG, Sist P. An experimental comparison of different harvesting intensities with reduced-impact and conventional logging in East Kalimantan, Indonesia. Forest Ecological Management. 1997;**94**(1–3):209-218

Chapter 5

Regeneration Dynamics of an African Tropical Forest Canopy Dominant Tree Species

Mugatsia H. Tsingalia

Abstract

The population structure and regeneration of Olea welwitschii in Kakamega differs in different forest blocks. There are differences in abundance between the forest blocks, that can be explained by past human-forest interactions – harvesting by mechanized loggers who clear-felled stands of desired merchantable species and the skilled pitsawers who selectively harvested desired tree species. There is a clear absence of seedlings/saplings pointing to a lack of regeneration inside the forest over the last 40 years. Intense seed predation by rodents and attack by fungal pathogens account for up to 99% mortality of fruits and seeds under the parent crowns. *Olea welwitschii* seedlings/saplings are evident in forest glades. These glades have been maintained by grazing and burning. Fires discourage the growth of trees, while grazing encourages the establishment of mound-building termite species upon which, grassland species such as *Combretum molle* colonize. Grazing appears to inhibit rodent predators while *Combretum molle* serves as perch and nesting sites for avian frugivores. Avian frugivores drop Olea seeds under Combretum's crowns, which germinate and establish under reduced predation and fungal attacks germinate and establish. Patches inside the glades in which Olea regenerates become centres from which the forest continues to invade the glades.

Keywords: Olea welwitschii, Kakamega forest, seed predation, seed dispersal, regeneration, canopy dominants, glades

1. Introduction

The Elgon teak, *Olea welwitschii* (Knobl) Gilg. & Schelleneb is a canopy emergent tree in the family Oleacea [1]. This tree species is indigenous to sub-Saharan Africa ranging from Cameroon in the west to Ethiopia and Kenya in the east, and south to Angola, Zambia, and Mozambique [2]. It is typically a forest tree species, that grows from lowland tropical rainforests to evergreen montane forests [3]. The tree can grow to a height of 25 m with a straight bole and a small crown. It has a pale gray to white bark, that is grooved vertically. The flowers are small and white, in sprays to 8 cm long. The fruits are narrow, oval, and have small drupes. Its wood is extremely good, and stable with esthetic timber used in shipbuilding. It is also used to make fine furniture, door and window frames, and any other applications that require strong, durable, and stable wood. It is termite resistant and its branches are also used for firewood while its bark is medicinal. This species dominates the western Kenya forests, especially the Kakamega and Mt. Elgon forests. In these forests, *Olea welwitschii* does not show notable evidence of natural regeneration under the parent crowns [4]. This paper examines the factors that may be responsible for the lack of apparent natural regeneration in the Kakamega forest, Kenya.

The family Oleacea has been prized since the dawn of civilization. For instance, its most famous member, the olive (Olea Europa L.) was cultivated 3000 years B.C [5]. Today, in a large part of the world, the olive still represents life and plenty in the minds of men. The European ash (*Fraxinus excelsior*), another member of the Oleacea, was used by Greeks to make cupid bows. The most familiar members of the family are the jasmine species (*Jasminum* spp.) which have been favorites in home gardens for centuries [2].

Three Oleacea genera -*Schrebera*, *Olea,* and *Linociera*, occur in Africa [6]. Of these, *Olea* is by far the most common and widespread. In Kenya, it is represented by five species -*Olea africana* Milla, *Olea hochstetteri* Baker, the East African olive, *Olea kilimandscharica* Knobl, *Olea mildbraedii* (Gill & Schelleneb) and *Olea welwitschii*.

Olea welwitschii is typically a rainforest tree species endemic to Mt. Elgon [7, 8]. It also occurs in Kakamega, North, and South Nandi forests which appear to be its easternmost limit. Information on the pollination ecology of this species is scant. In Kakamega and the neighboring Kisere forests, Olea welwitschii is canopy dominant comprising some 7.5 and 49 percent respectively of the total volume of merchantable species [9]. Prior to 1966, this species comprised 69% of the total utilizable timber in the Kakamega forest and by 1980, this percentage had dropped to a mere 7.5% due to uncontrolled harvesting by commercial loggers.

In the Kakamega forest, it grows in localized patches of very few senescing adults -except for those in enrichment plantations. Kisere forest dominates the canopy but shows a skewed population distribution in the 500-ha forest [10]. For instance, it occurs in an almost pure stand in the south, east and north-eastern parts of this forest, and in neither forest does Olea welwitschii exhibit any apparent natural regeneration.

2. Study sites and methods

2.1 Study sites

Three study sites were selected from different parts of the forest. Two sites were selected in the southern part of the forest while another site was in the northern part of the forest (**Figure 1**). Each of the three sites was contiguous with forest glades (grassy areas that are surrounded by forest). The northern study site was located in the Kisere forest which is part of the National Reserve and a long-time nature reserve. The southern study sites were located in the newly established Kakamega Forest National Reserve and at the Kakamega Forest Station. All three study sites had large mature *O. capensis* adults. The southern study sites have been logged selectively in the 1970s and early 1980s. During logging, many large *O. capensis* adults were selectively harvested [9].

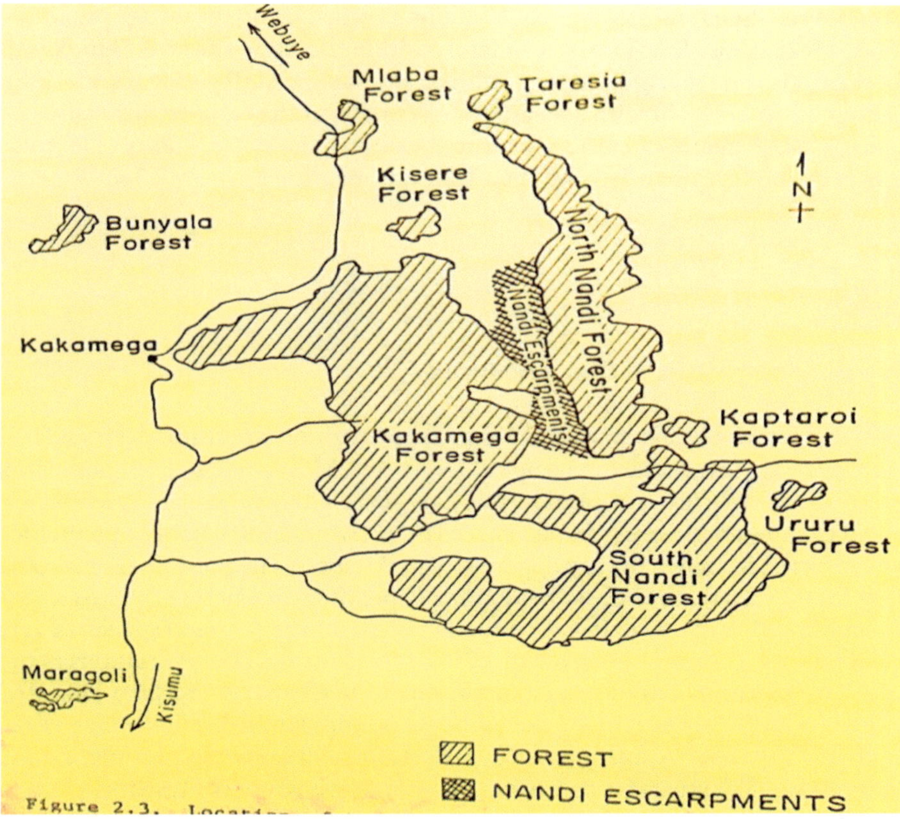

Figure 1.
Map of the Kakamega forest complex. Source: [9].

2.2 Study methods

Plotless methods were used throughout the sampling regime [10–14]. Density and size-class distribution were measured using a wedge prism B.A.F. 4.6 m^2 which allowed the discrimination between trees of different size classes. Distances between sampling points were established using a tape measure. Distances from this tree to the nearest closest first, second, and third nearest neighbors were measured [10, 11, 15]. A wedge prism was used to establish of variable plots that were evenly and randomly distributed within the study sites along established transects.

Spacing was determined by comparing the mean distances to the nearest neighbors with the distance from random points to the nearest tree [14, 15]. The mean observed distance was calculated using the formula,

$$rA = \breve{E}r / N \qquad (1)$$

where N is the number of measurements of distance in the observed population or sample and r is the distance in any specified units from a given individual to its nearest neighbor.

Because seedlings and saplings could not be sampled using the wedge prism method, 25×25 m quadrats located randomly within the study sites were employed. These quadrats were inventoried for seedlings and saplings. All young plants not exceeding 0.5 m in height with a measurable diameter at the base were classified as saplings. Their heights were also measured using a clinometer. Adults of all tree species present were identified and their Dbh was measured using a 2.5 m Dbh tape measure.

3. Results

3.1 Population structure of *Olea welwitschii*

Olea welwitschii in Kakamega and Kisere forests exhibit a multimodal population structure that is unique for a tropical forest tree. Most studies of the population structure of tree species in tropical forests have revealed that most of these species will tend to show an inverse J-shaped curve [16, 17]. The inverse J-shape population structure has been shown to vary depending on the species in question. For instance, [18, 19] found variations in the J-shape curve depending on whether the species in question was a climax tree that was permanently established or seral species of type I or II of either invading or unstable types in a mixed forest. These variations in the J-shape have also been demonstrated in forest trees in Ghana, west Africa [20, 21].

There are two explanations for the multimodal population distribution in Olea welwitschii. The first explanation invokes past human disturbances mainly, extractive harvesting. Kakamega forest has been logged in the past (for 50 years) [22]. In principle, this logging was supposed to be selective but there appears to be no evidence of this practice. If implemented, selective logging would have minimized overexploitation of certain preferred species of trees more so in size class distribution. But selective logging is difficult to implement because it requires strict supervision and manpower.

To date, there are no reported cases in tropical forests where supervised selective logging has been achieved. Besides, selective logging would be easier if trees in tropical forests were uniformly distributed or were in monospecific stands, but the spacing of trees in tropical forests is not uniform [23, 24]. It is therefore clear that logging cannot explain the multimodal population structure of Olea given the similarity of the population structure in logged and unlogged sites in the forest. Thus, careful logging can only explain differences in abundance.

The second explanation of the multimodal population structure is based on regeneration dynamics. Olea welwitschii regenerates in waves such that groups of trees are clustered in various age groups that correspond to periods of population recruitment. These age groups appear to be a function of the existence of conditions that favor germination and establishment. But for such a pattern to occur, requires that conditions for regeneration occur in waves or bursts. The differences in age groupings when calculated in years, would provide information on time periods between different regeneration events and whether these events are cyclical or random in nature. Such an explanation would imply that the largest individuals of Olea in the forest are not relics of a colonizing species, rather, they represent a different older age group.

The population structure of regeneration that Olea welwitschii in Kakamega forest is typically a northern temperate that has been observed among maple-basswood forests [19], oak-hickory forests [20], the new forests in Pennsylvania [21] and the hardwoods in New England (USA) [23]. These forests share one common phenomenon, disturbance, which has been shown to be a major cause of wave patterns of regeneration. A high correlation was reported between disturbance and population recruitment in their study of age-class distributions among various tree species in the New Forest of Great Britain [24]. The regeneration pattern in the species was not continuous and the successful establishment of seedlings was dependent on disturbance by fire and hurricanes.

3.2 Size-class, density, and density distribution

The density and spatial distribution of *Olea welwitschii* vary in Kakamega and Kisere forests (**Figure 2**). The overall size-class structure is significantly different in both forests. In addition, the mean Dbh in the Kakamega forest varies between 56.1 ± 49.6 cm −97.3 ± 65.4 cm while in the Kisere forest, it is 107.7 ± 58.2 cm. In essence, the population structure of *Olea welwitschii* shows a multimodal distribution (**Figure 2**) instead of the inverse J-shaped population structure typical of most tropical tree species [25, 26].

The density of *Olea welwitschii* varies depending on the forest site in question. **Table 1** summarizes the density in different blocks of forest. In all three sites, the values of R indicate that *Olea welwitschii* has a varied degree of clumped distribution.

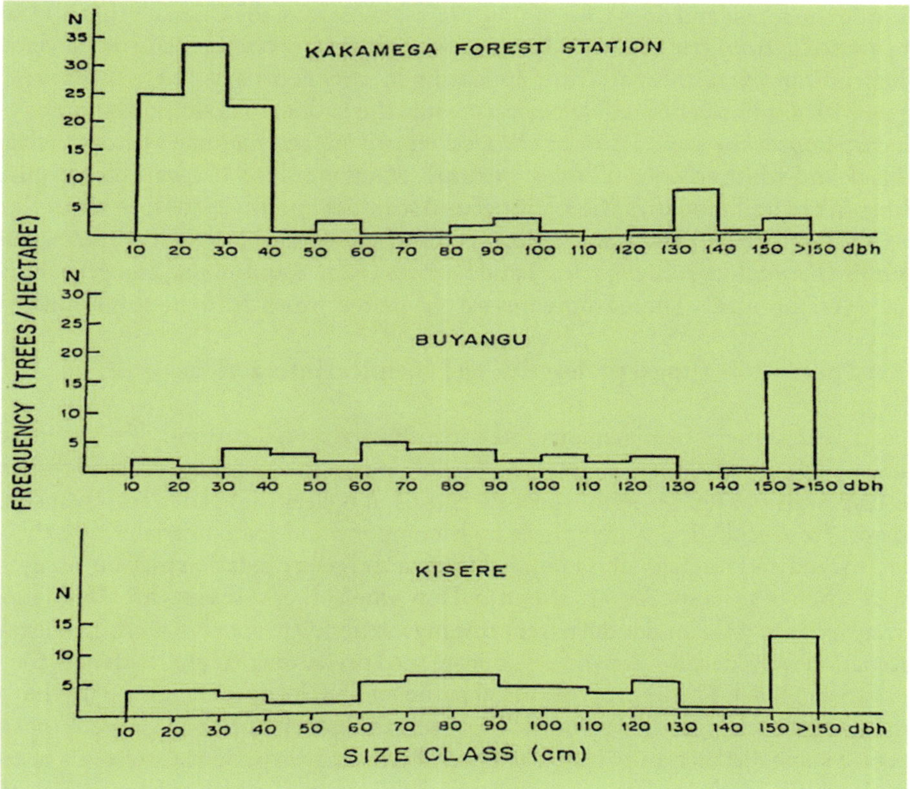

Figure 2.
Population structure of Olea welwitschii in different sites in Kakamega forest.

Forest Sites	Density/ha	Spatial contagion (R)	Z value
Kakamega Station	8.4	94.6	0.012
Buyangu	6.3	143.6	0.014
Kisere	425.2	368.0	0.002

Table 1.
Density of Olea welwitschii in three different forest sites together with values of the spatial contagion and the significance level.

The species is highly clumped in the Kisere forest. Variation in clumping appears to correspond to the degree of exploitation in the past [9].

The presence of the multimodal distribution pattern suggests periods of regeneration that are interspersed with periods of no regeneration [27, 28]. In fact, it has now been established that very small intensities of disturbances will produce a non-J-shaped population structure. According to [29, 30], non-J-shaped size-class distribution structures are relatively common and may represent stable population structures. The modal size-class distribution at the Kakamega forest station is probably due to enrichment planting rather than natural regeneration.

3.3 Natural enemies of natural regeneration

Perhaps the question to ask at this point is "why does *Olea welwitschii* not regenerate inside the forest and does *Olea welwitschii* regenerate at all? One fact that is clear is that *Olea welwitschii* fruits every other year with flowering commencing in November of the fruiting year with small fruits appearing in early February. These fruits are drupes with a thin layer of edible pericarp while the endocarp is stony. The stony endocarp houses the seed. Fruits of *Olea welwitschii* attract mammal visitors such as the black and white colobus (*Colobus guereza*), Sykes monkeys (Cercopithecus mitis stuhlmani) redtail monkeys (Cercopithecus Ascanius) and the giant tree squirrels (*Protoxerus stangeri*). The are also avian visitors that include black and white-casqued hornbills (*Bycannistes subcylindricus*) and barbets (Fam. Capitonidae and greenbuls (Fam. Pycnonotidae). These frugivores eat the pericarp and drop the stony endocarp.

3.4 Seed predation, fungal pathogens, and chemical interactions

Seed predation has been demonstrated to influence spatial patterns of regeneration of many plant populations [31, 32]. Seed predation may take place before seed dispersal (pre-dispersal) or after dispersal (post-dispersal). It is clear that plants and their seed predators from ecological systems have high atemporal and spatial variability with regard to seed and predator abundance. For instance, seed predation may be intense in years when other resources are scarce and low when other resources are abundant. Besides, seeds that fall under the parent tree may suffer disproportionately high levels of predation from density-dependent obligate seed predators that are resident at the parent tree [33–35]. The impacts of fungal pathogens are likely to increase with the ongoing climate change with a tendency for increased precipitation [36]. Seeds that are dispersed some distance away from such parent trees may experience low levels of seed predation [36]. In *Olea welwitschii*, smaller fruits (70%) suffer high predation rates than the larger ones (30%). In addition, more small fruits fall below the parent plant (89%)

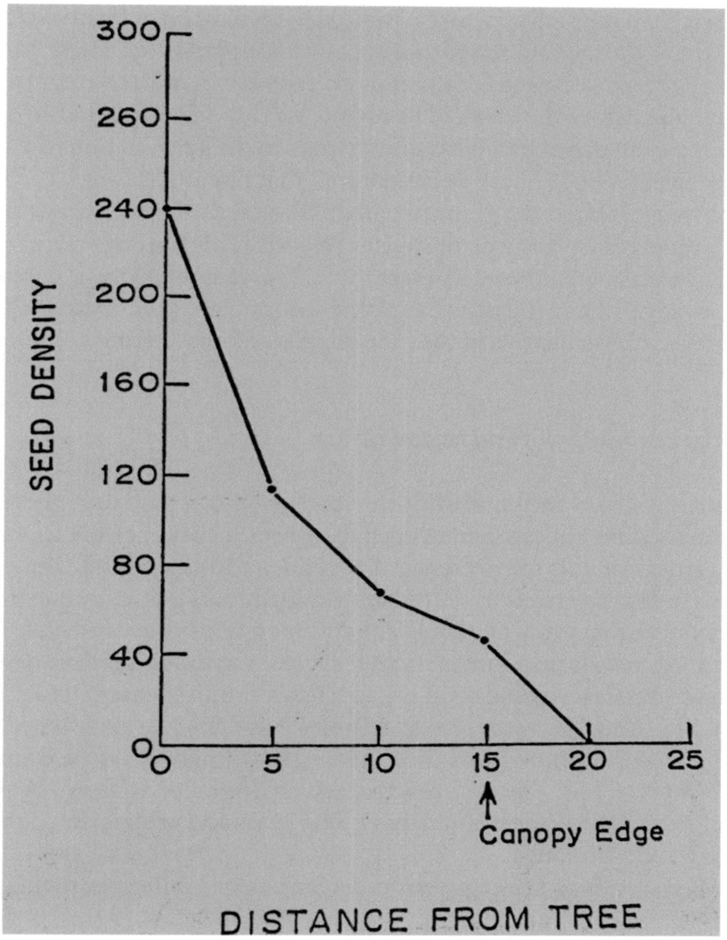

Figure 3.
Seed density as an inverse function of the distance from the parent crown.

than large ones (11%). Fruit and seed density is a decreasing function of distance from the parent trees and tends to be highly leptokurtic (**Figure 3**). Data on predation rates on fruits and seeds placed along transects away from the parent trees reveal that density has no influence.

Distance from the parent tree appears to have a significant effect on the rates of seed and fruit predation. Regardless of density, seeds tend to be eaten within hours of falling on the ground below the parent tree while fruits tend to be ignored. In fact, predation rates can be ranked with seeds within 10 m of the parent crown inside the forest (away 95%), fruits in the forest (>10 m) from the parent crown (68%), seeds in the forest away (>10 m) from the parent crown (48%) and fruits under the parent crown (7%). Clearly, while seed predation is distance-dependent, fruit predation is distance independent. The major seed predators are rodent species *Praomys jacksoni* whose population increases significantly under the *Olea welwitschii* crowns during fruiting. In addition, *Olea welwitschii* seeds under the parent tree suffer a high proportion of fungal attacks. Interestingly, mold infection tends to be influenced by seed density majority suffering an infection in the first 24 hours after falling from the parent tree. Two mutually exclusive mold species have been identified that are responsible for seed attack -*Cercosporella* sp. and *Gloesporium* sp. These two fungal species are obligate parasites of Olea that live on the leaves of the parent tree whose spores mature just before the onset of rains and are dispersed by raindrops [37, 38].

There is also evidence of chemical interactions in the germination of Olea *welwitschii* seeds. Studies by [9] have demonstrated that phytotoxin leachates from shoots of the parent tree inhibited the germination of Olea seeds. This apparent allelopathy and has been reported in many plant species [39–41]. This finding confirms the observation by the forest nursery workers that Olea seeds take longer to germinate (30 days on average) while the majority of the seeds do not germinate at all. In fact, shoot leachates additionally retard the growth of seedlings making seedling establishment difficult.

3.5 Factors that enhance natural regeneration

The absence of Olea seedlings inside the forest would appear to suggest that this canopy dominant does not regenerate at all. However, a survey of the nearby forest grasslands (glades) reveals the presence of seedlings and saplings of *Olea welwitschii*. These glades are characterized by abundant termite mounds and unique tree species (*Combretum molle*) that grow on them. The common termite species were *Cubitermes montanus* and *Macrotermes* sp. Interestingly, in glades without *Combretum molle*, there were fewer termite mounds and the species of termite present was different (*Odontotermes* sp. And *Sphaerotermes* sp. (**Tables 2** and **3**). The glade with *Combretum molle* and *Cubitermes montanus* and Olea seedlings/saplings are referred to as the Olea Regeneration Sites (ORS). The ORS tend to be heavily grazed with little evidence of burning while non-ORS glades tend to have long grass and are subjected to regular burning, sometimes annually.

In ORS, Olea seedlings/saplings tend to be found on Cubitermes montanus mounds.

3.6 Olea regeneration sites

Studies by [9] have demonstrated a strong association between termites and Combretum. For instance, of the 582 sampled in one of the glades, 422 (75.5%) had Combretum molle growing on them while 160 (27.5%) did not have *Combretum* on them. In addition, mounds in ORS tend to be highly aggregated. Clumping appears to be strong in areas where Combretum molle is permanently established but less clumping in areas with establishing Combretum trees. Mounds in non-ORS tend

Glade	Type	Mounds per Ha
Kisere	ORS	627.2
Kalunya	Non-ORS	206.4
Khavega	Non-ORS	123.2
Khasali	Non-ORS	131.2
Miyao	Non-ORS	112.0

Table 2.
Cubitermes distribution in ORS and Non-ORS glades in Kakamega forest.

Glade	*C. montanus*	Odontotermes sp.	Sphaerotermes sp.	Macrotermes	Total
Kisere	627.2	0	0	1.6	628.8
Kalunya	206.4	32	4	0	242.4
Khavega	123.2	40	0	0	163.2
Khasali	131.2	8	0	0	139.2
Miyao	112.0	2.8	0	0	114.8

Table 3.
Termite's species found in different glades in Kakamega forest.

to be regularly or randomly spaced pointing to the strong influence of Combretum on mound distribution. It appears like Combretum cannot establish strongly in areas without termite mounds. The mechanism by which Combretum establishes on mounds remains unclear and needs further investigations. What is clear is that establishment of *Combretum molle* on mounds appears to be a necessary first step in Olea regeneration. And the establishment of Combretum on mounds appears to inhibit the growth of the mounds forcing termites to construct yet another mound a short distance from the dying mound; with the process repeating itself.

3.7 Agents of seed dispersal

Olea seeds weigh a little over 1gm making it hard for them to be dispersed by wind. In addition, most Olea seedlings tend to occur at a distance from the parent trees inside the forest. This points to the animal dispersal of Olea seeds. Olea has two potential fruit predators -forest mammals and birds. For mammals, the home ranges would limit them from long-distance dispersal leaving birds as the best candidates for dispersal. Such avian Olea seed dispersers tend to be non-forest and reside in ORS and are capable of flying up the canopy, are highly frugivorous and non-territorial. Observational determination of Olea seed dispersers revealed that the yellow-whiskered greenbul (*Andropadus importunis*), the joyful greenbul (*Chlorocichla laetissima*), and the common yellow-vented bulbul (*Pycnonotus goiavier*) as the most efficient dispersers of Olea seeds from the forest into the ORS. **Table 4** clearly shows that *P. goiavier* makes far more visits to fruiting Olea than the other two, making it the most probable transporter of Olea seeds into the ORS. But these dispersers also facilitate germination. Seeds collected from their droppings germinated earlier and faster than those that had not been eaten.

Figure 4 shows the flight paths of the three dispersers. It is clear that the common yellow-vented bulbul is the only one that flies directly from the ORS into the fruiting Olea in the forest and back.

Disperser	Before Fruiting	During Fruiting	After Fruiting
Common yellow-vented bulbul	156	2344	115
Yellow-whiskered Greenbul	0	85	1
Joyful Greenbul	0	3	0

Table 4.
Number of visits to fruiting Olea by three possible dispersers.

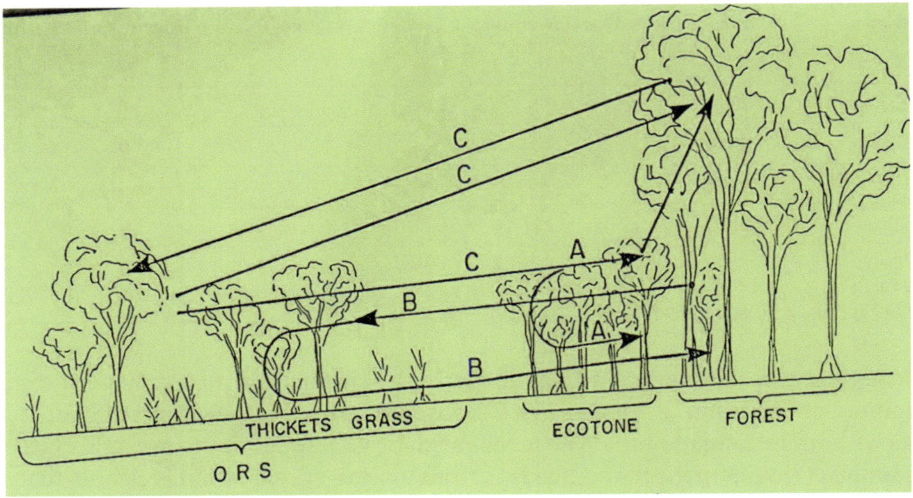

Figure 4.
Flight path of dispersers from the forest to the ORS. A is the flight path for the joyful greenbul, B for the yellow-whiskered greenbul and C for the common bulbul.

4. Conclusion

Does *Olea welwitschii* in Kakamega forest regenerate in response to disturbances that can explain its population structure? One common disturbance in tropical forests is the canopy openings resulting from tree falls [42]. And the Olea response to such a disturbance would be demonstrated by the presence of a large pool of seedlings/saplings inside the forest. Unfortunately, such pools do not exist [43]. What kind of disturbance could Olea regeneration be responding to? Regeneration of Olea appears to respond to disturbances that are unique in nature [44]. The formation of glades (open grassy areas in the forest) is one such unique disturbance that allows the regeneration of this species. Glades are by definition, disturbed sites that differ from openings in the canopy that last longer before they are transformed into a forest. Faunal and floral communities in these glades differ from those in the forest. The closest resemblance of glades to the forest lies in the species composition of canopy forest trees that are in their early and intermediate stages of succession.

Glades have two attributes that are seen to determine whether Olea will regenerate or not, low rates of seed predation, given their distance from the fruiting trees, and low incidences of attack by fungal pathogens, given their low humidity. Low seed predation rates are a function of the reduced population of rodents that destroy seeds and reduced numbers of rodents in these sites result from intense grazing which reduces ground cover that would otherwise provide habitats for the seed-eating rodents and fires that are used to stimulate the growth of new grass. Fire and grazing combined reduce ground cover. But fires affect the second component of Olea regeneration in the Olea Regeneration Sites (ORS) -termites. These glades are dominated by mound-building *Cubitermes montanus* which are absent in non-ORS. Studies by [45] revealed that fire reduces the density of mound-building termites.

The presence of mounds seems to encourage the establishment of a tree species, *Combretum molle* under whose crown, Olea seedlings can be found. The interplay between termites and Combretum on one hand and the common bulbul (*Pycnonotus barbatus*) appear to drive the regeneration of Olea in the Kakamega forest (**Figure 5**).

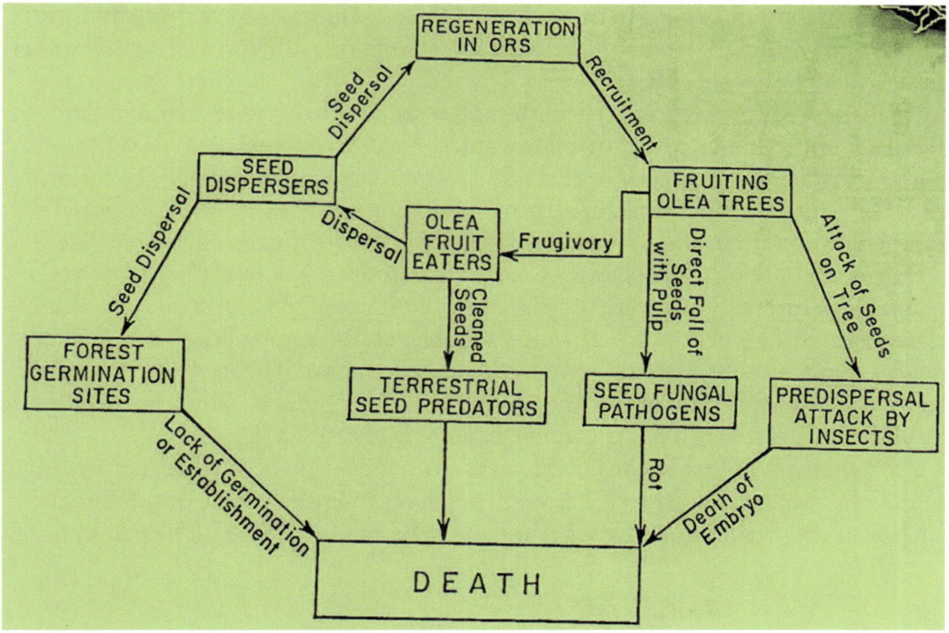

Figure 5.
Summary of destination and fate of Olea seeds in Kakamega forest. Successful seeds are those that are dispersed to the ORS.

A similar observation has been reported in Budongo forest, Uganda by [46] where it was observed that in sample plots where *Olea welwitschii* was regenerating, there were many grassland-type termites' mounds of *Bellicosus aurivilli* Sjost.

The absence of grass on the termite mounds keeps away grazing herds enabling Combretum to establish itself. Established Combretum ultimately shades the mounds preventing grass from growing. But Combretum is an important perch tree for the common bulbuls (and other bird species) during swarming of termites which tend to congregate above these tree species. The common bulbuls are strictly frugivorous and feed on Olea fruits during the fruiting season, even though they are open-country species. During fruiting, seeds are then dropped by the bulbuls under the crowns of *Combretum* and the microclimates under the crowns facilitate the germination of Olea seeds [44, 47]. Contrary to reports by [48, 49], Olea is not a colonizing species in view of its regeneration dynamics.

Survival of Olea seedlings in these ORS is good despite attacks by grazing domestic herds of the forest-adjacent communities. Survival of the saplings is also subject to harvesting by forest-adjacent communities. Olea poles tend to be harvested for use in the construction of grain storage units [50]. The diagram below summarizes the important processes in the regeneration dynamics of Olea in the ORS in the Kakamega forest. Disturbances that cause glades are the most important processes in the chain of events that will ultimately facilitate regeneration not only of Olea but also of other small-fruited canopy tree species inside the forest.

Regeneration of *Olea welwitschii* outside the forest is unique for a tropical forest tree species that is not a colonizer. Such a strategy of regeneration has not been reported for any other forest canopy dominant in the tropics. A number of studies have reported species with regeneration characteristics similar to those of Olea. Studies by [51] found that *Pithecellobium saman* and *Enterolobium cycloparm* were incapable of regenerating under their own crowns even though there were sufficient viable seeds to colonize gaps in their vicinity. Successful seeds were those that were dispersed to germination sites outside the forest in pasture fields where cattle were the possible dispersal agents [46].

Small-fruited canopy trees species like *Olea welwitschii*, *Prunus Africana*, and *Sapium ellipticum* tend to be the first forest tree species to establish in the ORS, representing a new category of forest tree species regenerating outside the forest. They, however, lack the characteristics of colonizing species such as rapid growth, short generation time, and a short lifespan that is typical of early colonizers like *Trema guinensis, Polyscias kikuyensis, Croton megalocarpus and Macaranga kilimanjaris* in Kakamega forest. Neither are they large gap colonizers given that they are absent in large gaps inside the forest [52]. Once established, the composition of the canopy remains the same contrary to the predictions of the Mosaic Theory [53]. What does change is the composition of the under-canopy species brought about by the different sets of dispersers that inhabit the under-canopy in the forest [54].

The canopy structure in the ORS derives from the origin of patches in different parts of the regeneration sites. Each small patch represents a refuge or safe site in which seeds of these canopy trees germinate and establish [55–57]. The patchy nature

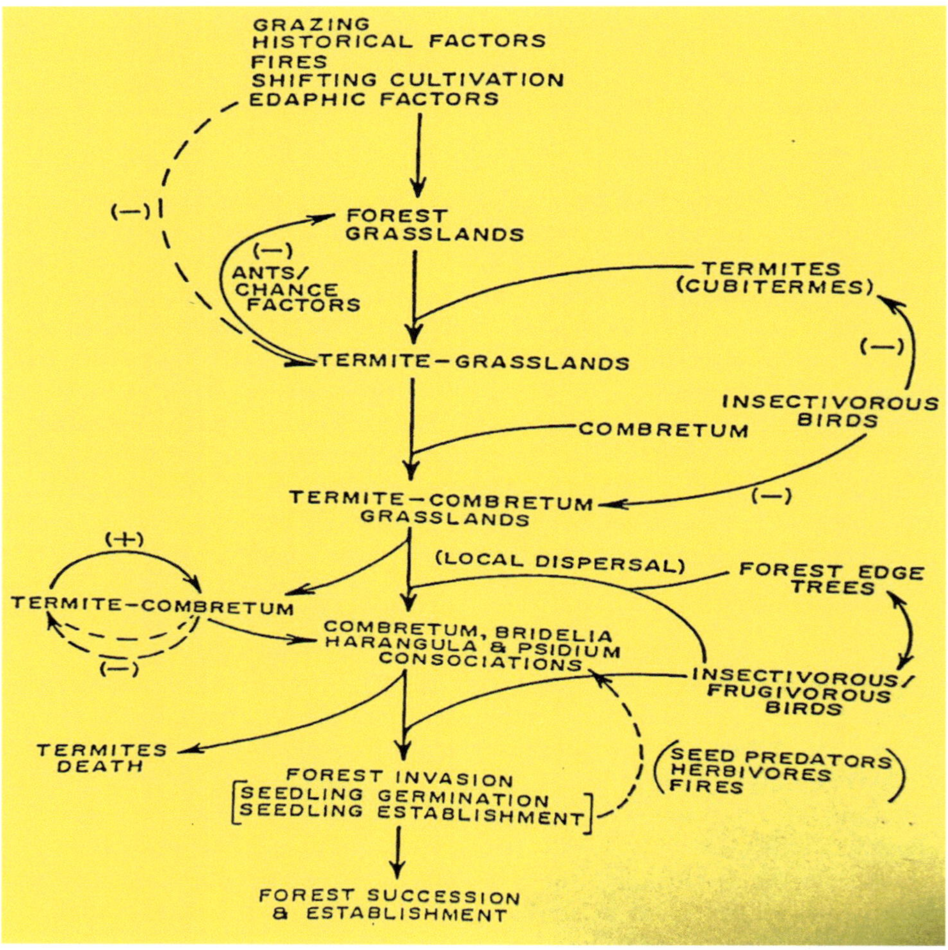

Figure 6.
Summary of olea regeneration cycle in Kakamega forest.

of these safe sites means that patches of forest develop independently and in parallel to each other. Expansions of these patches over time lead to coalescence which ultimately joins the larger forest block through the forest edge to form a continuous forest [58, 59]. In essence, the forest regenerates by shifting (a moving forest) the spatial distribution of the canopy dominants. Conditions necessary for the regeneration of the canopy dominants are rather specific requiring mound-building termites, a tree species that specialize in growing on these mounds, a transport agent, and later, a pool of the small-fruited forest canopy dominants (**Figure 6**). It is clear that canopy trees in the Kakamega forest initiate regeneration outside the forest.

Studies by [60, 61] have shown that the regeneration of a few canopy tree species such as *Prunus africana* can take place inside the forest successfully if the process is managed [62, 63]. In the Kakamega forest, past disturbances have resulted in the domination by the understory shrub, *Brillantaisia* sp., a light-demanding perennial herb [60, 63].

Author details

Mugatsia H. Tsingalia
Department of Biological Sciences, Masinde Muliro University of Science and Technology, Kakamega, Kenya

*Address all correspondence to: mugatsi2005@gmail.com

References

[1] Aerts R. *Olea capensis* L. In: Lemmens RHMJ, Louppe D, Oteng-Amoako AA, editors. PROTA (Plant Resources of Tropical Africa/Ressources Végétales de l'Afrique Tropicale). Netherlands: Wageningen; 2011

[2] Stevens, PF. Oleaceae at Angiosperm Phylogeny Website. At: Botanical Databases. St Louis: University of Missouri; 2001

[3] Turrill WB. Flora of Tropical East Africa: Oleaceae. Vol. 31. Royal Botanical Gardens, Kew; 1952

[4] Tsingalia HM, Nyongesa H. Inherently delayed germination of Elgon Teak (Olea capensis) seeds in Kakamega forest, Kenya. Journal of Biodiversity. 2010;**1**(2):71-75

[5] de Jussieu AL. ORDO IV Jasmineae. In: Genera Plantarum: Secundum Ordines Naturales Disposita. Parisii: Apud Viduam Herissant et Theophilum Barrois; 1789. pp. 104-106

[6] Palmer E, Pitman N. Trees of Southern Africa: Covering All Known Indigenous Species in the Republic of South Africa, South-West Africa. Vol. 1 & 2. Botswana, Lesotho & Swaziland, Cape Town: A.A. Balkema; 1972

[7] Dale IR, Greenway PJ. Kenya Trees and Shrubs. London: Hatchards; 1964

[8] Kokwaro JO. Ecological facets of the unique vegetation types of tropical Africa, with special reference to East Africa. [PhD thesis]. West Germany: Universität Bayreuth; 1978

[9] Tsingalia HM. Animals and the regeneration of a canopy tree in an African Tropical Forest. [PhD thesis] University of California, Berkeley. 1988

[10] Greig-Smith P. Quantitative Plant Ecology. 3rd ed. New York and London: Academic Press; 1983

[11] Kershaw KA. Quantitative and Dynamic Plant Ecology. 2nd ed. Edward Arnold; 1973

[12] Evans DA. The experimental evidence concerning contagious distributions in ecology. Biometrika. 1953;**40**:186-211

[13] Pielou EC. Association tests versus homogeneity tests: their use in subdividing quadrats into group's vegetation. Vegetatio. 1969;**18**:4-18

[14] Pielou EC. Mathematical Ecology. New York: Wiley; 1977

[15] Clark DA, Evans FC. Distance to nearest neighbour as a measure of spacing relationship in populations. Ecology. 1954;**35**:445-453

[16] Tsingalia HM. Seed predators, drupe predation rates and their effects on the regeneration dynamics of *Olea welwitschii* in Kakamega forest, Kenya. Asian Journal of Plant and Soil Sciences. 2016;**1**(1):9-16

[17] Swaine MD, Lieberman D, Putz FE. The dynamics of tree populations in tropical forest: A review. Journal of Tropical Ecology. 1987;**3**:359-366

[18] Lieberman D, Lieberman M. Forest tree growth and dynamics at La Selva, Costa Rica (1969-1982). Journal of Tropical Ecology. 1987;**3**(4):347-358

[19] Eggeling WJ. Observation on the ecology of Budongo rain forest, Uganda. Journal of Ecology. 1947;**34**:20-87

[20] Muvengwi J, Chisango T, Mpakairi K, Mbiba M, Witkowski ETF. Structure, composition and regeneration of miombo woodlands within harvested and unharvested areas. Forest Ecology and Management. 2020;**458**:117792

[21] Swaine MD, Whitmore TC. On the definition of ecological species groups in tropical forests. Plant Ecology. 1988;**75**(1):81-86

[22] Kasenene JM, Murphy P. Post-logging tree mortality and major branch losses in Kibale Forest, Uganda. Forest Ecology and Management. 1991;**46**:295-307

[23] Borah N, Athokpam FD, Garkoti SC, Das AK, Hore DK. Structural and compositional variations in undisturbed and disturbed tropical forests of Bhuban hills in south Assam, India. International Journal of Biodiversity Science, Ecosystem Services & Management. 2014;**10**(1):9-19. DOI: 10.1080/21513732.2013.873823

[24] Hubbell SP, Foster RB. In: Whitmore TC, Chadwick AC, Sutton AC, editors. Tropical Rain Forest: Ecology and Management. Oxford: The British Ecological Society; 1983. pp. 25-41

[25] Thorington, R.W, Tanuebaum, B., Tarak, A. & Rudran, R. (1982). Distribution of trees on Barro Colorado Island: A five-hectare sample. In the Ecology of a Tropical Forest: Seasonal Rhythms and Long-term Changes, edited by Egbert, G., Leigh Jr, Rand, S.A. & Windsor, D.M. Smithsonian Institutional Press, Washington DC.

[26] Eggler WA. The maple-bass wood forest types in Washburn County Wisconsin. Ecology. 1938;**19**:243-263

[27] Oosting HJ. An ecological analysis of plant communities of Piedmont, North Carolina. The American Midland Naturalist. 1942;**28**(1):1-126

[28] Keever C. Mechanisms of plant succession on old fields of Lancaster County, Pennsylvania. Bulletin of the Torrey Botanical Club. 1974;**106**(4):299-308

[29] Henry and Swan. Reconstructing forest history from live and dead plant material–An approach to the study of forest succession in southwest New Hampshire. Ecology. 1974;**55**(4):772-783

[30] Peterken GF, Tubbs CR. Woodland regeneration in the New Forest, Hampshire, Since 1650. Journal of Applied Ecology. 1965;**2**(1):159-170

[31] Goncalves F, Revermann R, Gomes A, Aidar PM, Finckh M, Jürgens N. Tree species diversity and composition of Miombo Woodlands in South-Central Angola: A chronosequence of forest recovery after shifting cultivation. International Journal of Forestry Research. 2017;**71**:13

[32] Mengich EK, Macharia JM, Mitloehner R, Too DK, Muturi FM. The dispersion distribution of indigenous trees is an indicator of adapted species in the semi-arid rangelands of Kenya. Journal of Tropical Forest Science. 2020;**32**(2):144-153

[33] Janowiak M, Nagel LM. Minimum tree size and interpretation of stand structure in uneven-aged northern hardwoods. Northern Journal of Applied Forestry. 2010;**27**(1):34-37. DOI: 10.1093/njaf/27.1.34

[34] Kayombo CJ, Ndangalasi HJ, Giliba RA, Kikoti I. Assessment of natural regeneration potential of tree species in image forest reserve, Tanzania. International Journal of Advanced Research. 2022;**5**(1):49-59

[35] Kand D, Guo Y, Ren C, Zhao F, Feng Y, Han X, et al. Population structure and spatial pattern of main tree species in secondary *Betula platyphylla* forest in Ziwuling mountains, China. Scientific Reports. 2014;**4**:6873. DOI: 10.1038/srep06873

[36] Velásquez AC, Castroverde CDM, He SY. Plant-pathogen warfare under changing climate conditions. Current Biology. 2018;**28**(10):R619-R634. DOI: 10.1016/j.cub.2018.03.054

[37] Elmugheira MIM, Hassan TT, Idris EA, Abdel-Magid TD. Tree population structure, diversity, regeneration status, and potential disturbances in Abu Gadaf natural reserved forest, Sudan. Environmental Challenges. 2021;**5**:1-16. DOI: 10.1016/j.envc.2021.100366

[38] Kollman J, Schill HP. Spatial patterns of dispersal, seed predation, and germination during colonization of abandoned grassland by Quercus petraea and Corylus avellana. Plant Ecology. 1996;**125**(2):193-205

[39] Calvino-Cancela M. Spatial patterns of seed dispersal and seedling recruitment in *Corema album* (Empetraceae): The importance of unspecialized dispersers for regeneration. Journal of Ecology. 2002;**90**(5):775-784

[40] Janzen DH. Seed predation by animals. Annual Review of Ecology and Systematics. 1971;**2**:465-492

[41] Comita JFA. Density-dependent pre-dispersal seed predation and fruit set in a tropical tree. Oikos. 2010;**119**:1841-1847

[42] Xiao Z, Mi X, Holyoak M, Xie W, Cao K, Yang X, et al. Seed–predator satiation and Janzen–Connell effects vary with spatial scales for seed-feeding insects. Annals of Botany. 2017;**119**(1):109-116

[43] Bustamante RO, Simonetti JA. Seed predation and seedling recruitment in plants: The effect of the distance between parents. Plant Ecology. 2000;**147**:173-183. DOI: 10.1023/A:1009852400814

[44] Nakamura RR, Mitchell-Olds T, Manasse RS, et al. Seed predation, pathogen infection and life-history traits in *Brassica rapa*. Oecologia. 1995;**102**:324-328. DOI: 10.1007/BF00329799

[45] Williams PJ, Ong RC, Brodie JF, et al. Fungi and insects compensate for lost vertebrate seed predation in an experimentally defaunated tropical forest. Nature Communications. 2021;**12**:1650. DOI: 10.1038/s41467-021-21978-8

[46] Turk MA, Tawaha AM. Allelopathic effect of black mustard (*Brassica nigra* L.) on germination and growth of wild oat (*Avena fatua* L.). Crop Protection. 2003;**22**(4):673-677

[47] Tsingalia HM, Auma JM. Allelopathic Interactions of Parthenium hysterophorous with Kidney bean, *Phaseolus vulgaris*. Asian Journal of Biology. 2020;**10**(3):25-33

[48] Zhao J, Yang Z, Zou J, Li Q. Allelopathic effects of sesame extracts on seed germination of moso bamboo and identification of potential allelochemicals. Scientific Reports. 2022;**12**:6661. DOI: 10.1038/s41598-022-10695-x

[49] Eilu G, Hafashimana DLN, Kasenene JM. Density and species diversity of trees in four tropical forests of the Albertine rift, Western Uganda. Diversity and Distributions. 2004;**10**:303-312

[50] Kasenene JM. Post logging structural changes and regeneration

of *Olea welwitschii* (Knobl) Gilg. & Schellemb. in the Kibale National Park, Uganda. African Journal of Ecology. 2007;**45**(3):109-115. DOI: 10.1111/j.1365-2028.2007.00866.x

[51] Denslow JS, Hartshorn GS. Tree falls gap environments and forest dynamic processes. In: Mcdade LA, Bawa KS, Hespensheide HA, Hartshorn GG, editors. La Selva; Ecology and Natural History of a Neotropical Rainforest. Chicago, IL: University of Chicago Press; 1994

[52] Benzie JAH. The distribution, abundance and the effect of fire on mound-building termites (*Trinervitermes*) and *Cubitermes* sp. (Isoptera: Termitidae) in northern Guinea savannaj, West Africa. Oecologia. 1986;**70**:559-567

[53] Egler FE. Vegetation concepts I. Initial floristics composition - A factor in Oldfield vegetation development. Vegetatio. 1954;**4**:412-417

[54] Latja P, Valtonen A, Malinga GM, Roininen H. Active restoration facilitates bird community recovery in an Afrotropical rainforest. Biological Conservation. 2016;**200**:70-79

[55] Brokaw N, Busing T. Niche versus chance and tree diversity in forest gaps. Trends in Ecology & Evolution. 2000;**15**:183-188

[56] Hamilton AC, Taylor D, Vogel J. Early clearance and environmental degradation in Southwest Uganda. Nature. 1986;**320**:164-167

[57] Tsingalia HM. Impacts of selective logging on population structure and dynamics of a canopy tree (*Olea capensis*) in Kakamega forest. African Journal of Ecology. 2010;**48**(3):569-575

[58] Tsingalia HM. Wave regeneration in an Africa Tropical Forest canopy tree. Journal of Tropical Ecology. 1990;**10**:133-145

[59] Sangsupan HA, Hibbs DE, Withrow-Robinson BA, Elliott S. Effect of microsite lighton survival and growth of understory natural regeneration during the restoration of seasonally dry tropical forest in upland northern Thailand. Forest Ecology and Management. 2021;**419**:91-100

[60] Hatheway WH, Baker HG. Reproductive strategies in *Pithecolobium* and *Enterolobium*-further information. Evolution. 1970;**24**:253-254

[61] Pickett STA. Different adaptation of tropical tree species to canopy gaps and its role in community dynamics. Tropical Ecology. 1983;**24**:69-84

[62] Weber W, White JT, Vedder A, Naughton-Treves L. African Rain Forest Ecology and Conservation. New Haven, CT: Yale University Press; 2001

[63] Tsingalia HM. Variation in seedling predation and herbivory in *Prunus africana* in the Kakamega Forest, Kenya. African Journal of Ecology. 1989;**27**:207-217

Chapter 6

Biodiversity of Fabaceae in the Brazilian Amazon and Its Timber Potential for the Future

Luiz Augusto Gomes de Souza

Abstract

The vegetation of Neotropical forest of Amazonian region is one representative part of global biodiversity in Fabaceae, with numerous species for wood production, oil, fruits, or other forests products or same ecosystemic services for agricultural production. This chapter gathers information on the biodiversity of Fabaceae of the Amazon, highlighting the timber species in the set of existing plant genetic resources. Systematic information emphasizes the importance of basic forestry information, and potential for cultivation in agroecosystems or for management, forestry and reforestation. The product of the highest economic value of the Fabaceae of the Brazilian Amazon is wood, which is sold in local, regional markets and exported to other countries. Also noteworthy is the ability acquired evolutionarily by many Fabaceae: the symbiotic fixation of N_2 with soil bacteria of the Rhizobia group, character particularly important for soil recuperation or reforestation.

Keywords: biodiversity, Fabaceae, Amazon, forestry, timber

1. Introduction

The Amazon rainforest is known as one of the hot spots of biodiversity in the world, and has, in its geographical area, numerous zones of detachable biological variety, centers of diversity, with incalculable stock of genetic resources for current and future use. Recent estimates of plant biodiversity in the Amazon indicate that of the 270,000 terrestrial plant species, 32,000 are present in Brazilian biomes (11.8% of global biodiversity) and, of these, ± 14,000 species in the Brazilian Amazon representing 5.2% of global biodiversity [1].

Territorially the continental Amazon occupies 50% of the surface of South America, distributed geographically in nine countries: Bolivia, Colombia, Ecuador, Guyana, French Guiana, Peru, Suriname, Venezuela and Brazil (**Figure 1**). In Brazil, the ecological domain of the Amazon biome has 3.68 million km^2, which added to the ecotone zones with the Caatinga and Cerrado biomes total 4.24 million km^2, which corresponds to 49% of the Brazilian territory [2].

The need for natural wood managed or reforestation to meet local and global market demand is a consensus among reforestation and, in the future, natural and

Figure 1.
Geographical distribution of the Amazon biome in northern South America. Fonte: https://news.files.bbci.co.uk/include/vjamericas/304-countries-amazon-rainforest/portuguese/app/embed.

cultivated forests will have great economic and social importance, in addition to contributing to carbon sequestration and global climate stability. Among the natural riches of the Amazon, its timber potential stands out, which are estimated at tens of millions of m^3, and which can, in unprotected areas, be rationally exploited with sustainable forest management practices.

2. Biodiversity of Fabaceae in the Amazon region

In the Amazon, the Italian botanist Adolpho Ducke (1876–1959) pioneered the diversity of the Fabaceae family. In 1925, Ducke published "The legumes of the state of Pará," and, with the progress of his research in the region, in 1949, presented the basic treatise: "The Legumes of the Brazilian Amazon," relating 867 species of Fabaceae [3, 4]. Subsequently, the botany Marlene Freitas da Silva (1937–2005) related the Fabaceae of the Brazilian Amazon with estimates of 1241 species distributed in its three subfamilies: Caesalpinioideae, Mimosoideae and Faboideae [5].

Standing out in the tropical forest in number of species and frequency in the listing of the main timbers exported from the Amazon region, the trees of the Fabaceae family it's adapted in the different ecological environments of this biome. In the world, the geographical distribution of Fabaceae stands out in tropical and subtropical regions and numerically represents the second largest family of Magnoliopsid plants (surpassed by Orchidaceous), estimating global biodiversity in 19,325 species classified in 727 botanical genera [6]. In the different Brazilian biomes, which include the Amazon, Cerrado, Caatinga, Atlantic Forest, Pantanal and Pampas, the diversity of Fabaceae is currently estimated at 2964 species (1458 endemic), 53 subspecies (16 endemic) and 731 varieties (419 endemic), sheltered in 210 genera [7].

The Fabaceae of the Amazon are represented in arboreal, shrubby, liana, herbaceous, underwater and aquatic forms. This detachable plasticity of growing habits includes trees of different sizes, occupying the niches of understory to the upper canopy, but also diversified in shrubs, lianas and herbs. In Fabaceae phylogeny, the tree component forms the most primitive group with tropical origin and the most evolved herbaceous group, whose evolutionary differentiation occurred mainly in subtropical zones and even in temperate zones [8].

The family Fabaceae (Leguminosae) is one of the most natural of the Division Magnoliophyte and belongs to the Fabales order. As for phylogeny, Fabaceae are divided into three subfamilies: Caesalpinioideae, Mimosoideae and Faboideae and each has branches consisting of classification tribes. In each subfamily, the distinctions between genera and species are related to floral morphology. Thus, the main differential characteristics between the three subfamilies are [9]:

Caesalpinioideae: These are trees, shrubs and rare herbaceous or scanning plants, distributed mainly in tropical and subtropical regions, with ±180 genera and between 2500 and 3000 species. The leaves are usually pined or sometimes pecked. The flowers are slightly zygomorphic, and the side petals (wings) cover the banner on the button. The stamens are numbered 10 or less (free or monoadelphal). It is considered the most primitive subfamily within the Fabaceae.

Mimosoideae: They are trees, shrubs, woody vines, with few perennial herbaceous, in tropical and subtropical regions, with approximately 56 genera and ± 3000 species. The leaves are usually bipinated, with glands in the raque. The petiole usually presents pulvino and the presence of thorns is common, in those species differentiated in the dry environment, and these may originate from stipulations or occur along the branches. The flowers are regular or actinomorphs, androgynous or unisexual, with valve petals on the bud, with 10 stamens or more. In evolutionary terms, the species of this subfamily are positioned between the Caesalpinioideae and Faboideae.

Faboideae: Trees, shrubs and mainly annual or perennial herbs, distributed in temperate, tropical and subtropical zones, with 400–500 or more genera and > 10,000 species. The leaves are usually pined, but sometimes simple unifolioladas or trifolioladas. The flowers are hermaphrodite, typically zygomorphic, with side petals (wings) covered by the banner, on the button; 10 theses, usually diadelfs (9 + 1), but sometimes monoadelphal or free. The species of this subfamily, especially herbaceous ones, are considered more evolved among the Fabaceae.

The differentiation of species in subfamilies is mainly made by describing their floral morphology (**Figure 2**). Caesalpinioideae species have flowers in ascending pendulums, with five petals and 10 free stamens, e.g. *Aldina heterophylla*, *Campsiandra laurifolia* and *Senna multijuga*. Mimosoideae has flowers consisting of multiple and numerous stamens, defining a particular floral morphology in the set of Fabaceae, e.g. *Hydrochorea marginata*, *Inga cinnamomea* and *Parkia igneiflora*. The Faboideae, the most numerous group, have flowers with five petals and 10 stamens united at the base, or nine united stamens and one free, for example. *Clitoria fairchildiana* and *Dipteryx odorata*.

The most accepted evolutionary theory for the differentiation of Fabaceae attributes the origin of the family in the Mesosolic era, in the Late Cretaceous period, ±97 million years ago [10]. In the Amazon the genus *Hymenaea* (Caesalpinioideae) was researched with an evolutionary and ecological perspective, evidencing the African origin of neotropical species, possibly by migration across the Atlantic occurred at the beginning of the Tertiary (lower Paleocene), when the continents were closer and the tropical rain forest had a larger geographical distribution than the current one [11].

Figure 2.
Floral morphology of tree species of Fabaceae from the Amazon region of woody habit and wood interest. Photos: Souza LAG.

In primitive terms, the subfamily Caesalpinioideae originated the other ones and it is currently considered that the representatives of Faboideae form a more evolved group within the family, with 441 genera and ± 12,300 species with worldwide geographical distribution [8]. Considering phylogeny, the primitive Fabaceae have tropical origin and almost exclusively arboreal growth habit [12]. In fact, 95% of

Caesalpinioideae and Mimosoideae are woody species, frequent and abundant in tropical forests, as well as in the Brazilian Amazon.

3. Timber importance of Fabaceae of the Amazon

In Brazil, Fabaceae are classified into 210 genera, 160 of them represented in the Amazon [7]. In this geographical area, the biological diversity of Fabaceae is higher for tree and/or woody habit species (98 genera - 61.2% of the total) (**Table 1**). In ten of the genera listed (6.2% of the total), there are similar species with mixed growth habits, because they harbor tree plants, but also of a gaping, non-woody, shrubby and herbaceous habit. In 32.6% of the genera present in the region, there are no species with tree or woody growth.

The physiognomy of the rainforest is a mosaic of different landscapes that, in the Amazon, include the meadows and peasants, forests and lowland fields, seasonal igapós, streams, tidal forests, land fields, dense forests of soil no hydromorphic, dry or deciduous forests, transition forests, open forests of palm trees, vine forests and bamboo forests [13]. In these different physiognomies the richness in Fabaceae is numerically well represented and documented in floristic and phytosociological surveys already performed.

Although the number of woody genera is almost twice as high as non-woody ones, there is a lot of variation in the intraspecific number. The most representative woody genera are *Inga*, with 93 species, *Swartzia* with 80 and *Tachigali* with 44 species. In *Mimosa*, a genus that has mixed habit species, there are 36 species. In the flora of Brazil, some genera of Fabaceae are unique to the Amazon, among them *Eperua* and *Cynometra* (with 12 species each), *Aldina* (10 species), *Elizabetha* (9), *Crudia* and *Campsiandra* (6), *Alexa* (5), *Taralea* and *Heterostemon* (4), *Samanea*, *Hydrochorea* and *Clathrotropis* (3) and *Vouacapoua* and *Cenostigma* (two species) (Lima, 2010). There are also many monotypic genera, where some species such as *Cedrelinga cateniiformis* and *Dinizia excelsa*, which produce wood of economic value and there are also numerous genera that are represented by a single taxon.

There are other genera of Fabaceae very well represented, with diversity in species greater in the Amazonian flora, but which also has similar species in Brazilian biomes. Among them are: *Macrolobium* (34 species in the Amazon), *Ormosia* (23), *Dimorphandra* (19), *Abarema* and *Parkia* (16), *Peltogyne* (15), *Andira* (13), *Hymenolobium* (12), *Diplotropis* (9), *Dipteryx* (8), *Albizia*, *Erythrina* and *Hymenaea* (7), *Enterolobium*, *Vatairea* and *Platymiscium* (6), *Martiodendron* (3) and *Bowdichia* (2 species) [7].

Fabaceae trees of wood importance grow in native forests in economically unexplored areas and also in those altered by regional development. The size of the trees, preferred environment, trunk diameter and wood volume are diversified, as well as their technological properties and market value. In wood, there is a wide variety of coloration, density, texture, resistance to curving and other technological properties, suggesting multiple forms of utilization. For the species with the highest market value, there is an advance in the knowledge of the technological properties of wood supporting quality indicators that expand its potential for use.

In the identification of wood species in field conditions, to help identify the trees, the timber collector and parabotanists observe the morphological characteristics of the trunk, such as shape, color and texture of the bark and wood, presence or not of latex or resin and characteristic smell after small cut [14]. It is not always possible to

Growing habit	Genera of Fabaceae of the Amazon
Woody (trees)	*Abarema*, *Albizia*, *Aldina*, *Alexa*, *Amburana*, *Anadenanthera*, *Andira*, *Androcalymma*, *Apuleia*, *Ateleia*, *Barnebydendron*, *Batesia*, *Bocoa*, *Bowdichia*, *Brownea*, *Browneopsis*, *Caesalpinia*, *Calliandra*, *Campsiandra*, *Cassia*, *Cedrelinga*, *Cenostigma*, *Centrolobium*, *Chloroleucon*, *Clathrotropis*, *Cojoba*, *Copaifera*, *Crudia*, *Cyclolobium*, *Cynometra*, *Dialium*, *Dicorynia*, *Dimorphandra*, *Dinizia*, *Diplotropis*, *Dipteryx*, *Diptychandra*, *Dussia*, *Elizabetha*, *Enterolobium*, *Eperua*, *Erythrina*, *Etaballia*, *Guianodendron*, *Guilandina*, *Heterostemon*, *Hydrochorea*, *Hymenaea*, *Hymenolobium*, *Inga*, *Jacqueshuberia*, *Lecointea*, *Leptolobium*, *Leucaena*, *Macrolobium*, *Martiodendron*, *Monopteryx*, *Mora*, *Myrocarpus*, *Myroxylon*, *Ormosia*, *Paloue*, *Paloveopsis*, *Panurea*, *Paramachaerium*, *Parkia*, *Peltogyne*, *Pentaclethra*, *Petaladenium*, *Pithecellobium*, *Plathycyamus*, *Plathymenia*, *Platymiscium*, *Platypodium*, *Poecilanthe*, *Poeppigia*, *Poincianella*, *Pseudopiptadenia*, *Pterocarpus*, *Pterogyne*, *Recordoxylon*, *Samanea*, *Schizolobium*, *Stryphnodendron*, *Swartzia*, *Tachigali*, *Tamarindus*, *Taralea*, *Tipuana*, *Trischidium*, *Ulenthus*, *Vachellia*, *Vatairea*, *Vataireopsis*, *Vouacapoua*, *Zapoteca*, *Zollernia* e *Zygia*. (n = 98)
Mixed (woody and non-woody)	*Bauhinia*, *Chamaecrista*, *Clitoria*, *Dalbergia*, *Lonchocarpus*, *Machaerium*, *Mimosa*, *Piptadenia*, *Senegalia* e *Senna*. (n = 10)
Non-woody (shrubs, herbaceous and lianas)	*Abrus*, *Aeschynomene*, *Amphiodon*, *Arachis*, *Barbieria*, *Cajanus*, *Calopogonium*, *Camptosema*, *Canavalia*, *Candolleodendron*, *Centrosema*, *Chaetocalyx*, *Cleobulia*, *Collaea*, *Coursetia*, *Cratylia*, *Crotalaria*, *Cymbosema*, *Deguelia*, *Derris*, *Desmodium*, *Dicymbe*, *Dioclea*, *Discolobium*, *Entada*, *Eriosema*, *Exostyles*, *Fissicalyx*, *Galactia*, Harpalyce, *Indigofera*, *Macroptilium*, *Macrosamanea*, *Milletia*, *Mucuna*, *Neptunia*, *Nissolia*, *Pachyrhizus*, *Periandra*, *Phanera*, *Phaseolus*, *Psophocarpus*, *Pueraria*, *Rhynchosia*, *Sesbania*, *Sommeringia*, *Sophora*, *Stylosanthes*, Tephrosia, *Teramnus*, *Vigna* e *Zornia*. (n = 52)

Source: [7], adapted.

Table 1.
Diversity and classification regarding the habit of growth for 160 Amazonian genera of Fabaceae, regarding the growth habit of the species.

access the canopy of the plant when it develops in the upper canopy of the forest. Some characteristics of the trunk of amazon wood species are illustrated in **Figure 3**. Field identification recognizes the individual at the family, gender or even species level, and may be more accurate with sampling canopy, leaves and/or flowers or fruits. In addition to the knowledge of high-level botanical systematics, new technologies can help in the reliable recognition of species.

The use of wood of the species exploited by this potential is also quite varied, from civil and shipbuilding, carpentry, furniture, tacos, cutlery, adornment objects, and

Figure 3.
Aspect of the trunk and bark of Fabaceae species from the Amazon region of wood interest. Photos: Souza LAG.

numerous other applications. Thus, the main economic product of the Fabaceae of the Amazon is wood, notified by those used for noble purposes.

In the group of species that has been most used for its timber value, those whose uses and mechanical properties have already been investigated are classified, revealing their economic and industrial potential. In the research already carried out, it was

Scientific name	Characteristics and use of wood
Cedrelinga cateniiformis	Wood easy to saw, plan, nail and screw. It gets good finish. Used in light construction, furniture, general works of carpentry, joinery and finishing, floors, partitions, boxes, plywood and for the construction of vessels (internal).
Dalbergia spruceana	Very heavy wood, with violaceous core, which receives attractive treatment. Used in decoration coatings, joinery, segeria, ebanisteria, marquetry, turned and adornment objects, knife cables, brushes, boxes or notched cases. Good for offices.
Dicorynia paraensis	Hard wood and imputable, rich in silica. It is used in civil construction, shipbuilding, carpentry, segeria, cuttings and hydraulic works, dormant. It's flexible, elastic.
Dinizia excelsa	Wood very heavy and hard to cut, high resistance to the attack of xylophages and very resistant. Suitable for external uses such as poles, bridges, piles, sleepers, mains, for construction such as rafters, beams, slats, such as clubs and boards, stops, bodywork, wagons and shipbuilding.
Diplotropis martiusii	Very hard and resistant wood in contact with water, used in shipbuilding, in parts of boats such as hull, also in furniture, carpentry, for frames, doors, turned parts, crosses, poles, floorboards, clubs and civil construction in general.
Dipteryx odorata	Wood very heavy, hard to cut, very resistant to the attack of termites and rotting fungi. Indicated for civil construction, such as beams, rafters, slats, floor boards and clubs, door stops, liners, for external use as poles, mains, piles, crosslets, bodywork, shipbuilding and furniture.
Enterolobium schomburgkii	Heavy wood that receives good finish with high chandelier, so even employed in joinery, franchisees, construction in general, board for floors, door stops, tacos, good quality furniture and sleepers.
Hymenaea courbaril	The wood is hard and sturdy, employed in grinders, firewood, mains, portals, etc. Heavy construction, hydraulic works, bodywork, mills, poles, beams, barrel, etc.
Macrolobium acaciifolium	Reddish wood, compact, imputable. Used in carpentry, boxes, pulp and paper, good quality boards and joinery. It is very flexible, used in red furniture taking various shapes.
Peltogyne catingae	Hard and compact wood, with intense purple heartwood or purplish violet, resistant to decomposition, glossy polishing. Used in segeria, dormant, construction and shipbuilding, sculptures, fine joinery, ebanisteria, floor clubs and carpentry.
Vatairea guianense	The wood is heavy, with light brown heartwood, resistant to fungi and termites, easy to work and receives good finish. It is used in carpentry, crate, construction, industrial boxes, poles, etc.
Zygia racemosa	Very heavy wood, light yellow heartwood, resistant to fungi. Used in ebanisteria, luxury joinery, parquets, floor clubs, and construction in general, plywood, decorative laminates cutlery, walking sticks, dormant sands and adornment objects.

Source: [15–18].

Table 2.
General characteristics of wood of economic value and its jobs, for some species of Fabaceae of the Amazon.

also found that there is availability of wood-producing species for secondary use, such as firewood and coal, and that they can be used for less noble uses, such as azimbre, plywood, crate, etc. However, the high diversity in species demands the advancement of wood research of those little used by the lack of knowledge of its technological properties.

Table 2 contains information on some general characteristics of Amazonian tree wood, with information on its current use, among them some of the most valuable. Among them are *Cedrelinga cateniiformis*, *Dalbergia spruceana*, *Dinizia excelsa*, etc., expanding its potential for identification and commercialization in the timber, national and foreign markets. Most of the wood that reaches foreign markets is hard, heavy and resistant, indicating that the density of wood is an important indicator to define its potential for use.

4. Random characteristics of Fabaceae of timber importance in the Amazon

In opposite in the advance of knowledge of the technological properties of wood that defines its potential form of use, forestry research aims to support the recovery of the landscape and the potential of supplying raw material for the future. Historically, the exploitation of native trees is the source of raw material for numerous products and by-products, such as wood, medicines, cellulose and paper, food for fauna and man, fibers, oils and resins, gums, etc. There is a need for research on the silvicultural and economic potential of the species, adding value to biodiversity, especially in planting and reforestation actions, contributing to the conservation of forest resources. Some specific information for 45 random tree species of the Amazon region, with the popular name in Brazil, phylogeny, size characteristic, wood density and N_2 fixation ability are presented in **Table 3**.

Species	Popular name	Phylogeny (Tribe)*1	Tree height	Wood density ($g\ cm^{-3}$)*2	N_2 Fixation
Caesalpinioideae					
Caesalpinia ferrea	Jucá	Caesalpinieae	Small	1, 19	No
Campsiandra laurifolia	Acapurana	Caesalpinieae	Medium	0,90–1,10	Yes
Cassia fastuosa	Chuva-de-ouro	Cassieae	Medium	0,60–0,70	No
Cassia grandis	Mari-mari sarro	Cassieae	Big	0,65–0,77	No
Cassia leiandra	Ingá-mari-mari	Cassieae	Small	0,86	No
Copaifera multijuga	Copaíba	Detarieae	Big	0,75–0,85	No
Crudia oblonga	Iperana	Detarieae	Big	0,70	No
Cynometra bauhiniifolia	Jutairana-preta	Detarieae	Medium	0,81	No
Dialium guianense	Jutaí-café	Cassieae	Medium	1,20	No

Species	Popular name	Phylogeny (Tribe)*1	Tree height	Wood density ($g\ cm^{-3}$)*2	N_2 Fixation
Dicorynia paraensis	Angélica-do-Pará	Cassieae	Big	0,90	No
Dimorphandra parviflora	Faveira-de-anta	Caesalpinieae	Big	1,00	Yes
Hymenaea courbaril	Jatobá	Detarieae	Big	0,88–1,00	No
Macrolobium acaciifolium	Arapari	Detarieae	Big	0,65–0,75	No
Macrolobium angustifolium	Apeu	Detarieae	Medium	0,65–0,75	No
Mora paraensis	Pracuúba	Caesalpinieae	Big	0,83–0,96	No
Peltogyne paniculata	Mulateiro	Detarieae	Big	1,20	No
Sclerolobium hypoleucum	Tachi-preto	Caesalpinieae	Medium	0,55–0,75	Yes
Tachigali paniculata alba	Tachi-branco	Caesalpinieae	Big	0,55–0,75	Yes
Faboideae					
Acosmium nitens	Taboarana	Sophoreae	Medium	1,00	Yes
Andira inermis	Manga-brava	Dalbergieae	Big	0,70	Yes
Clitoria fairchildiana	Palheteira	Phaseoleae	Medium	0,51–0,54	Yes
Dalbergia spruceana	Jacarandá	Dalbergieae	Small	1,00–1,10	Yes
Dipteryx odorata	Cumaru	Dipterygeae	Big	1,15–1,19	No
Erythrina fusca	Mulungu	Phaseoleae	Medium	0,31–0,33	Yes
Hymenolobium pulcherrimum	Angelim-da-mata	Dalbergieae	Big	0,79	Yes
Ormosia excelsa	Tento-amarelo	Sophoreae	Medium	0,70–0,72	Yes
Platymiscium trinitatis	Macacaúba	Dalbergieae	Big	0,95	Yes
Swartzia laevicarpa	Saboarana	Swartzieae	Medium	0,80	Yes
Swartzia polyphylla	Arabá	Swartzieae	Big	0,64	Yes
Taralea oppositifolia	Cumaru-da-beira	Dipterygeae	Big	0,82–0,95	No
Vatairea guianense	Fava-mutum	Dalbergieae	Big	0,65–0,80	No
Mimosoideae					
Albizia saman	Bordão-de-velho	Ingeae	Big	0,45–0,60	Yes
Anadenanthera peregrina	Angico-vermelho	Mimoseae	Medium	0,93–0,95	Yes

Species	Popular name	Phylogeny (Tribe)*1	Tree height	Wood density ($g\ cm^{-3}$)*2	N_2 Fixation
Dinizia excelsa	Angelim-pedra	Mimoseae	Big	0,98–1,15	No
Enterolobium maximum	Fava-tamboril	Ingeae	Big	0,60	Yes
Enterolobium schombugkii	Orelha-de-negro	Ingeae	Big	0,75–0,85	Yes
Hydrochorea corymbosa	Faveira-do-igapó	Ingeae	Medium	0,55	Yes
Inga alba	Ingá-turi	Ingeae	Medium	0,50–0,70	Yes
Inga edulis	Ingá-cipó	Ingeae	Small	0,76	Yes
Inga splendens	Ingá-açu	Ingeae	Small	0,55	Yes
Parkia multijuga	Pinho-cuiabano	Parkieae	Big	0,50	No
Parkia nitida	Faveira-benguê	Parkieae	Medium	0,40–0,55	No
Parkia pendula	Visgueiro	Parkieae	Big	0,80–0,85	No
Pentaclethra macroloba	Paracaxi	Parkieae	Medium	0,61–0,88	Yes
Stryphnodendron guianense	Faveira-camuzé	Mimoseae	Medium	0,48	Yes

**1[19] www.ildis.org; *2 Values compiled from the technical literature.*

Table 3.
Scientific and popular name, subfamily, tribe, size, wood density and symbiotic n2 nodulation ability of 45 species of Fabaceae native to the Amazon.

The Fabaceae of wood interest are classified in all subfamilies and many have large size, but also medium or small, varying with the taxon. There are large trees that occupy the upper canopy of the tropical forest in areas of no hydromorphic soil as *Dinizia excelsa* that can reach 55 m high or in flooded forests such as *Mora paraensis*, which grows up to 50 m, but also species such as *Dalbergia spruceana* which, despite the small size, has economic value in the timber market. Some of the largest trees in the Amazon are very old and have an estimated age of up to 1200 years, as verified for *D. odorata*, by dating ^{14}C, in a tree that had trunk diameter of ±120 cm, with an estimated growth rate of ±0.1 cm $year^{-1}$, while *Parkia pendula*, with trunk diameter of 110 m corresponded to the age of ±200 years with an estimated growth rate of ±0.65 cm $year^{-1}$ [20].

The timber trade is hampered by the little knowledge of the wood itself and low identification at the species level by the high variation of popular names, yet in the listings of timber exported from the Amazon, angelim, jacarandá, jatobá, sucupira, macacaúba, etc., are frequent presences. The density values of the wood are directly related to its potential for use, so the light woods have density < 0.50 g cm^{-3}, medium density woods have between 0.50 and 0.70 g cm^{-3} and the stiffest and heavier woods >0.70 g cm^{-3}. The woods of higher density are harder and heavier viable for uses already described, but also for energy production (firewood, alcohol and coal) and the woods with low density are much lighter and able for the production of pulp and paper, light woods and rich in thin-walled fibers.

For the sustained exploitation of timber resources, it is necessary to consider that in tropical forests, some ecological and economic characteristics can determine the increase or decline of forest species populations. The commercial value, limited geographical distribution, little dispersal ability, slow growth and reduced number of juveniles, will affect the decline of populations if the species is exploited by severe logging [21]. Obtaining more forestry and autecology information of species with greater potential management and cultivation is an important step for the preservation of these natural resources and the main challenge is the great biodiversity and numerical grandeur of wood-producing species.

Additionally, there is a property that also characterizes Fabaceae: the ability to associate with soil bacteria, from the rhizobium group, producing small root nodules, an effective site for biochemical and enzymatic processes that allow the dynamics of biological fixation of N_2 - BNF. Numerous tree legumes, but not all, can develop root nodules and fix N_2, adding to these species another role besides the productive, that of providing services to increase the entry of NH^{4+} into ecosystems in imbalance [22]. In soils altered in its organic matter stocks, BNF has a strategic agroecological function. N_2 fixer trees in symbiosis with Rhizobia are designated nitrogen-fixing trees – NFTs.

Among the timber species, several with nodulation species of economic importance stand out among the species, they are *Anadenanthera peregrina*, *Campsiandra laurifolia*, *Hymenolobium pulcherrimum*, *Platymiscium trinitatis*, etc. On the other hand in several Fabaceae of wood importance the nodulifiable ability has not been proven, e.g. *Dialium guianense*, *Dinizia excelsa*, *D. odorata*, *Peltogyne paniculata*, etc. It is noted that the fixing ability of N_2 in Fabaceae evolved mainly in species of the subfamilies Mimosoideae and Faboideae, and has less representation among Caesalpinioideae, where only species cited for the Caesalpinieae tribes had this natural ability (**Table 3**).

The N_2-fixing Fabaceae have advantages over other plants, due to the natural absorption ability of atmospheric N in the form of NH^{4+}, and BNF is an economically and biologically rewarding process, contributing to the productivity of the soil-plant-animal system. The current concepts of Agroecology have emphasized the use of NFTs in forestry or agricultural production systems with particular importance in the composition of species in Agroforestry Systems that seek sustainability in nitrogen. Thus, the Fabaceae of the Amazon emerge as an important element for sustainable production systems in the future.

5. Silvicultural bases for the cultivation and management of woody Fabaceae for the future

As for their use and practical use, tree legumes can be classified into three distinct groups: (1) trees that produce wood and miscellany of by-products such as firewood, oil, resin, tannin, coal and cellulose; (2) forage and human food trees; and (3) trees that aid soil fertility in agricultural practices such as the establishment of retention ranges, shelter areas, erosion control and increase of N status in the soil in agricultural rotation models or in rotation systems whose benefit is the service performed by trees to the environment.

Regardless of the choice of species, basic forestry knowledge is indispensable. Among the Fabaceae there is a wide variety of fruit forms, being possible to define

14 different morphological types [23]. An illustration of the morphological aspect of the fruits and seeds of Fabaceae from the Brazilian Amazon is shown in **Figure 4**. For the advance in basic forestry research of Fabaceae of the Amazon, there is a need to ensure the correct identification of taxa in existing herbaria. The following is the application of drying techniques, cleaning and fruit processing, germination studies, seedling production and sequential forest plantations contributes to the generation of

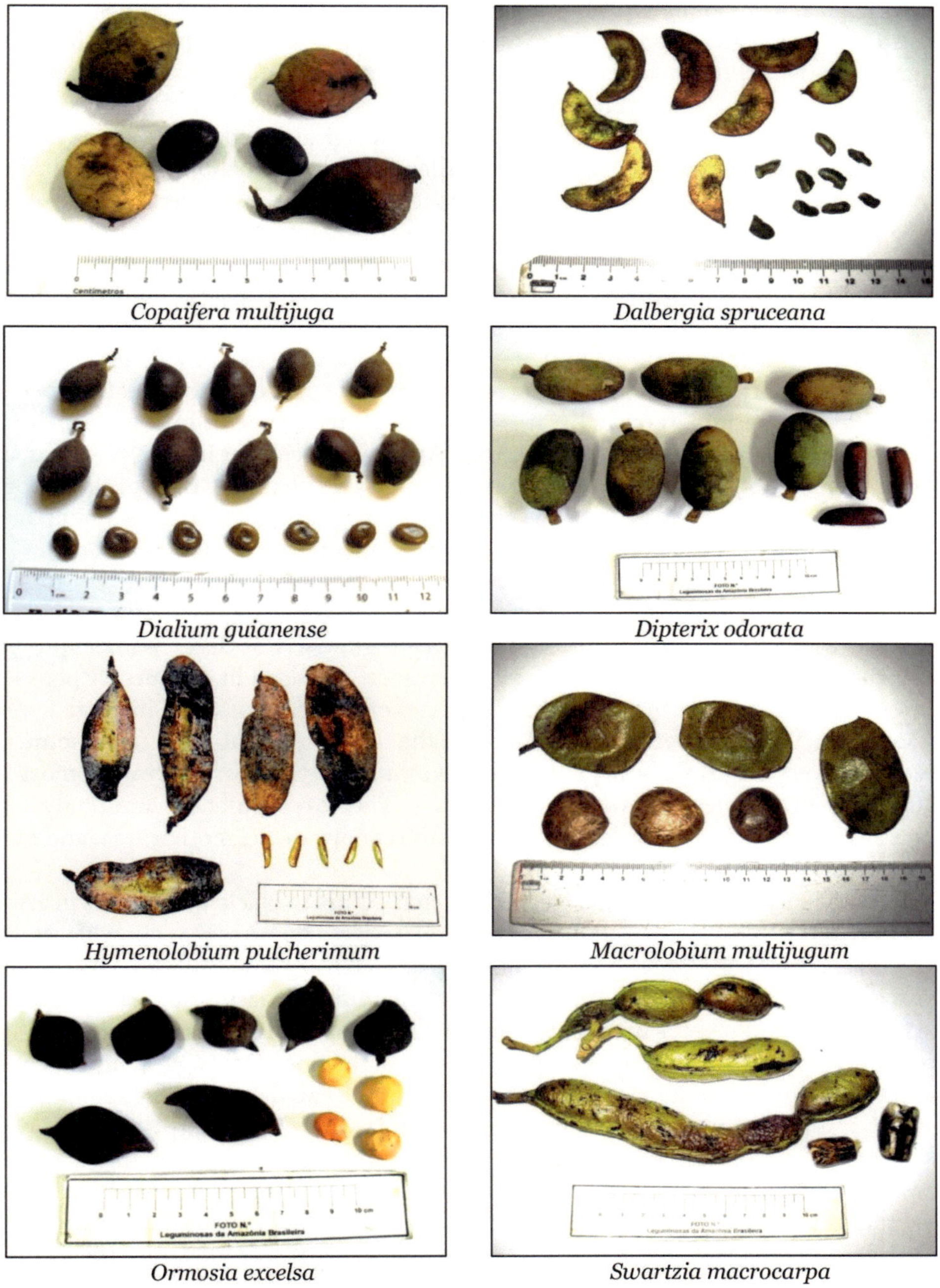

Figure 4.
Morphology of fruits and seeds of Fabaceae species from the Amazon region of wood interest. Photos: Souza LAG.

basic information to identify the species with the greatest potential, followed by their silvicultural use.

In the tropical forest, tree populations with higher potential for regeneration or management should present the following ecological parameters: (1) Seed dispersal facility; (2) Abundance of rods in regenerating forests; (3) Good growth rate; (4) Good regrowth capacity after cutting/burning; (5) Fire resistance; (6) Good breadth of geographical distribution; and, (7) Plenty of adults [21].

In Brazil, at the National Institute of Amazonian Research - Inpa, in Manaus, Amazonas, the research actions follow sequential steps to identify the species with higher potential for management and planting. In bioprospecting in natural or altered areas throughout the region, after fruit collection, the studies follow the following flow: → correct botanical identification of the taxon and its ecological group; → evaluation of fruit and seed production; → seed resistance to desiccation germination of seeds quickly and homogeneity → evaluation of seedling production in nurseries → identification of N_2 fixing ability → phytosanitary aspects and survival after planting → planting definitive to identify rapid growth, rusticity, high biomass production and adaptation to dystrophic soils→ progeny assays, genetic improvement and more advanced stages of research.

Basic forestry studies prioritize the generation of information that can help the planting of species contributing to their conservation and value of biodiversity resources. To contribute to the planting on a larger scale of little-known native species, the basic determinations in the fruit processing phase provide information for seed acquisition or marketing of propagules. Among the relevant information scans are the average weight of the fruits and in the processing phase the fruit purity test provides data on the investment of the fruit in seed formation, and can be an important indicator to evaluate the quality of a batch of intraspecific fruits.

The number of seeds per kilo and the weight of 1000 seeds are auxiliary information for seed acquisition, and, considering their size, provide indicators for the later forestry stages such as seedling production, defining the size of the container and total amount of substrate to be used for each species, managed in the nursery phase. Some of these basic determinations for 45 species of Fabaceae of wood importance are presented in **Table 4**. There is great variety in the fruit size of Fabaceae trees, some developing very large and heavy fruits such as *Inga edulis* whose fruits reach almost 400 g or *Swartzia polyphylla* with just over 200 g. The heavy fruits are commonly dispersed by weight, and are deposited under the canopy of the mother tree, and can then be distributed by water or wildlife.

In the flooded forests of Igapó, vegetation that occurs in black water rivers in areas with nutrient scarcity, large and heavy fruits may have structures that allow floating, and some species with large and voluminous fruits such as *Cassia grandis*, *Mora paraensis* and *Vatairea guianense* form their own comparatively mass group. In the Igapó, the fruits containing seeds with high biomass compensate for the loss of nutrients in the initial phase of seedling establishment and in the floodplains, in sedimentary basins with high nutrient availability, the seeds may be small, but produce elongated seedlings, suitable for survival in regular flood cycles [24].

There are also species with small fruits with weight < 1 g as observed for *Acosmium nitens*, *D. guianense*, *Ormosia excelsa* and others and also a large group of intermediate weight defined in the range of the median fruits. The size of the fruit, therefore, can sometimes be associated with vigor and growth capacity of seedlings and, therefore, development in juvenile stages. For this, each self-ecological species the investment in the fruit in the formation of seeds, also defined as fruit purity index, which is very varied and is not related to its size (**Table 4**).

Specie	Fruit weight	Weight of 1000 seed	Purity (%)	Nº of seeds kg^{-1}
	g			
Caesalpinioideae				
Caesalpinia ferrea	7,51	197,7	17,87	5.051
Campsiandra laurifolia	80,92	9.629,3	59,57	104
Cassia fastuosa	39,29	49,3	15,72	20.408
Cassia grandis	194,41	629,7	17,49	1.608
Cassia leiandra	385,60	540,86	11,84	1.885
Copaifera multijuga	7,77	2.562,6	21,90	372
Crudia oblonga	28,07	10.330,2	64,88	118
Cynometra bauhiniifolia	1,44	1.225,5	87,78	816
Dialium guianense	0,47	144,3	37,54	6.977
Dicorynia paraensis	0,87	257,6	45,50	3.944
Dimorphandra parviflora	11,09	241,64	22,25	4.141
Hymenaea courbaril	25,52	3.816,9	32,79	262
Macrolobium acaciifolium	8,91	5.506,4	63,90	190
Macrolobium angustifolium	10,85	4.558,4	52,38	222
Mora paraensis	126,92	91.033,0	67,04	12
Peltogyne paniculata	0,74	445,8	58,72	2.243
Sclerolobium hypoleucum	0,99	178,0	15,96	5.620
Tachigali paniculata alba	0,52	173,6	24,40	5.720
Faboideae				
Acosmium nitens	0,43	203,8	43,88	4.927
Andira inermis	24,61	9.573,5	38,93	104
Clitoria fairchildiana	18,84	604,8	22,89	1.657
Dalbergia spruceana	2,03	1.285,3	74,69	782
Dipteryx odorata	19,40	2.222,4	9,87	450
Erythrina fusca	4,30	368,6	41,02	2.724
Hymenolobium pulcherrimum	0,76	341,3	4,09	4.298
Ormosia excelsa	2,65	1.468,8	55,42	680
Platymiscium trinitatis	0,61	493,9	58,20	2.029
Swartzia laevicarpa	54,99	17.399,2	47,00	58
Swartzia polyphylla	203,53	66.967,7	44,56	15
Taralea oppositifolia	10,77	4.069,3	38,10	247
Vatairea guianense	93,17	44.448,1	47,30	23
Mimosoideae				
Albizia saman	13,22	203,9	26,50	4.930
Anadenanthera peregrina	2,30	158,2	29,80	6.345

Specie	Fruit weight	Weight of 1000 seed	Purity (%)	Nº of seeds kg^{-1}
	g			
Dinizia excelsa	10,31	66,8	31,96	17.681
Enterolobium maximum	75,97	1.020,0	18,60	953
Enterolobium schombugkii	16,53	54,7	5,75	18.289
Hydrochorea corymbosa	1,64	64,88	18,28	15.470
Inga alba	4,50	203,72	46,04	4.909
Inga edulis	395,05	3.594,0	17,18	279
Inga splendens	104,90	1.710,0	20,91	588
Parkia multijuga	177,42	6.625,8	44,70	155
Parkia nitida	97,81	1.746,3	33,70	573
Parkia pendula	11,58	170,50	60,00	9.870
Pentaclethra macroloba	172,37	6.333,4	26,31	158
Stryphnodendron guianense	3,91	74,2	22,81	13.491

1 – Determinations made at Inpa, in Manaus, AM, Brazil.

Table 4.
Fruit weight, percentage of purity, number of seeds per kilo and weight of 1000 seeds of Amazonian legumes of wood interest. [1]

Among the species the greatest investment in seed formation in relation to the fruit was verified in *Cynometra bauhiniifolia*, where the purity indexes (%) of the fruits reach 87.78%, but also *Campsiandra laurifolia* and *Macrolobium acaciifolium* all of the flooded forests, which possibly relate to their dispersal syndrome, which in this environment is by hydrocoral process. There is a group of species with low purity in fruits, with little investment in seed formation, which in *Hymenolobium pulcherrimum* is <5%, and there are others with <20% such as *Cassia fastuosa*, *D. odorata*, *Enterolobium maximum*, etc. Many species have intermediate values, reflecting their variety and diversity.

Basic forestry determinations of the number of seeds per kilo and/or the weight of 1000 seeds are important for the acquisition and/or marketing of germplasm for planting or reforestation. A kilo of *Cassia fastuosa* seeds can contain >20,000 seeds, because they are small and light seeds. For other species, in 1 kg of seeds there are >10,000 seeds such as for *Cassia leiandra*, *Dinizia excelsa*, *Enterolobium schomburgkii*, *Stryphnodendron guianense*, etc.

On the other hand, in 1 kg of *Mora paraensis* seeds is the weight of only 12 seeds and the high individual weight is a hindrance to its acquisition, regardless of its timber value. Thus also, in 1 kg of *Swartzia polyphylla* seeds there are only 15 seeds, 23 in *Vatairea guianense*, and 58 of *Swartzia laevicarpa*. They are large fruits and seeds, characteristic of the flooded areas of Igapó.

The acquisition of 1000 seeds for planting trees of various species of trees in the Amazon exceeds 10 kg. In this group, for the species that has heavier seeds, *Mora paraensis*, to gather 1000 seeds is necessary more than 90 kg of seeds. In contrast, to dispose of 1000 seeds of *Cassia fastuosa* are <50 g of seeds.

There is a sequence of research steps generating applied silvicultural technology, which will not be explored at this time. In seed germination there are large differences between species regarding the classification for desiccation tolerance and also a predominant characteristic among Fabaceae: the existence of dormancy mechanisms in seeds, associated with the impermeability of the integument, which can be overcome by the application of scarification methods.

They follow the experimental research on seedling formation in forest nurseries followed by definitive planting for selection of the best-developing species. The oldest experimental plantations with native species are more than 60 years old and over time new groups of species have been evaluated to identify their potential for cultivation and favor reforestation actions and forest production. In more advanced phases of forest research, progeny tests are performed for selection of superior germplasm in forest breeding programs.

6. Conclusions

Seen from above, the Amazon rainforest often appears some floristic and relief uniformity. Conversely, the large extent of continuous natural forests is a mosaic of different ecosystems and overlapping ecotone zones, constituting different environments, and biodiversity in fauna and flora, on irregular topography. The botanical inventories conducted in the Central Amazon have pointed out, in general, a high diversity of Fabaceae in its floristic composition [13], which leads us to propose strategies that can provide instruments for the planned use of these resources. The assessment of the economic potential of biodiversity becomes a priority in areas of high diversity.

In practical terms, the results of these studies could potentially reach the following groups of beneficiaries: the industrial chemical sector of wood products; reforestation companies for timber production or for species suitable for agroforestry systems and for the recovery of degraded areas; small farmers' communities, generating production alternatives; the scientific community, through the dissemination of specialized information; and, the general population, through the dissemination of technical information made accessible to a larger audience.

Recognizing that wood is the most valuable product of Amazonian Fabaceae, there is no doubt that the role of many species in reforestation and recovery of eroded soils may be greater than the current one. Reforestation with legumes inoculated with bacteria from the rhizobium group can be a promising alternative in Amazonian agroecosystems contributing to nitrogen sustainability. The planting of NFTs has been successful in areas disturbed by poor land use, such as those resulting from abandoned pastures and mining activities [25].

In the Brazilian Amazon, even with the maintenance of protection and conservation areas distributed in this biome, such as National Parks, Biological Reserves and Ecological Stations, Natural Protection Areas, etc., the economic potential of wood has been exploited for economic purposes. Despite their economic value for the timber potential, Fabaceae are important in the recovery of degraded areas and in the reforestation of marginal soils to reduce the costs of wood produced as well as obtaining raw material for industrialization [26]. For the products and services that fabaceae species offer is certainly an important resource for exploration, preservation and conservation of biodiversity in the Amazon and, in this third millennium, its valorization to the world.

Author details

Luiz Augusto Gomes de Souza
Instituto Nacional de Pesquisas da Amazônia: Inpa, Manaus, Amazonas, Brazil

*Address all correspondence to: souzalag@inpa.gov.br

References

[1] Perez EL. A diversidade vegetal da Amazônia e o Patrimônio Genético Mundial. In: Borém A, Lopes MTG, Clement CR, editors. Domesticação e melhoramento: espécies amazônicas. Viçosa: Editora UFV; 2008. pp. 39-52

[2] Meireles Filho JC. O livro de ouro da Amazônia: mitos e verdade sobre a região mais cobiçada do planeta. Rio de Janeiro: Ed. Ediouro Publicações; 2004. p. 398

[3] Ducke A. As leguminosas do estado do Pará. Rio de Janeiro: Ministério da Agricultura Indústria e Comércio; 1925. p. 133

[4] Ducke A. Notas sobre a flora neotrópica - II As leguminosas da Amazônia Brasileira. Boletim Técnico do IAN, Belém. 1949;**1949**(18):249p

[5] Silva MF, Carreira LMM, Tavares AS, Ribeiro IC, Jardim MAG, Lobo MGA, et al. As leguminosas da Amazônia Brasileira. Lista prévia. Acta Botânica Brasílica. 1989;**2**(1):193-237. DOI: 10.1590/S0102-33061988000300017

[6] Lewis GP, Schrire B, Mackinder B, Lock M. Legumes of the Word. Kew Publishing; 2005. p. 592. DOI: 10.1017/S0960428606190198

[7] Lima HC. Fabaceae. In: Forzza RC et al., editors. Catálogo de plantas e fungos do Brasil. Vol. 2. Rio de Janeiro: Instituto de Pesquisas Jardim Botânico do Rio de Janeiro; 2010. pp. 989-1102. DOI: 10.7476/9788560035083

[8] Corby HDL. The systematic value of leguminous root nodules. In: Pohlhill RM, Raven PH, editors. Advances in Legume Systematics. In: Proceedings of the International Legume Conference. Kew, Surrey, England; 1981. pp. 657-670

[9] Allen ON, Allen EK. The Leguminosae. A Source Book of Characteristics, Uses, and Nodulation. USA: The University of Wisconsin Press; 1981. p. 812. DOI: 10.1007/BF02858721

[10] Norris DO. Lime in relation to the nodulation of tropical legumes. In: Hallsworth EG, editor. Nutrition of Legumes. Butterwoths, London: Sci. Publ; 1958. pp. 164-182

[11] Langenheim JH, Lee YT, Martin SS. An evolutionary and ecological perspective of Amazonian Hylaea species of *Hymenaea* (Leguminosae: Caesalpinioideae). Acta Amazonica. 1973;**3**(1):5-38. DOI: 10.1590/1809-43921973031005

[12] Tutim TG. Classification of legumes. In: Hallsworth EG, editor. Nutrition of the Legumes. Academic Press: New York; 1958. pp. 3-14

[13] Nelson BW, Oliveira AA. 2001. Estado de conhecimento florístico da Amazônia. In: Capobianco JPR, Veríssimo A, Moreira A, Sawyer D, Santos I, Pinto LP, editors. Biodiversidade da Amazônia Brasileira. São Paulo: Estação Liberdade; 2001. pp. 132-176

[14] Procópio LC, Secco RS. A importância da identificação botânica nos inventários florestais: o exemplo do "tauari" (*Couratari* spp. e *Cariniana* spp. - Lecythidaceae) em duas áreas manejadas no estado do Pará. Acta Amazonica. 2008;**38**(1):31-44. DOI: 10.1590/S0044-59672008000100005

[15] Loureiro AA, Silva MF, Alencar JC. Essências madeireiras da Amazônia. Suframa, Manaus. 1979;**1**:187

[16] Inpa/CPPF. Catálogo de Madeiras da Amazônia, Área da Hidrelétrica de Balbina. Manaus.: Inpa; 1991. p. 163

[17] Chimelo JP. Controle de qualidade de madeiras da região amazônica. Acta Botânica Brasílica, Belo Horizonte. 1989;**2**(1):43-53. DOI: 10.1590/S0102-33061988000300005

[18] Lorenzi H. In: Odessa N, editor. Árvores brasileiras: Manual de identificação e cultivo de plantas arbóreas do Brasil. 2nd ed. São Paulo: Editora Plantarum; 2002. p. 394

[19] Roskov YR, Bisby FA, Zarucchi JL, Schrire BD, White RJ. ILDIS World Database of Legumes: draft checklist, version 10 [published June 2006, but CD shows November 2005 date]. Reading, UK: ILDIS; 2006

[20] Chambers J, Higuchi N, Schimel J. Ancient trees in Amazonia. Nature. 1988;**391**:135-136. DOI: 10.1038/34325

[21] Martini A, Rosa NA, Uhl C. Espécies de árvores potencialmente ameaçadas pela atividade madeireira na Amazônia. In: Capobianco JPR, Veríssimo A, Moreira A, Sawer D, Santos I, Pinto LP, editors. Biodiversidade da Amazônia Brasileira. São Paulo: ISA; 2001. pp. 339-347

[22] Andrews M, Andrews ME. Specificity in legume-rhizobia symbioses. International Journal of Molecular Sciences. 2017;**705**:18. DOI: 10.3390/ijms18040705

[23] Barroso GM, Morin MP, Peixoto AL, Ichaso CLF. Frutos e sementes: morfologia aplicada à sistemática de dicotiledôneas. Viçosa: Editora UFV; 1999. p. 443

[24] Parolin P. Submergence tolerance vs. escape from submergence: Two strategies of seedling establishment in Amazonian floodplains. Environmental and Experimental Botany. 2002;**48**:177-186. DOI: 10.1016/S0098-8472(02)00036-9

[25] Franco AA, Faria SM. The contribution of N_2 fixing tree legumes to land reclamation and sustainability in the tropics. Soil Biology and Biochemistry. 1997;**29**(5/6):897-903. DOI: 10.1016/S0038-0717(96)00229-5

[26] Fortes JLO, Balieiro FC, Franco AA. Leguminosas arbóreas como agentes de recuperação de áreas degradadas. In: Moura EG, editor. Agroambientes de transição entre o trópico úmido e o semiárido do Brasil. Atributos, alterações; uso na produção familiar. São Luiz: Uema; 2004. pp. 101-132

Chapter 7

Perspective Chapter: Traditional African Vegetables Agrobiodiversity – Livelihood Utilization and Conservation in Tanzania Rural Communities

Michael Kazyoba Benedict, Frenk M. Reuben, Luseko Amos Chilagane and George Muhamba Tryphone

Abstract

The agricultural sector is challenged to fulfill the nutritional requirements of ever-increasing population. Rates of growth in crop productivity have declined, among many reasons is due to climate change and the decline of water and land resources. Dependence by 50% on few cereal crops like wheat, maize and rice for total food intake and calorie requirements has countersigned loss of biodiversity and decline in crop productivity. The genetic diversity of traditional crop varieties offers resilience to environmental risks, socioeconomic shocks, adaptation and mitigation to climate change which is crucial for crop production. Traditional African Vegetables (TAVs) are an integral constituent of the diets of many rural and urban communities. They are important sources of essential macro and micro-nutrients. In addition, they offer a source of livelihood when marketed, and also contribute to crop biodiversity. Tanzania needs to conserve the Traditional African vegetables and their genetic resources against stressful conditions and increased selection pressures which causes loss of genetic variation and a decrease in fitness by a process called genetic erosion. Conservation and use alleviate genetic drift and inbreeding depression, then, is critical to guarantee TAVs persistence in rural areas. This review explores agrobiodiversity of traditional African vegetables (TAV) from livelihood of Tanzanian rural communities' perspectives and how the country has managed to conserve these species.

Keywords: traditional African vegetables (TAVs), agrobiodiversity, genetic erosion, genetic drift, inbreeding depression, conservation, utilization

1. Introduction

Agriculture and food systems are both drivers and recipients of major changes in global health and socioeconomic systems in this era of intense human interaction with the planet "Anthropocene" [1]. The agricultural sector is challenged to fulfill the food and nutritional requirements of ever-increasing population [2–4]. For instance, feeding 9.7 billion people in 2050, and 11 billion in 2100, is one of if not the most important challenge facing mankind during the remaining years of the century. Global population, which was only 1.7 billion at the turn of the century in 1900, is now 7.3 billion. The world has added approximately 1 billion people in the span of the last 12 years. It is expected to climb to 9.7 billion by 2050, and to 11 billion at the end of this century in 2100. Globally, 870 million people are currently chronically hungry and 2 billion are malnourished. Rates of growth in crop productivity have declined, among many reasons is due to climate change and the decline of water and land resources.

In an attempt to safeguard food security, the homogenization of the agricultural system through the green revolution led to a calorie-based food system [5]. It is only less than 20 plant species out of 5000 that are key source of food [4]. Such scenario has led to 40% of arable land globally be dedicated to only three crops which are wheat, maize, and rice. These crops are the main focus of plant breeding programmes, account for the largest share of the global seed market, they are depended for calorie requirement by over 50% of total food intake and are excessively prioritized [2]. Although these crops have contributed to resilience and long-term sustainability of human wellbeing, but the dependence on such few crops has countersigned loss of biodiversity and decline in crop productivity [6].

Agrobiodiversity which is defined as the variety and variability of animals, plants and micro-organisms that are used directly or indirectly for food and agriculture, including crops, livestock, forestry and fisheries as the key driver of the ecological transition of agriculture. Agrobiodiversity is significant to resilience to environmental risks, socio-economic shocks, adaptation and mitigation to climate change. It delivers valuable and response diversity to agroecosystems, potentially making them more robust, and may offer solutions to prevent pests and diseases and it plays an important role in nutrients recycling, soil organic matter maintenance and reduces the need for chemical pesticides and fertilizers. It also creates more favorable habitats for useful macro and micro-organisms and then improves soil quality. Its interactions with farmers provide key dimension of human and nature in agro-systems [7]. The genetic diversity of traditional varieties of crop in agrobiodiversity is the most economically valuable part which is crucial for future world crop production [8]. Therefore, this review intends to explore the contribution of Traditional African Vegetables (TAVs) as a biodiversity component in the livelihood of Tanzania rural communities and how these communities can conserve TAV species.

2. Agrobiodiversity of traditional African vegetables

2.1 Terminologies and definition of African traditional vegetables

Traditional African Vegetables (TAVs) have been in the spotlight in recent years in academic papers for agricultural biodiversity, nutritional and dietary diversity and society livelihood at large. Many other names have also been assigned to refer to these species in different contexts of studies including Indigenous African Vegetables (IAV); African Indigenous Vegetables (AIV) Traditional Leafy Vegetables (TLV); African Leafy Vegetables (ALV); Traditional African Leafy Vegetables (TALV or TLV), orphan and neglected or underutilized vegetables and all are subject to contested meanings [9–11]. In the context of this review, traditional vegetables follow the FAO definition [12], and refers to all categories of plants whose leaves, fruits or roots are acceptable and used as vegetables by urban and rural communities through custom, habit and tradition. It also refers to the review by Towns and Hackleton [13] who proposed the name Traditional African Vegetables (TAVs) and defined them as plant species that are indigenous or naturalized to Africa, well adapted to or selected for local conditions, and whose plant parts are used as a vegetable, and, preparation, and consumption is deeply embedded in local cuisine, culture, folklore, and language.

2.2 Typology and diversity of African traditional vegetables

Towns and Hackleton [13] reported that, 207 species out of 275 TAVs generalized as Traditional Vegetables are grown in Africa [14]. In their review, [13] most academic papers as a result it renders it difficult to focus on options and possibilities to manage, use, conserve, cultivate, market and disseminate knowledge specific to particular TAV species since different species will require distinct strategy from the other. Therefore, the need to establish a well-documented typology for TAVs based on botany, uses, useful parts, mode of cultivation and management practices of a particular species with respect to the context in which they are utilized is vital.

2.3 An overview of traditional vegetables of utilization in Tanzania

Traditional vegetables are important sources of nutrition which have potential to sustainably address malnutrition, a growing problem in sub-Saharan Africa [14, 15]. The TAVs hold excellent potential to improve nutrition and increase the dietary diversity of rural households however their consumption is limited by negative perceptions and lack of awareness on nutritional benefits [14]. In their study, Lotter et al. [15] and Afari-Sefa, et al. [16] found that, traditional vegetables have been known among Tanzanians for a long time, and utilization is affected by the difference in socioeconomic background, food and eating habits of one community to another. The same author noted that TAVs have been initially preferred for their taste, medicinal properties, availability and access while in recent years increased awareness of their nutritional value is the main focus. In Africa and Tanzania in particular, most of the traditional vegetables are used as folk medicine due to their nutritional properties and medicinal value [17]. Despite this fact, most of these have not been satisfactorily studied to substantiate their medicinal properties besides studies on diversity and conservation [17]. The medicinal value of these indigenous vegetables is an important element in the African traditional health systems. In some of African countries, it has been estimated that over 80% of the population consult traditional healers using indigenous plant species [18]. This shows high demand for these vegetable species that necessitates their deeper studies on their use and conservation. Bottlenecks in the production of TAVs are among many others, poor quality and availability of seeds, lack of market information and high postharvest losses which discourage farmers [19, 20]. Furthermore, Afari-Sefa et al. [16] noted that, the low adoption of indigenous vegetables is the inability of formal and centralized seed systems to meet their complex and diverse seed requirements.

3. The livelihood utilization of traditional vegetables in Tanzanian rural communities

The livelihood of Tanzania's rural communities depends on agricultural activities. Jackson et al. [21] used indicators of livelihoods capitals namely; natural, physical, human, social and financial capitals to determine status of agrobiodiversity. There is a wide range of vegetable crops grown in different parts of Tanzania [22]. The poor households rely mostly on TAVs for production and consumption. Approximately 40% of farmers are involved in cultivation of TAVs, of which 25% are relatively large scale farmers [23]. The share of both marketed and non-marketed TAVs in total house- hold income is on average, nearly 13% indicating significant contribution to overall household incomes Weinberger and Msuya [23], ATVs have rapid canopy expansion rates which allow them to accumulate vegetative biomass within a short period. This guarantees food and nutrition security. The ATVs have ability to grow all year round in warm subtropical environments and have co-evolved adaptive mechanisms to ensure broad adaptation to drought. Some species has a remarkable recovery rate after exposure to a prolonged drought. Most TAVs come into production within a short time after the first rains, and harvesting commences three to six weeks after emergence. Chepkoech et al. [24] found a high correlation between adaptive capabilities to climate change and five livelihood capitals. In their study, Berg et al. [25] noted that human capital was built by critical thinking, innovation, confidence and quality of life; social capital by mutual trust, bonding, collective action, networking and emancipation; natural capital was enhanced by improved field practices, food production, agricultural diversification and food security while financial capital by increased income and profits, savings and loan schemes. The framework described by Kissoly et al. [26] (**Figure 1**) show that livelihood assets directly affect agrobiodiversity and food security status of a particular community. Understanding how livelihood assets affects utilizations of traditional vegetable agrobiodiversity will enable establishment of policies that will be community specific thus result to expected outcomes.

3.1 Natural capital

The livelihood of rural populations of farmers and pastoralists relies on natural resources and ecosystem services to a greater extent [25]. Subsistence farmers in environments too adverse to support high yielding crop and livestock varieties rely on wide range of crop and livestock types [27]. Barbier, [28] reported that rural communities which depend on ecologically fragile environments face a vicious cycle of declining livelihoods, increasing ecological degradation and loss of resource commons as well as declining ecosystem services on which they depend. Therefore, in the context when yields for major crops go rancid globally, marginal crops could increase yields and food security and encouraging environmental sustainability [29].

Tanzania is among the regions rich in diversity of traditional vegetables particularly of wild species in Africa [30]. The study by Keller et al. [31] found considerable variation in richness of traditional vegetable diversity across different agroecologies of Tanzania although analysis showed even distribution.

3.2 Social capital

Social capital yields a flow of mutually beneficial collective action which serves as the basis for enhancing sustainable resource management to improve productivity, equity and the environment in farming communities [32]. A study by Jackson et al. [21] suggest that although social capital is not strongly associated with natural capital in agrobiodiversity management, it goes hand in hand with human capital such that individual knowledge of social–ecological system by maintaining dynamic set of social norms and institutions that support ecological intensification. Kessy et al. [33]

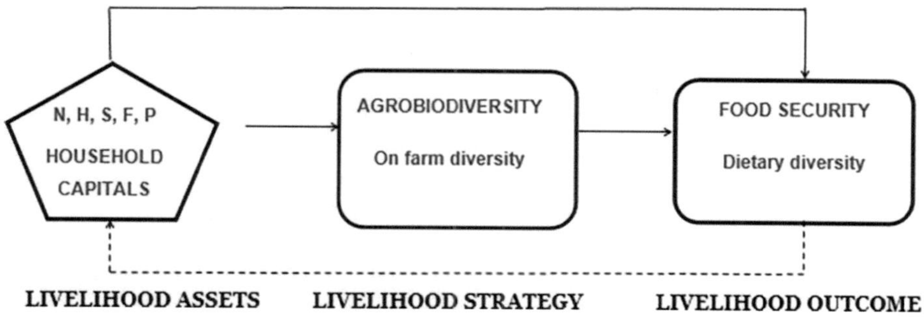

Figure 1.
Conceptual framework of relationship between livelihood asset, agrobiodiversity and food security. Source: Kissoly et al. [26].

reported that socio-demographic characteristics such as age, sex and educational level of head of household affect perception of traditional vegetables and thus utilization, other factors being size of household. Lambrou and Laub, [27] proposed understanding gender sensitivity in livelihood roles as an important approach of devising solution for conservation of biodiversity and alleviate poverty. The authors argued that, depletion of natural resources and decreasing agricultural productivity place additional burden on women's work and health while impede their participation in decision making processes and income generating activities.

3.3 Physical capital

High physical capital coupled with financial capital supported agricultural intensification hence food security [21]. TAV species despite having the potential to improve dietary requirements of households in Tanzania, 30% - 40% of the produce is lost after harvest due to inadequate handling skills, poor infrastructure, lack of appropriate processing technology and poor market information, lack of storage facilities and supporting system [33, 34]. Musebe *et al.*, [35] recommended that it is necessary to provide processing and market infrastructure to the processors of African indigenous vegetables. Owusu and İşcan, [36] estimated the likelihood of market participation of farms in Tanzania to be 41% and probability of a farm transition from subsistence to commercial is 30%, the authors argued market participation of a farm is highly influenced by land size, labour use and mechanization such that a unit increment increase likelihood of market participation by 5%.

3.4 Human capital

Training farmers in modern production techniques, quality control and standardization of selling unit as well as linking them to the market will open up market outlet of traditional vegetables [11]. Chepkoech et al. [24] associated high indicator of human capital such as farming experience, educational attainment of the household head, percentage of adults with primary education, type of training on farming, technical assistance, and access to climate information with capabilities to cope with environmental changes in crop production. Women dominate the traditional African value chain in sub – Saharan Africa from production to marketing thus improving traditional African vegetables is improving and the livelihood of resource-poor women [37].

3.5 Financial capital

Developing countries in general and Tanzania in particular exhibit inconsistence relationship between economic growth, rate of poverty reduction and nutrition outcome despite positive agricultural growth [38]. **Figures 2** and **3** show that the rate of declining poverty headcount ratio is relatively slow compared to the rate of increasing GDP per capita in Tanzania. This may be attributed to the need to shift in labour out of agriculture and into industry and service in order to diversify sectoral production [39], the author noted the fall of agricultural employment share from 72% in 2015 to 46.9% by 2030. However, agricultural sector will continue to be the backbone of many developing countries, hence, there is an urgent need for financial support to vulnerable small scale farmers whose livelihood depends on subsistence farming.

Growing vegetables for sale have been a reliable source of alternative income in many households in both rural and urban areas while offer dietary diversity from wide range of nutrients. Traditional African vegetables have been equally important nonetheless, however, in addition to many other challenges, lacking financial support at most if not any point in value chain of traditional vegetables limit growth of the sector and in turn household livelihood. A study by Ambikapathi et al. [40]

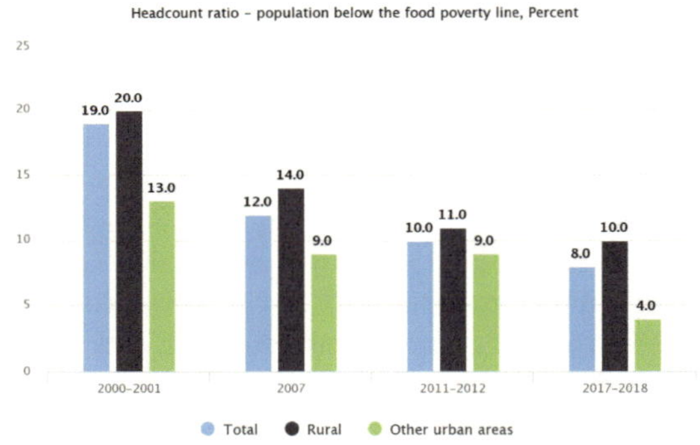

Figure 2.
Poverty headcount ratio.

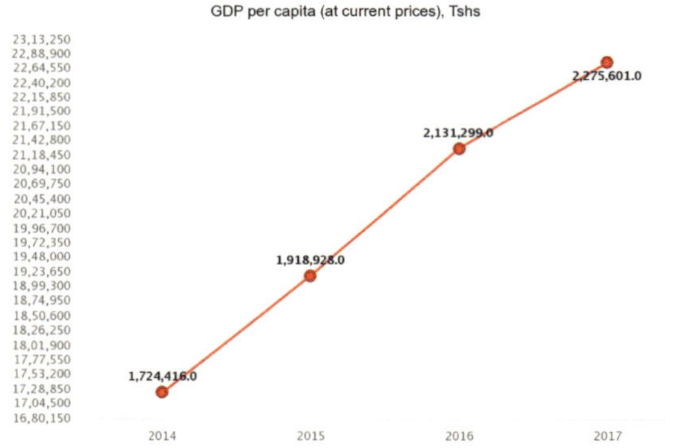

Figure 3.
GDP per capita from 2014 to 2017. Source: NBS [22].

showed that only 3.3% of green leafy vegetables are sold in formal market and 26.8% and 58.0% are sold semi-formal and informal respectively in urban regions of Tanzania suggesting even lower percentage of formal vegetable marketing in rural areas.

4. Conservation of traditional African vegetables in Tanzanian rural communities

Traditional African vegetables and their genetic resources need to be conserved against threats as they have high potential to contribute to healthy diets, climate resilience and food systems in Tanzania. TAVs are highly nutritious, easy to incorporate into crop rotation, and can generate more climate-resilient systems suitable for food production. Therefore, conservation of TAVs will support sustainably on food production, diversification and promote wider range of healthier foods Zonneveld et al. [30] especially in some areas of Sub-Saharan Africa which has some parts with acute nutritional deficiencies and highest level of hidden hunger in the world. Thus, world's sustainable food future depends on the response of plant biodiversity to global changes in climate and environmental impacts, market integration, demographical and nutritional transition and human health and disease Zimmerer and Haan [41] hence the need for integrative approach in conservation of agrobiodiversity. The conflicts between agricultural produce or and biodiversity are by no means inevitable if farming practices are not sustainable [42]. However, until recently relatively little effort on exploration of traditional vegetables as an important component of agricultural biodiversity have been done by researchers both ecologically and socioeconomically, Dinssa et al. [37] acknowledged the fact that farmers in sub-Saharan Africa have been the sole curators and developers of the neglected and underutilized traditional vegetables.

5. Erosion of diversity of traditional African vegetables and socio-economic implications in Tanzania rural community

Predominant pattern of agricultural growth has been at expense of eroded biodiversity of plant genetic resources, livestock, insects and soil organisms [21, 42]. Keller et al. [31] found that wild TAVs are under threat of genetic erosion due to increasing urbanization, encouraged the cultivation of exotic vegetables for sale and the disappearance of indigenous knowledge on where and how to collect, cultivate and prepare traditional vegetables. In their review Chaudhary et al. [6] noted that agrobiodiversity loss can be categorized into natural drivers for example specie's ability to tolerate biotic and abiotic stresses, disturbed habitat render inhabitable and anthropogenic drivers such as change in food preferences, market forces, gene manipulation, human response to climate changes and lack of transfer of traditional knowledge. Farmer's in a survey conducted by Keller et al. [31] in Tanzania mentioned threats of traditional vegetables to be the introduction of new vegetable species or cultivar replaced indigenous vegetables varieties, influence of politics on commercialization of exotic vegetables, loss of indigenous knowledge making young generation unable to utilize existing traditional vegetables, changes in food habits, loss of habitat of wild vegetables and climate changes led to excessive drought.

5.1 Natural drivers

Biodiversity is made up of ecosystems, species diversity, and genetic diversity which sustain our lives and to by preserving our culture, well- being, and to economic prosperity. Biodiversity is increasingly subjected to human-induced changes to the environment. To persist, populations continually have to adapt through natural selection to stressful changes including pollution and climate change. Natural selection is the process through which species adapt to their environments. When faced with new stressful conditions and increased selection pressures, organisms can respond in several ways. If they are not able to adapt, they will either go extinct or they have to avoid the stressful conditions: through changes in local behavior [43]. Small relatively isolated populations become increasingly subject to genetic drift and inbreeding, resulting in loss of genetic variation and a decrease in fitness by a process called genetic erosion. The genetic erosion in small populations, owing to fragmentation of natural habitats, is expected to obstruct such adaptive responses through genetic drift which causes a decrease in the level of adaptive genetic variation, thereby limiting evolutionary responses and inbreeding depression which reduces individual fitness and, consequently, the tolerance of populations to environmental stress. Genetic drift causes allele frequencies to fluctuate, which over time leads to random loss and fixation of alleles and an increase in homozygosity. Significantly, inbreeding mostly increases the sensitivity of a population to stress, thereby increasing the extent of inbreeding depression. As adaptation to stress is utmost frequently convoyed by augmented mortality (cost of selection), the rise in the 'cost of inbreeding' beneath stress is anticipated to harshly hinder evolutionary adaptive processes. Inbreeding therefore plays a fundamental part in this process and is anticipated to bound the chance of genetically eroded populations to effectively adapt to stressful environmental situations. Thus, the dynamics of slight fragmented populations may vary substantially from large non-fragmented populations. The resilience of fragmented populations to altering and deteriorating environments is expected to be greatly decreased. Alleviating inbreeding depression, then, is critical to guarantee population persistence.

5.2 Anthropogenic drivers

Human activities are considered to be the main driver of agrobiodiversity erosion by population numbers, use of land, and of peoples' lifestyles, causes damage to habitats for various species. The global biodiversity decline by 30% is driven by land use through food production followed by 20% overexploitation of natural resources such as overharvesting, overhunting and overfishing for food, timber and medicines. The activities by rural communities of Tanzania in similar context affect agrobiodiversity. For instance, the study conducted in 1997 in the Udzungwa Scarp Forest Reserve of Tanzania, found human activities threatening the biodiversity [44]. This tells the fact that, biodiversity takes convert if progressively is exposed to human alterations of natural habitats, and abiotic and biotic environments are both changing swiftly, often unpredictably, and species and populations are progressively more subjected to stressful environmental situations [43]. Industrial pollution and the use of agro-chemicals have revealed to interrupt biodiversity dramatically. Large-scale destruction of natural habitats has triggered large populations of various species to become fragmented, resulting in small 'remnant' populations that become increasingly isolated. Subdivision of large populations in combination with limited gene flow between the fragments has significant ecological and genetic consequences. Ecologically, habitat fragmentation will have demographic effects as small populations are progressively more affected by demographic and environmental stochasticity greatly increasing their extinction chance. The biodiversity needs to be explored and documented due to the fact that it consists of some of the species which are rare, endemic and threatened in such a way that they need protection and conservation measures [44].

6. Factors influencing conservation of traditional vegetables in Tanzania rural communities

Traditional African vegetables represent a diverse and widespread set of vegetables that are consumed across Tanzania. However, the knowledge base of traditional African vegetable conservation remains truncated due to a deficiency in research and policy support and impedes efforts to promote widespread cultivation, consumption and commercialisation. Therefore, even the conservation of traditional vegetable among rural farmers is generally low. Conservation is the wise utilization of natural resource that involves maintaining essential ecological processes and life support systems, preserving genetic diversity and ensuring sustainable utilization of species and ecosystems. Thus, conservation involves elements of protection, preservation, utilization and sustainable management of plant genetic resources that satisfy the increasing or matches with increasing efficiency, current and future needs of humanity and other organisms. Farmer's efforts in conservation, improvement and utilization of traditional vegetables currently under production serve as the basis of understanding different aspects of these crops by research community [37]. Dweba and Mearns [45] argued that it is vital to conserve indigenous knowledge on traditional vegetables to ensure the availability and utilization of these important food sources for resource-poor rural communities. Mpasiwakoma et al. [46] study found that most rural household in Tanzania depended on wild food plants during the periods of food crisis, this suggest that conservation by utilization is an effective method. Keller et al. [31] argued that socioeconomic status of rural communities has influence in the conservation of traditional vegetables such that a community with alternative sources of income and thus higher food security is more likely to forego utilization of traditional vegetables compared to a community whose members relies of on these species as source of their dietary requirements. Therefore, there is a need of Government and Non-Governmental organizations to encourage conservation of traditional vegetables in order to increase TAVs in natural habitats.

6.1 The effort to safeguard the biodiversity of traditional vegetables in Tanzania

Biodiversity in general and agrobiodiversity in particular are crucial for adaptation to climate change, for resilience and for human health as related to dietary diversity [47]. Chaudhary et al. [6] proposed utilization of crop wild relatives, farmer's fields, community seed bank, participatory crop improvement, and value addition of underutilized crops to be methods of in situ conservation while research stations, botanical garden, and national and international gene bank to be methods of ex situ conservation. Muhanji et al. [48] noted contribution made by institutions such as AVRDC in the conservation of traditional vegetables in Tanzania including awareness creation, capacity building to farmers, development of improved TAV varieties, and establishing seed multiplication and distribution systems for few TAV species, business support and marketing of TAVs.

6.2 International agreements to safeguard agrobiodiversity particularly of TAV

Considerable efforts have been made by the international community to commit itself towards international agreements and treaties on the Convention on Biological Diversity (CBD), the International Treaty on Plant Genetic Resources for Food and Agriculture (ITPGRFA), and the Global Plan of Action for Animal Genetic Resources. These translate international obligations to conserve biodiversity into national laws and policies. These also increase the internationally agreed and effective mechanisms of access and fairly sharing the benefits from the use of the genetical resources which can contribute to the creation of a fair, more equitable economy and support sustainable agricultural development. The secretariats of the Convention on Biological Diversity (CBD) and the FAO International Treaty on Plant Genetic Resources for Food and Agriculture (FAO International Treaty) developed synergies implementing a joint initiative for on-farm conservation; working on sustainable use of plant genetic resources for food and agriculture and protected areas; promoting the importance of biodiversity and plant genetic resources for food and agriculture, food security and nutrition under a changing climate. Since 1996 Tanzania became a party to the Convention on Biological Diversity (CBD) and it adheres to its international obligation to protect and conserve its biodiversity as a global resource [49] (URT, 2015). The country Tanzania became a signatory to the Convention on Biological Diversity (CBD) in 1992 (URT). Tanzania has taken number of initiatives in order to conserve its biodiversity and to achieve the Article 6 of the CBD. The country formulated her National Biodiversity Strategy and Action Plan (NBSAP) 2015–2020 which sought to address national biodiversity targets based on the national priorities that contribute to the global targets on number of emerging issues such as climate change and variability, invasive species, and genetically modified organisms (GMOs) among others [49].

6.3 National legislature to safeguard agrobiodiversity particularly of TAV

Tanzania is one of the twelve mega-diverse countries of the world endowed with different natural ecosystems that harbor a massive wealth of biodiversity [49]. By then, the country hosted 6 out of the 25 world renowned biodiversity hotspots hosting more than one-third of the total plant species on the continent and about 20% of the large mammal population [49]. In Tanzania, the Biodiversity wealth contributes significantly to the socio-cultural, economic and environmental goods and services to the country and peoples` livelihood. The public policies, market dynamics, wholesale and retail distributors in agriculture have led to legislation of varietal registration and seed quality, and for supporting the increase in productivity. The legislation has conditioned agriculture, shifting it towards specialization and monoculture. This decreases diversity in farming methods. This review recommends for developing legislative mechanisms to support conservation of genetic biodiversity under in situ and ex situ conditions through the establishment of field gene banks and cryo-preservation centres and establish a mechanism for capacity building, participation and empowerment of farmers through legislative measures for conservation and utilization of such genetic resources.

7. Participation of different stakeholder of TAV

7.1 Farmers and traders

The global market of vegetables is still predominantly local because only about 5% of vegetables grown worldwide are marketed internationally [50] (Sumalan et al., 2021). According to Sumalan et al. [50] "the total volume of exports of vegetables reached at 47 million tonnes with a value of 42.3 billion US$ in 2018." In the same year, 64% of total vegetable exports were contributed by Netherlands (6.1 million tonnes), Mexico (5.8 million tonnes), Spain (5.1 million tonnes), China (4.3 million tonnes), France (3.5 million tonnes), Germany (2.7 million tonnes) and the United States (2.4 million tonnes). It was noted the exports were increasing at an average annual rate of 1.7%. Vegetable imports have also kept upward trend due the emergency of some countries with high vegetable requirements. In Tanzania, the Traditional African vegetables are mainly produced for subsistence and few for cash. The study carried out in northern and central part of Tanzania to investigate participation of farmers and traders in relation to gender in vegetable production and trading, found farmers reporting more balanced intra-household labour arrangements paired with less-balanced income and expenditure shares, while traders indicated less-balanced labour contributions which went hand in hand with more balanced shares of benefits [51]. The participation of farmers and traders in vegetables shows opportunities to farmers who benefit from trade by selling surplus produce and purchasing needed goods and services [52]. Benefits and challenges to market participation faced by vegetable farmers include lack of information about the markets and high transportation and transaction costs, among others, which do not allow traditional vegetable farmers to efficiently participate in markets [52]. African traditional vegetables if given interest can offer an important entry point which provides an important economic pillar upon which women's livelihood can be supported to improve welfare in rural areas.

7.2 Consumers

Consumers in many parts of the world derive a major portion of their diets from vegetables of which 70% are sold as fresh (unprocessed) hortcultural products [50]. Traditional African Vegetables (TAVs) are an affordable and relatively inexpensive source of vitamins, essential minerals, dietary fibber, and various phytochemicals which have a role in reduction of malnutrition, stunting growth and poor health among consumers in different areas of Sub-Saharan Africa. However, these vegetables are often neglected and underutilized. The food basket regions of Tanzania lack of dietary diversity which is key factor causing hidden hunger in because the population consumes mostly maize. Women and children can use the traditional African vegetables to get an excellent means to complement maize for better nutrition. In order to increase consumption of traditional African vegetables, there is a significant role to be played in order to promote attitudes of consumers towards increasing the amounts of TAVs in their diets.

7.3 Research institutes and seed banks

The Traditional vegetables have received low priority of research and production, poor seed distribution and availability, and lack of awareness on their values (URT 2015). Researches on TAV in Tanzanian context are mainly in the fields of food and nutrition potential [19, 20, 37, 53], livelihood [14, 25, 35, 48], diversity and conservation [30, 31, 54], environmental conservation and resilience [37]. The work of institutions such as AVRD and affiliates in collaboration with stakeholders such as Universities, TARI, farmers and consumers must not go unnoticed, collection of germplasms of traditional vegetables [37] (Dinssa et al., 2016) is important for conservation and establishment of seed systems and marketing of TAV [16, 48] is vital for commercialization. This looks the same as other most of Sub Saharan African countries which have not prioritized these TAVs in their crop research, training and development programs. There are innovative ways required for conducting researches on TAV' varieties and promotion through posters, campaigns and other online and printed publications. Increased research and production of TAVs will offer multiple economic opportunities, reduce unemployment and poverty.

The gene bank is a reservoir of biodiversity that acts as a source of genes which are used in breeding programs [50]. The World Vegetable Center (WorldVeg) genebank of traditional African vegetables currently has 2500 accessions which originated from germ-plasm-collecting mission with national partners across Africa in early 2000s [55]. This genebank conserves, screens and distributes the germplasm to support plant breeding to agricultural research institutes and seed companies. Between 2013 and 2017, Tanzania, Kenya and Uganda received 42,514 seed kits (containing an average of four seed samples of different vegetable crops and varieties) and 183,193 seed samples from WorldVeg [55]. The genebank offers a huge diversity and variability of vegetables to researcher for breeding and genetic modifications.

8. Conclusion

Traditional African Vegetables (TAVs) are among the horticultural crops whose natural habitat originated in Africa. The prospect of TAVs lies on the effort of researchers to revive the once perceived as an important group of food for cultural identity of rural communities which in turn were utilized, conserved and the knowledge passed from one generation to another. Several studies have shown that TAVs have high market potential and can contribute substantially to food and nutrition security, and climate change adaptation. Their production in Tanzania has the potential to be highly profitable, provide employment opportunities and generate income. There are number of challenges that hinder the production of the traditional vegetable in Tanzania including but not limited to high perishability; unavailability of quality seeds; high price and post-harvest losses, and lack of will-structured and organized market linkages. The traditional vegetables constitute a valuable genetic pool of gene diversity, which can be exploited both in breeding programs for obtaining new commercial varieties with targeted traits. Thus, more effort should be kept in characterization of existing TAV germplasms both morphologically and genotypically so as to first understand the extent of genetic diversity existing within the species but also to exploit genetic potentials of these species especially wild relatives so as to maximize production. Emphasize must also be on the marketing of TAV in order to elevate the livelihoods of rural families that grows these species. To realize the potential of TAVs, the value chain actors must improve the competitiveness of their vegetable production and marketing commodities to increase market share and profits.

Conflict of interest

No conflict of interest.

Author details

Michael Kazyoba Benedict[1], Frenk M. Reuben[2], Luseko Amos Chilagane[1] and George Muhamba Tryphone[1*]

1 Department of Crop Science and Horticulture, College of Agriculture, Sokoine University of Agriculture, Morogoro, Tanzania

2 Tanzania Agricultural Research Institute (TARI)-Ilonga, Morogoro, Tanzania

*Address all correspondence to: muhamba@sua.ac.tz

References

[1] Zimmerer KS, de Haan S, Creed-kanashiro H, Carrasco M, Meza M, Amaya FP, et al. The biodiversity of food and agriculture (agrobiodiversity) in the Anthropocene: Research advances and a conceptual framework. Anthropocene. 2019;**45**:100192

[2] Ebert AW. Potential of underutilized traditional vegetables and legume crops to contribute to food and nutritional security. Income and More Sustainable Production Systems. 2014;**6**:319-335. DOI: 10.3390/su6010319

[3] Kahane R, Hodgkin T, Jaenicke H, Padulosi S, Looney N. Agrobiodiversity for food security, health and income. Agronomy for Sustainable Development. 2012;**33**:671-693. DOI: 10.1007/s13593-013-0147-8

[4] Sogbohossou EOD, Achigan-dako EG, Maundu P, Solberg S, Deguenon EMS, Mumm RH, et al. A roadmap for breeding orphan leafy vegetable species: A case study of Gynandropsis gynandra (Cleomaceae). Horticulture Research. 2018;**5**:1-15. DOI: 10.1038/s41438-017-0001-2

[5] Meldrum G, Padulosi S, Lochetti G, Robitaille R, Diulgheroff S. Issues and prospects for the sustainable use and conservation of cultivated vegetable diversity for more nutrition-sensitive agriculture. Agriculture. 2018;**8**:112. DOI: 10.3390/agriculture8070112

[6] Chaudhary P, Bhatta S, Aryal K, Joshi BK, Gauchan D. Threats, drivers, and conservation imperative of agrobiodiversity. The Journal of Agriculture and Environment. 2020;**21**:44-61

[7] Mariel J, Carrière SM, Penot E, Danthu P, Rafidison V, Labeyrie V, et al. Exploring farmers' agrobiodiversity management practices and knowledge in clove agroforests of Madagascar. People and Nature. 2021;**3**:914-928. DOI: 10.1002/pan3.10238

[8] Wood D, Lenne JM. The conservation of agrobiodiversity on-farm: Questioning the emerging paradigm. Biodiversity and Conservation. 1997;**129**:109-130

[9] Ambrose-Oji B. Urban food systems and African indigenous vegetables: Defining the spaces and places for African indigenous vegetables in urban and Peri-urban agriculture. African indigenous vegetables in urban agriculture. 2009;**1**:33

[10] Maseko I, Id TM, Tesfay S. African leafy vegetables: A review of status, production and utilization in South Africa. Sustainability. 2017;**10**:1-16. DOI: 10.3390/su10010016

[11] Akinola R, Pereira LM, Mabhaudhi T, Bruin FM, Rusch L. A review of indigenous food crops in Africa and the implications for more sustainable and healthy food systems. Sustainability. 2020;**12**(8):34-93

[12] FAO. Traditional Food Plants, Food and Nutrition. Rome, Italy: FAO; 1998 Paper 42

[13] Towns AM, Hackleton CS. Traditional, indigenous, or leafy ? A definition, typology, and way forward for African vegetables. Economic Botany. 2019;**X**:1-17

[14] Kansiime MK, Ochieng J, Kessy R, Karanja D, Romney D, Afari-Sefa V. Changing knowledge and perceptions of African indigenous vegetables: The role of community-based nutritional outreach. Development in Practice. 2018;**28**(4):480-493

[15] Lotter DW, Marshall MI, Weller S, Mugisha A. African indigenous and traditional vegetables In Tanzania: Production, post-harvest management and marketing. African Crop Science Journal. 2014;**22**(3):181-189

[16] Afari-Sefa V, Chagomoka T, Karanja DK, Njeru E, Samali S, Katunzi A, et al. Private contracting versus community seed production systems: Experiences from farmer-led seed Enterprise development of indigenous vegetables in Tanzania. Vegetable Breeding in Africa: Constraints, Complexity and Contributions toward Achieving Food and Nutritional Security. Acta Horticulturae. 2013;**1007**:671-680

[17] Nakaziba R, Anyolitho MK, Amanya SB, Sesaazi CD, Byarugaba F, Ogwal-Okeng J, et al. Traditional medicinal vegetables in northern Uganda: An ethnobotanical survey. International Journal of Food Science. 2021;**55**:88-196. DOI: 10.1155/2021/5588196

[18] Venter SL, Van den Heever E, Allemann J, Viljoen J. Potential vegetable and medicinal uses of traditional crops in South Africa. Acta. Horticulture. 2000;**523**:25-28. DOI: 10.17660/ActaHortic.2000.523.2

[19] Ochieng J, Afari-sefa V, Lukumay PJ, Dubois T. Determinants of dietary diversity and the potential role of men in improving household nutrition in Tanzania. PLoS One. 2017;**12**:1-18. DOI: 10.7910/DVN/INRWQA.Funding

[20] Ochieng J, Afari-sefa V, Karanja D. Promoting Consumption of Traditional African Vegetables and its Effect on Food and Nutrition Security in Tanzania. Cambridge University press; 2016

[21] Jackson LE, Pulleman MM, Brussaard L, Bawa KS, Brown GG, Cardoso IM, et al. Social-ecological and regional adaptation of agrobiodiversity management across a global set of research regions. Global Environmental Change. 2012;**22**:623-639. DOI: 10.1016/j.gloenvcha.2012.05.002

[22] NBS. National Sample Census of Agriculture 2019/20. Dar Es Salaam, Tanzania: NBS, National Report; 2021. p. 99

[23] Weinberger K, Msuya J. Indigenous Vegetables in Tanzania—Significance and Prospects. Shanhua, Taiwan: AVRDC—The World Vegetable Center, Technical Bulletin No. 31, AVRDC Publication 04-600. 70 pp; 2004

[24] Chepkoech W, Mungai NW, Stöber S, Lotze-campen H. Understanding adaptive capacity of smallholder African indigenous vegetable farmers to climate change in Kenya. Climate Risk Management. 2020;**27**:100204. DOI: 10.1016/j.crm.2019.100204

[25] Berg H, Phillips S, Dicke M, Fredrix M. Impacts of farmer field schools in the human, social, natural and financial domain: A qualitative review. Food Security. 2020;**14**:43-59

[26] Kissoly LD, Karki SK, Grote U. Diversity in farm production and household diets: Comparing evidence from smallholders in Kenya and Tanzania. Frontiers in Sustainable Food Systems. 2020;**4**(June):1-13. DOI: 10.3389/fsufs.2020.00077

[27] Lambrou Y, Laub R. Gender, local knowledge and lessons learnt in documenting and conserving agrobiodiversity. In: Food Insecurity, Vulnerability and Human Rights Failure. UK: Palgrave Macmillan; 2007. pp. 161-194. DOI: 10.1057/9780230589506_7.2007

[28] Barbier EB. Natural Capital, Ecological Scarcity and Rural Poverty. Washington DC: World Bank; 2012

[29] Boulay B, Khan R, Morrissey O. Under-utilised crops and rural livelihoods: Bambara groundnut in Tanzania. Oxford Development Studies. 2021;**49**(1):88-103. DOI: 10.1080/13600818.2020.1839040

[30] Zonneveld M, Kindt R, Solberg SØ, Danikou SN, Dawson IK. Diversity and conservation of traditional African vegetables: Priorities for action. Diversity and Distributions. 2020;**27**:216-232. DOI: 10.1111/ddi.13188

[31] Keller GB, Mndiga H, Maass BL. Diversity and genetic erosion of traditional vegetables in Tanzania from the farmer' s point of view. Plant Genetic Resources. 2006;**3**(3):400-413. DOI: 10.1079/PGR200594

[32] Meliyo JL, Mowo JG, Wickama JM, Shemdoe R. The role of social Capital in Enhancing INRM innovations in Lushoto, Tanzania. In: Enhancing Gender Inclusion, Equity and Social Awarenesss. World Vegetable center, Taiwan. 2016. pp. 192-198

[33] Kessy RF, Ochieng J, Afari-Sefa V, Chagomoka T, Nenguwo N. Solar-dried traditional African vegetables in rural Tanzania: Awareness , perceptions, and factors affecting purchase decisions. Economic Botany. 2019;**1**(X):1-13

[34] Marson M. Direct access to markets by farmers and the role of traders: Insights from Kenyan and Tanzanian leafy vegetables markets. Studies in Agricultural Economics. 2022;**124**(2):66-73. DOI: 10.7896/j.2287

[35] Musebe R, Karanja D, Rajendran S, Kessy R, Kansiime M, Marandu D, et al. Development of market opportunities through post- harvest processing of the African indigenous vegetables in Tanzania. AJBM. 2017;**11**(17):426-437. DOI: 10.5897/AJBM2017.8286

[36] Owusu O, İşcan TB. Drivers of farm commercialization in Nigeria and Tanzania. Agricultural Economics (United Kingdom). 2021;**52**(2):265-299. DOI: 10.1111/agec.12618

[37] Dinssa FF, Hanson P, Dubois T, Tenkouano A, Stoilova T, Hughes JA, et al. AVRDC–the world vegetable center's women-oriented improvement and development strategy for traditional African vegetables in sub-Saharan Africa. European Journal of Horticultural Science. 2021;**81**(2):91-105. DOI: 10.17660/eJHS.2016/81.2.3

[38] Pauw K, Thurlow J. Agricultural growth, poverty, and nutrition in Tanzania. Food Policy. 2011;**36**(6):795-804. DOI: 10.1016/j.foodpol.2011.09.002

[39] Meade CG. Strategy for traditional African vegetables in sub-Saharan Africa. European Journal of Horticultural Science. 2015;**81**(2):91-105

[40] Ambikapathi R, Shively G, Leyna G, Mosha D, Mangara A, Patil CL, et al. Informal food environment is associated with household vegetable purchase patterns and dietary intake in the DECIDE study: Empirical evidence from food vendor mapping in peri-urban Dar es Salaam, Tanzania. Global Food Security. 2021;**28**:100474. DOI: 10.1016/j.gfs.2020.100474

[41] Zimmerer KS, Haan SD. Agrobiodiversity and a sustainable food future. Nature Plants. 2017;**3**:17047. DOI: 10.1038/nplants.2017.47

[42] Thrupp LANN, Thrupp LA. Linking agricultural biodiversity and food security: The valuable role of agrobiodiversity for sustainable agriculture. International Affairs. 2020;**76**(2):265-281

[43] Bijlsma R, Loeschcke V. Genetic erosion impedes adaptive responses to

stressful environments. Evolutionary Applications. 2012;**5**(2):117-129

[44] Shangali CF, Mabula CK, Mmari C. Biodiversity and human activities in the Udzungwa Mountain forests, Tanzania. 1. Ethnobotanical survey in the Uzungwa scarp Forest reserve. Journal of East African Natural History. 1998;**87**(1):291-318

[45] Dweba TP, Mearns MA. International journal of information management conserving indigenous knowledge as the key to the current and future use of traditional vegetables. International Journal of Information Management. 2011;**31**(6):564-571. DOI: 10.1016/

[46] Mpasiwakoma RA, Nyomora AMS, Gimbi A. Diversity and utilization of wild edible plant species from the Uvinza woodlands, Tanzania. Huria Journal. 2017;**24**(3):150-168

[47] Ceccarelli S, Grando S. Return to agrobiodiversity: Participatory plant breeding. Diversity. 2022;**14**:126

[48] Muhanji G, Roothaert RL, Webo C, Stanley M. African indigenous vegetable enterprises and market access for small-scale farmers in East Africa African indigenous vegetable enterprises and market access for small-scale farmers in East Africa. International Journal of Agricultural Sustainability. 2011;**9**:194-202. DOI: 10.3763/ijas.2010.0561

[49] URT. National Biodiversity Strategy and Action Plan (NBSAP) 2015-2020. Dar es Salaam, Tanznia: URT; 2015

[50] Sumalan R, Ciulca S, Sumalan R, Popescu S. Vegetable landraces: The "gene banks" for traditional farmers and future breeding programs. In: Landraces. London, UK: InTech Open; 2021

[51] Fischer G, Patt N, Ochieng J, Mvungi H. Participation in and gains from traditional vegetable value chains: A gendered analysis of perceptions of labour, income and expenditure in producers' and traders' households. European Journal of Development Research. 2020;**32**:1080-1104

[52] Amao IO, Egbetokun OA. Market participation among vegetable farmers. International Journal of Vegetable Science. 2018;**24**(1):3-9

[53] Powell B, Kerr RB, Young SL, Johns T. The determinants of dietary diversity and nutrition: Ethnonutrition knowledge of local people in the east Usambara Mountains, Tanzania. Journal of Ethnobiology and Ethnomedicine. 2017;**13**:1-12. DOI: 10.1186/s13002-017-0150-2

[54] Hallensleben M, Polreich S, Heller J, Maass B. Assessment of the importance and utilisation of cowpea (Vigna unguiculata L. Walp.) as leafy vegetable in small-scale farm households in Tanzania–East Africa. In: Biophysical and Socio-Economic Frame Conditions for the Sustainable Management of Natural Resources. Vol. ii. Witzenhausen DITSL Gmbh, Germany: 2009. pp. 1-4

[55] Stoilova T, Zonneveld M, Roothaert R, Schreinemachers P. Connecting genebanks to farmers in East Africa through the distribution of vegetable seed kits. Plant Genetic Resources. 2019;**17**(3):306-309

Section 3

Animal Diversity and Conservation

Chapter 8

Changes to Health Parameters of White-Tailed Deer during a Drought in the US Virgin Islands

Suzanne L. Nelson, Nicola Justice, Kaitlynn M. Apple, Aidan H. Liddiard, Madeleine R. Elias and Jon D. Reuter

Abstract

Resident white-tailed deer (*Odocoileus virginianus*) on St. John, US Virgin Islands offer a unique case study for understanding a population under pressure from climate change. During a 2015–2016 regional drought, deer health parameters including body condition, coat condition, tick prevalence, musculature, and stress hormones were tracked over three field seasons representing the onset, peak, and recovery phases of the drought. All health indicators showed significant change over the course of the drought, and post-hoc tests suggest some indicators (body condition, musculature, and ticks) were more sensitive during drought onset. High levels of cortisol during the peak period indicated substantial stress to the population, which normalized during recovery. The strongest correlations were between overall health/body condition and musculature and overall health/body condition and coat condition. The weakest correlations were between ticks and the remaining three variables. These results support the hypothesis that various measures of deer health are related. The frequency and intensity of droughts and environmental stressors are predicted to increase in the future due to climate change, which will further challenge this island deer population.

Keywords: stress hormones, climate change, body condition, musculature, coat condition, ticks

1. Introduction

Historically, the Caribbean region has been characterized by relatively predictable seasonal rainfall patterns and moderate fluctuations in annual temperatures. However, this stability is changing rapidly with climate change, and is projected to be highly variable as levels of greenhouse gases in the atmosphere continue to increase [1, 2]. The most recent climate change data predicts that climate change will bring extreme heat waves at greater frequency, droughts that will occur 2–3 times more often, stronger hurricanes with the trend toward hotter temperatures, and wet and dry extremes in local areas [2, 3]. The Caribbean is considered particularly vulnerable to the effects of climate change as weather events that occur in the region increase in both frequency and duration (www.drought.gov). As a result, this area will experience increasing

IntechOpen

weather variability and extremes, which will manifest as hotter temperatures, shifting rainfall patterns, more frequent water shortages following decreased annual rainfall, and longer dry seasons. The Virgin Islands have already experienced a series of droughts in recent years, the most severe of which occurred in 2015–2016.

Changing landscape conditions, including more frequent drought, can exert stressors on wildlife populations that have the potential to be detrimental to the health and fitness of individuals [4, 5] and populations over time [6]. For wildlife populations, drought presents multiple simultaneous environmental challenges such as high temperatures, low food availability, and low water availability [1]. Water shortages associated with drought can bring about reduced plant primary productivity and seed survivability, which causes food reduction and changes in water quantity and quality. As a result, suboptimal consumption of protein, vitamins, minerals, and other essential nutrients can lead to malnutrition. Subsequently, malnutrition, prolonged dehydration from heat stress, and parasitism can lead to the depletion of fat reserves, anemia, and poor body condition, and has the potential to result in immunocompromised individuals [7, 8]. These individuals are even more at risk for malnutrition, parasitism, or starvation [9]. Wildlife adapt to drought using physiological and behavioral adaptations, but the stress from prolonged drought can eventually overwhelm their resiliency [10].

Several studies have evaluated deer survival during drought events and demonstrated a clear connection between a weather event and population effects. During drought years, deer often overgrazed available flora and harmed plant species due to intense herbivory [11]. In addition, deer consumed fewer plants, and plants of lower forage quality, and often did not meet their nutritional requirements, which had the potential to limit lactation [12]. Reduced lactation can decrease deer numbers either because of neonate starvation or result in smaller and weaker fawns [13]. Bucks responded to reduced food quantity and quality by displaying smaller body size and antler growth, particularly in young males that were still growing [12]. Therefore, a single year of drought might have lifetime consequences for a cohort of both female and male deer [14]. Overall, the effects of climate change are multidimensional, and exacerbated by the stress of prolonged drought, and can be largely deleterious to the health of wildlife.

The goal of this study was to assess the resilience of a population of white-tailed deer on St. John, US Virgin Islands as they responded to a severe drought in 2015–2016. Health observations were collected through three successive field seasons representing drought onset, peak, and recovery. This work is the first to document the physiological changes observed during drought for a population of isolated residential island deer in the Caribbean region, and as a result, this work was largely exploratory. We hypothesized the lack of food and water resources associated with prolonged drought would have a negative effect on deer on St. John and we predicted that there would be an adverse change to their physiological condition as a result. However, there is little prior research alluding to chronic, pre-existing stress factors or the time points at which the changes would be demonstrated in the deer, including if and when they would resolve. We hypothesized that there would be strong associations between drought and musculature, coat condition, and body condition, due to limited island resources, and that these parameters might decrease due to direct and indirect effects of the drought. Similarly, we hypothesized that there might be an association between the values for cortisol, T-3 levels, and tick presence, and they might increase with drought-associated stress. However, we did not know the strength of the relationship between these different parameters, or which parameters would show differences within the three evaluation periods of this study. Our goal was to provide quantitative evidence for these changes for a protected island population at three distinct time periods of a drought. The monitoring of stress and nutrition in wildlife populations can provide researchers with valuable insight into the baseline stability and physiological impacts of environmental change on wildlife populations [6]. The deer of St. John represent a unique and intriguing case study for understanding a population facing highly altered future conditions due to climate change.

2. Methods

2.1 Field data collection

The island of St. John is located in the Caribbean Sea between 18°18′ and 18°22′N latitude and 64°40′ and 64°48′W longitude (**Figure 1**). It is 11 km long and 5 km across at its widest point [15]. St. John is part of the US Virgin Islands which includes St. John, St. Thomas, St. Croix, and Water Island. Virgin Islands National Park (VINP) lies on the island of St. John and comprises 60% of the landmass of the island and protects one of the largest tracts of secondary dry forest in the eastern Caribbean [16]. Dryland plant communities on St. John include forests, shrublands, coastal hedges, and a rare cactus community [17, 18]. The forests and forest structure are largely shaped by hurricanes and drought [16]. The climate is relatively dry, with an average temperature of 27°C with 75% relative humidity. Water is in limited supply because of high temperatures, high evaporation rates, and run-off from steep slopes [15, 19]. There are no permanent streams or rivers on the island and only a few intermittent streams [20]. Precipitation is highest from May through November with a range of 890–1400 mm/yr [15]. The driest months are February and March [20]. The region is prone to cyclical patterns of drought and excessive moisture and structural damage from seasonal hurricanes.

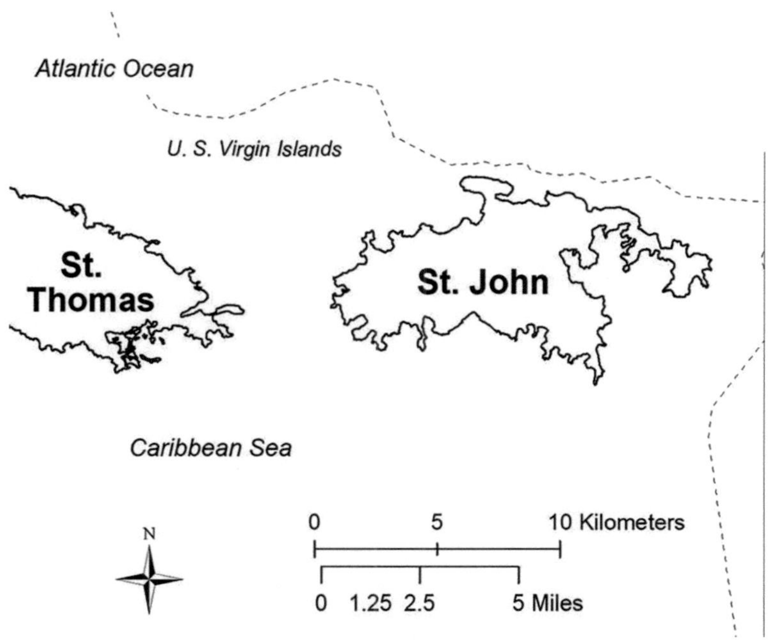

Figure 1.
A map of the study location on the island of St. John, U.S. Virgin Islands in the Caribbean.

This information builds upon a series of articles describing the natural history parameters of the deer of St. John. Despite their isolation, St. John deer demonstrated low inbreeding and average heterozygosity [21], were positive for antibodies for bluetongue and epizootic hemorrhagic disease but no clinical signs of the disease were observed during field observations [22]. A third study identified the ticks found on the St. John deer as a species of cattle tick and tropical horse tick [23].

Data were collected over three field seasons on St. John, USVI.; first during the onset of a drought (July 2015, henceforth "onset"); second at the peak of the drought (March 2016, henceforth "peak"); and third during recovery from the drought (July 2016; henceforth "recovery") (**Figure 2**). The drought occurred regionally throughout the Caribbean and was prevalent in 2015 on St. John [23].

Deer musculature, coat condition, body condition, and tick levels were recorded using two data collection methods: remote assessment and direct assessment of tranquilized deer. Remote assessments of nutritional conditions offer a noninvasive alternative when capture of deer is not possible, and can be used to evaluate changes at a population level [24]. Remote assessments were conducted at all three time points (onset, peak, and recovery of the drought), including collecting data on deer seen either on transects or opportunistically by trained technicians working in pairs and using binoculars at a distance of less than 25 m. Deer on the island are highly acclimatized to humans and did not move when observed by technicians, particularly near popular tourist trails or beaches. This allowed the research team to observe deer directly for accurate data collection. Technicians for

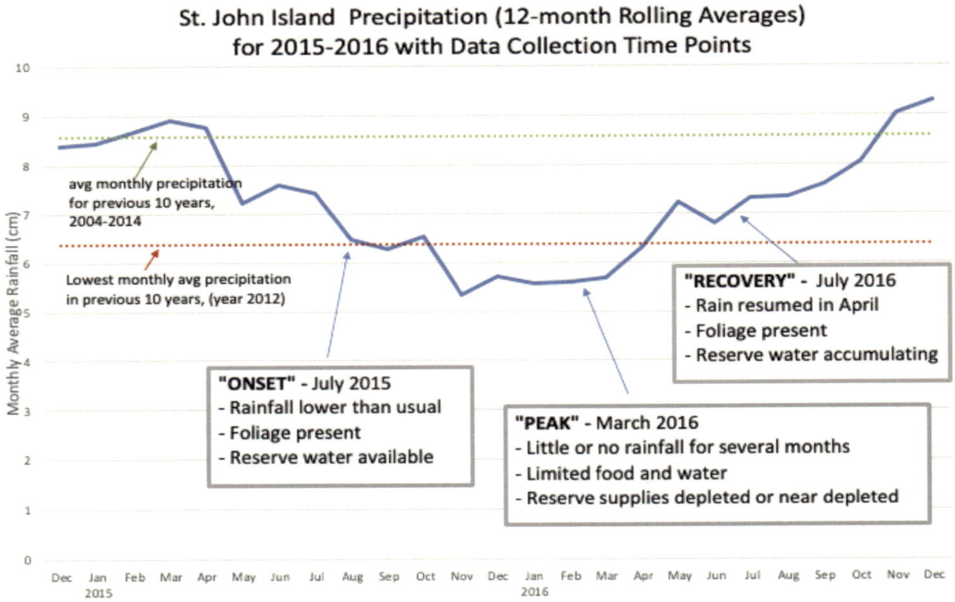

Figure 2.
The 12-month rolling average precipitation (calculated using the previous 12 months) on St. John Island for the years 2015–2016, with annotations for the three data collection time points and field notes regarding observed conditions in a study of white-tailed deer (Odocoileus virginianus) health indicators during the onset, peak, and recovery phases of a 2015–2016 drought. Initially (December–April 2015) the rolling average is on par with average precipitation for the previous 10 years. Precipitation decreases beginning in April of 2015. Data collected in July 2015 ("Onset") occur when the rolling average is low but reserve supplies are still available. By October 2015 the rolling average precipitation drops below the lowest annual precipitation in the previous 10 years. Data collected in March 2016 ("Peak" of the drought) occurred when the monthly average precipitation for the past 12 months has remained lower than usual for several months. Precipitation returned shortly after, bringing the rolling average up to more favorable conditions ("Recovery" phase of the drought).

each field season were trained to determine body condition, musculature, coat condition and tick levels by completing both pre-visit trainings and in-field calibration sessions to reduce the amount of inter-observer variability and to retain a high fidelity of rating. A body scoring system that uses anatomical landmarks provided an immediate evaluation that is non-invasive [25]. Data were collected in real time using binoculars, iPads, and a custom app for data storage that was specifically designed for this project. For examples of the different parameters measured, see **Figure 3** and **Table 1**. Pregnant deer and fawns were purposefully excluded.

The second method of data collection was conducted via direct assessment when deer were tranquilized. This method was only used during the recovery phase of the drought [22]. Deer data were collected throughout St. John [21] and were geographically representative of the deer population in all areas of the island. Pregnant or nursing does and deer that were one year old were not tranquilized. In total, twenty-three adult deer were tranquilized using butorphanol, azaperone, and medetomidine (BAM, Wildlife Pharmaceuticals, Windsor, Colorado, USA). Relative body size was used to determine the administered dose according to the manufacturer's guidelines. Most deer

Figure 3.
(A) A white-tailed deer (Odocoileus virginianus) showing excellent coat condition, musculature, and body condition, and an absence of ticks (B) showing good coat condition, musculature, and body condition, and no ticks (C) showing fair coat condition, musculature, and body condition, with some ticks and (D) showing a poor body condition, with a high degree of tick infestation in the ears. These conditions were observed in a study of deer health indicators during the onset, peak, and recovery phases of a 2015–2016 drought on St. John Island.

	Body composition	Coat condition	Musculature
Excellent	• Hip, rib and spine bones not visible • Rump rounded and possible extra fat seen	• Shiny and smooth hair with luster • No blemishes or scars	• Adequate/robust development • Toned and well-defined muscles • Abundant fat reserves
Good	• Hip, rib and spine bones starting to be visible • Rump not as rounded	• Mostly free of scars • Less shiny and smooth and losing luster • Some blemishes on hair	• Healthy muscle present • Not well defined • Adequate fat reserves
Fair	• Hip, rib and spine bones visible • Rump starting to flatten/concave.	• Some scars • No luster or shine • Small patches of missing hair	• Atrophy of muscle seen • No definition • Depleted fat
Poor	• Hip, rib and spine bones prominent • Rump concave and obvious absence of muscle	• Dullness to coat • Large clumps of missing or patchy hair • Large amounts of scars	• Severe atrophy • Underdeveloped • No fat reserves

Table 1.
White-tailed deer (Odocoileus virginianus) were scored as excellent, good, fair, and poor for body condition, coat condition, and musculature on St. John Island during the onset, peak, and recovery phases of a 2015–2016 regional drought. Ratings loosely follow [26].

were small to medium in size (41.1 ± 13.2 kg), receiving 1.0–1.5 mL intra-muscularly in the hind quarter by pneumatic dart gun (Pneu-Dart, Williamsport, Pennsylvania, USA). Vitals monitored included heart rate, respiratory rate, mucous membrane color, body temperature, time to recumbency, and recovery. Body temperatures were stabilized with a wet cooling blanket (Equi Cool-Down, Jacksonville, Florida, USA). After examination, the anesthesia was reversed with 2–3 mL of atipamezole (25 mg/mL) and 0.5 mL of naltrexone (50 mg/mL, Wildlife Pharmaceuticals). This work was conducted under Scientific Research and Collection Permit VIIS-2016-SCI-0026 for the Virgin Islands National Park to S. Nelson and IACUC (1602.01-15Mar2016) from the University of Colorado at Boulder and the National Park Service to S. Nelson.

Stress hormones were analyzed from fresh deer fecal material. Feces were collected opportunistically throughout St. John when we could directly attribute the fecal deposit to an individual deer, which allowed us to describe the age, sex, and health condition of the deer [27, 28]. Fecal samples were collected using sterile gloves and placed into labeled plastic bags, stored with ice packs, and given a unique identifying number [29]. The samples were frozen and sent to the Wasser lab at the University of Washington for stress hormone analysis. Samples were analyzed for both cortisol (ng/g) and T3 levels (ng/g). All samples were freeze-dried and homogenized, and then 0.1 g was extracted using 15 ml of 70% ethanol [30].

2.2 Statistical analysis

Deer musculature, coat condition, body condition, and tick levels were determined to be excellent, good, fair, or poor (**Figure 3** and **Table 1**). The initial four categories of excellent, good, fair, and poor were collapsed into binary outcomes: *Good/Excellent* vs. *Fair/Poor* to simplify interpretation and bolster sample sizes. Borderline cases were omitted (e.g., deer rated as "Good/Fair"). After collapsing, all four variables satisfied conventional conditions for sample size for Chi-Square tests of independence to assess the relationship between deer health and time point (onset, peak, and recovery phase, respectively). In post-hoc analyses, for each of these measures three pairwise two-proportion two-tailed z-tests were conducted to discern differences between time points: drought onset vs. recovery, peak vs. recovery, and onset vs. peak. No corrections for multiple testing were used so as to flag potential effects for future study.

To evaluate relationships between the four health indicators collected via visual assessment (musculature, body condition, coat condition, and ticks; both remote and direct), the phi coefficient was calculated for each of the pairwise relationships. Marginal distributions indicate that it is impossible for phi to reach a magnitude of 1; to avoid inflating the perceived relationships, no adjustment was made. The resultant correlations are conservative.

Two deer that had been tranquilized were also observed remotely in July 2016 following the drought. The remote observations were removed from the primary analyses, but used to check consistency between data collected remotely and data collected by tranquilizing. Observations of body condition and musculature were rated the same, however, ticks were observed more readily within the tranquilized deer when compared when they were viewed remotely. Thus, to promote consistency when comparing across time periods, observations of tranquilized deer were not included in the analysis for ticks. Data from tranquilized deer were included for the other variables because inconsistencies were not meaningful after the collapse into binary groups.

To quantify the stress from drought using fecal cortisol levels, two F-tests were conducted (one for average cortisol and one for average T3 levels). There were no extreme outliers and conditions for the test were otherwise adequately met. In post-hoc analyses three pairwise two-sample t-tests were conducted to discern differences between time points. All analyses were conducted using R statistical software (version 3.6.0).

3. Results

In total, 515 deer were observed remotely and 23 deer were tranquilized over three study seasons. In July 2015, 265 deer were observed, in March 2016, 189 deer were observed, and in July 2016, 61 deer were observed remotely, and 23 were tranquilized. At each time point, the samples were slightly more heavily represented by female deer; approximately 60–70% were female. The remote samples were heavily represented by mature deer, and ranged from approximately 75–90% mature deer at each time period.

3.1 Deer health indicators associated with drought conditions

The data confirm the hypothesis that there is an association between deer health and phase of drought conditions. For all six measures, a relationship between the stage of drought was statistically discernible (significant), which suggests that the onset, peak, and recovery phases were not all equally difficult for the deer. More specifically, a relationship was found between time points relative to drought and average cortisol levels ($F_{(2, 63)} \approx 18, p < .00001$) as well as average T3 levels ($F_{(2, 63)} \approx 5.8, p \approx 0.005$). Similarly, statistically discernible (significant) associations were found between time point relative to drought and proportion of deer with fair/poor rankings on each of the four measures observed visually (tick level, coat condition, body health, and musculature, with respective p-values of .013, .01, .005, and .001, respectively) (**Figures 4, 5** and **Table 2**).

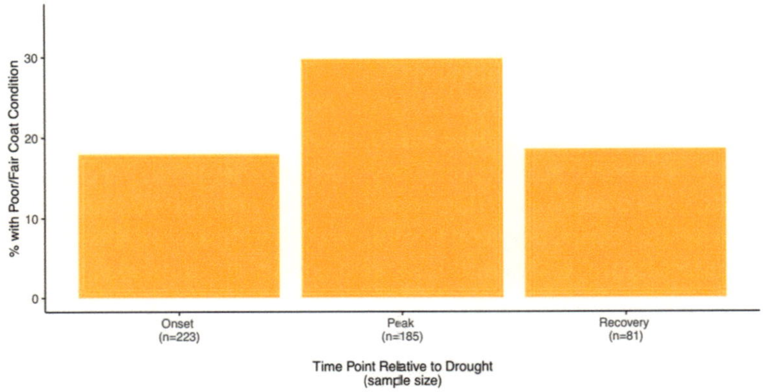

Figure 4.
Percentages of deer with coat condition issue (i.e., categorized as poor or fair condition) at each of the three time points (drought onset, peak, and recovery). This was the only indicator that presented higher prevalence of issues during the peak of the drought, while having lower prevalence both at the drought onset and recovery phases. Chi-square test indicates an association between time and prevalence of coat condition issues (χ^2 (2 degrees of freedom) = 8.9, p = .01).

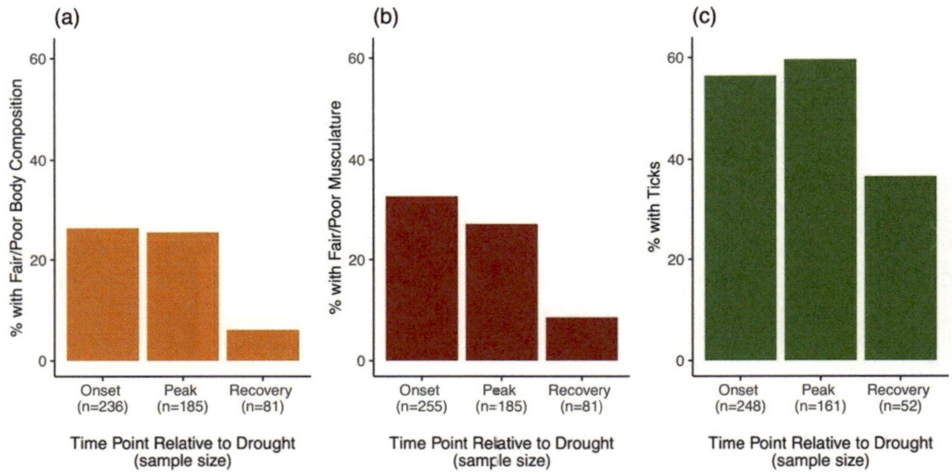

Figure 5.
Indicators that presented higher prevalence of issues both at the onset and peak of the drought, while having lower prevalence after drought in a study of white-tailed deer (Odocoileus virginianus) during a 2015–2016 drought on St. John Island. (a) Percentages of deer with body decomposition (i.e., categorized as poor or fair condition) at each of the three time points (onset, peak, and recovery of the drought). Chi-square test indicates an association between time and prevalence of body decomposition (χ^2 (2) = 15.1, p = .005). (b) Percentages of deer with muscle atrophy observed at each of three time points (onset, peak, and recovery of the drought). Chi-square test indicates an association between time and prevalence of muscle atrophy (χ^2 (2) = 17.9, p = .0001). (c) Percentages of deer with ticks observed at each of three time points (onset, peak, and recovery of the drought). Chi-square test indicates an association between time and prevalence of ticks (χ^2 (2) = 8.8, p = .013).

In summary, all six measures revealed an association between drought conditions and health of the deer, which does not support a hypothesis that the onset, peak, and recovery phases are all equally difficult on the deer.

Post-hoc analyses were conducted to explore various hypotheses about patterns of how each of the six health measures appeared to respond to the different stages of drought (**Table 2**). There are a few similarities; perhaps not surprisingly, all six measures had high levels of poor health at the peak of the drought. Another similarity is that all six measures revealed relatively improved health at the third time point

	P-value for Two-tailed Pairwise Test		
Health indicator	**Onset vs. peak**	**Peak vs. recovery**	**Onset vs. recovery**
1. Coat condition	.007	.08	1
2. T-3	.2	.02	.02
3. Body condition/overall health	.9	.0005	.0002
4. Musculature	.3	.001	.00004
5. Ticks	.6	.006	.01
6. Cortisol	.02	.0006	.0005

Table 2.
Post-hoc results (two-tailed p-values) of white-tailed deer (Odocoileus virginianus) health indicator comparisons across time points relative to drought (onset, peak, recovery) on St. John Island during a 2015–2016 regional drought. To flag potential effects, no correction for multiple testing was used.

(during the recovery phase after the drought). This suggests the deer population showed resilience in the recovery phase.

On the other hand, there were differences in the extent to which the health measures appeared to be sensitive to the onset stage of the drought: some measures indicated poorer health even in the onset, whereas other measures were less sensitive. Three patterns of behavior were observed and named according to relative health at each of the onset, peak, and recovery periods of the drought. The patterns are named: "Low-High-Low," "High-High-Low," and "Medium-High-Low" (where "High" indicates high levels of poorer health).

Coat condition displayed the Low-High-Low pattern. This measure had discernibly higher (statistically significant) prevalence of poorer health only at the peak of the drought (**Figure 4** and **Table 2**). In the drought onset and recovery, percentages were significantly lower than the peak. No statistically discernible (significant) differences were observed when comparing onset vs. recovery phases of the drought ($p > .5$). This supports a hypothesis that coat condition is not particularly sensitive to the onset stage of the drought.

In contrast, the T-3 levels, body condition, musculature, and tick indicators displayed the "High-High-Low" pattern. These three measures showed higher prevalence of poorer health both during the drought onset and peak (**Figures 5, 6** and **Table 2**). To be clear: for these indicators there were no statistically discernible (significant) differences in deer health when comparing the first two time points: onset versus peak of drought ($p = .2, .3, .6$, and $.9$, respectively, for T-3, musculature, ticks, and body condition measures). The prevalence of poor health on these indicators was only discernibly different (significant) at the third time point, after the drought. This supports the hypothesis that T-3 levels, body condition, musculature, and tick prevalence are sensitive to drought even in the onset stage.

Cortisol was the only measure to display a "Medium-High-Low" pattern. This measure had statistically discernible (significant) differences when comparing each of the three time points with each other (**Figure 6** and **Table 2**), with a mild sensitivity during the onset stage, increased severity at the peak of the drought, and of course the ability to recover in the third stage.

3.2 Relationships between health measures

Positive relationships were also present between each of the visually observed measures of deer health (**Table 3**). For example, a deer with fair/poor musculature was

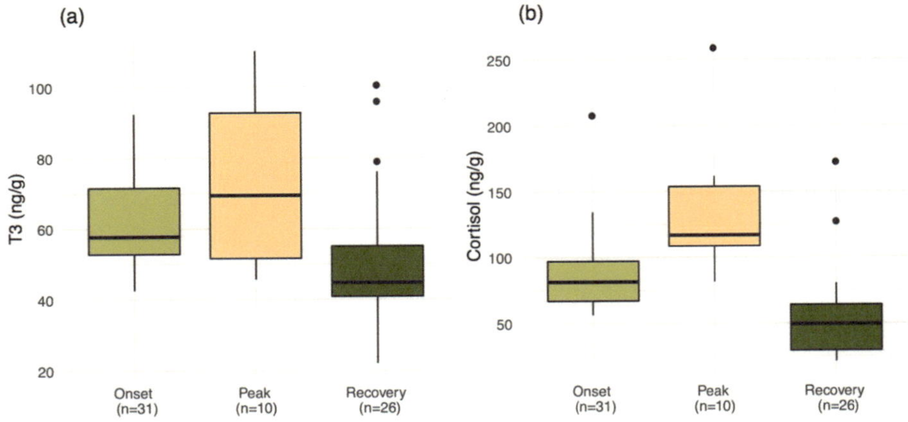

Figure 6.
Levels of fecal T3 (a) and cortisol (b) in white-tailed deer (Odocoileus virginianus) populations in a study of deer health indicators during the onset, peak, and recovery phases of a 2015–2016 drought on St. John Island. Sample sizes collected indicated below each period.

	1	2	3	4
1. Coat condition	1.00			
2. Body condition/overall health	0.48	1.00		
3. Musculature	0.34	0.64	1.00	
4. Ticks	0.31	0.26	0.2	1.00

Table 3.
Correlation (as measured by Phi coefficient) indicating strength of association between health indicators (coat condition, body condition, musculature, and ticks) for white-tailed deer (Odocoileus virginianus) scored as Excellent/Good vs. Fair/Poor on St. John Island during the onset, peak, and recovery phases of a 2015–2016 regional drought. To avoid inflation, no adjustment for binary outcomes was used (e.g., adjusted phi, etc.). Thus, the correlations given here are conservative.

more likely to also have fair/poor coat condition (φ = .34). Conversely, a deer with good/excellent musculature was more likely to also have excellent/good coat condition. Overall health/body condition was strongly correlated to musculature (φ = .64) and coat condition (φ = .48). Coat condition and tick infestation were moderately correlated (φ = .31), as were coat condition and musculature (φ = .34). Smaller correlations were observed between tick infestation and musculature (φ = .2), and between tick infestation and health/body condition (φ = .26). The fact that all the pairwise correlation coefficients between health/body condition, musculature, coat condition, and ticks are positive supports the hypothesis that that these four measures of deer health are related, and/or largely dependent on similar underlying reasons. For this work, that reason was thought to be the multifaceted effects of stress associated with prolonged drought on the island.

4. Discussion

Positive pairwise correlations were present between all four measures of individual deer: tick prevalence, musculature, coat condition, and body condition. That is, deer that had one health issue were more likely to also have another. This could simply reflect that some measures may be related (e.g., coat condition and overall body condition have some common sources of overlap). However, other measures (e.g., tick prevalence and musculature) do not naturally overlap. Therefore, our results suggest health factors are indeed associated perhaps because the underlying stress of deer represents a common etiology. The population of deer on St. John are largely disease-free [22], and demonstrate relatively high allelic diversity for an isolated island population [21] which may act as a mitigating factor for stress. However, this study demonstrates a strong response during peak drought for all of the variables.

Our observations suggest that deer on St. John were highly nutritionally deficient, particularly within the drought peak in July 2015, when deer faced severely limited feeding within a denuded landscape. During the peak drought period, deer appeared to be eating atypical forage, often called famine foods, in the absence of typical and available forage. Famine foods are eaten by animals when all known food sources have been exhausted. They often contain low levels of protein, are low in calories provided [31], and can be very energetically expensive to metabolize. Observations during the drought peak included a fawn eating dead leaves and adult deer spending considerable time and energy digging up tubers and roots. Dependable sources of water on St. John were also severely reduced in both quantity and quality, and deer were often seen near anthropogenic water sources such as pipes, foot baths, and shower areas. Their body condition during the height of the drought indicated both severe water and nutritional stress [23]. Below we analyze multiple indicators of deer health to show individual responses to drought onset, peak, and recovery.

4.1 Indicators of deer health

4.1.1 Body condition and coat condition

Results for body condition followed the High-High-Low pattern, demonstrating a higher prevalence of poor health during both the drought onset and peak. This indicates that the deer were in a state of stress at the drought's onset and were without the reserves needed from the lack of food and water availability during the drought peak. Body condition scoring represents an amalgam of multiple health parameters representing external markers of internal health. Many of the factors that were evaluated with body condition, including coat condition and musculature, are complementary data for assessing the health and nutritional state of an animal. These factors are influenced by both current and past food availability, and can therefore indicate health in both the present and recent past. Prolonged nutritional challenges such as drought can cause individuals to exhaust stored fat reserves which can result in a deterioration in body condition [6]. In contrast, coat condition demonstrated a Low-High-Low pattern during the drought. This suggests that coat condition is not as sensitive to the onset stage of the drought as body condition, and coat condition is able to return to original condition during drought onset within the drought recovery period. Coat condition represents a sub-category of the body composition assessment, and one that can further indicate nutritional health. Essential vitamins, as well as fats and oils in the diet, are needed to provide coat and hair luster [32, 33].

4.1.2 Muscle and muscle atrophy

Results for musculature followed the High-High-Low pattern, demonstrating a greater prevalence of poor health both during the drought onset and peak. The response of musculature is similar to body condition in that the pattern indicates that individual deer were stressed at drought onset, indicating a deficiency in both food availability and in high quality foods that contain adequate protein reserves to support muscle development and growth. Muscle is an energetically expensive tissue to build and maintain [34]. Muscle presence and appearance can be an indicator of overall health and an excellent external indicator of both health and protein reserves. However, with chronic and insufficient protein consumption, muscle catabolism can occur, resulting in muscle atrophy [33]. In addition, because more than 80% of protein in the animal body is dedicated to maintaining proper functioning of the immune system, disease state can indirectly result in muscle loss and atrophy. As a result, there can be a strong correlation between muscle appearance and the disease state of a deer. If muscle atrophy is present, this indicates that the animal body is using muscle as an energy source in the absence of any remaining fat reserves. More energy is released when a unit of fat is metabolized as compared to a unit of protein. Therefore, when a deer has metabolized most of its fat and is using muscle for energy, it will lose weight very quickly as subcutaneous fat is already depleted [35] and will appear gaunt and emaciated in appearance as a result [32]. In contrast, excellent musculature indicates that the deer is eating sufficient protein to both maintain muscle and to support the high protein demands of the immune system [36]. These changes in subcutaneous fat can be evident via visual assessment [24]. The results of this study add to evidence that both musculature and fat reservoirs are sensitive to drought even in the early onset stages within island populations.

4.1.3 Ticks

The dynamics of infection often depend on the host's vulnerability, as poor body condition is likely to predispose individuals to infectious and parasitic diseases [6, 7]. Tick prevalence in this study followed a High-High-Low pattern, indicating that ticks were already both prevalent during both the onset and peak of the drought. The additive effect of this continued high level of parasitism has negative consequences for individual deer health for blood loss and the potential for disease development. A common parasite found on the deer of St. John included ticks. The two tick species found on St. John include the southern cattle tick (*Rhinocephalus* (*Boophilus*) *microplus* (Canestrini) and the tropical horse tick, *Dermacentor* (*Anocenter*) *nitens* Neumann [23]. High tick densities could result in associated health problems, including pruritis, alopecia, anemia, and low weight gain. In addition, ticks can deplete the iron resources of the deer through each blood meal taken, which can result in the development of iron-deficiency anemia and deprive tissues of necessary oxygen [37, 38]. There can be a considerable energy requirement to replace daily blood loss which could result in further accelerated nutritional decline and weight loss. Additionally, the relative energetic cost associated with compensating for blood loss is higher for animals in poor condition who can experience more pronounced energy and protein deficits compared to healthier animals [37].

4.1.4 Stress hormones

Stress hormones help the body handle adverse conditions, and levels can vary by situation and species [39]. Cortisol is released by the body when there is either an acute or chronic stressor. Cortisol can suppress the conversion of T4 thyroid hormone into T3, and lower circulating T3 levels can be an indicator that cortisol levels are high. Together, both T3 and cortisol levels create a complementary data set that demonstrates stress levels in individual animals [28].

Cortisol was the only measure from this study to display a "Medium-High-Low" pattern. This measure had statistically discernible (significant) differences when comparing each of the three time points with each other. Our data for all three phases of the drought show that the deer were already stressed in July 2015 at the start of the drought, and to a degree that shows widespread stress across the population. This was further exacerbated at the height of the drought, in March 2016, when the deer were in acute stress. The stress hormone data ranged from 24.3 to 110.1 ng/g for T3 and 26.5 to 258.5 ng/g for cortisol.

The cortisol values for St. John deer at the onset and recovery of the drought were within published values for animals within ongoing stressful situations, but cortisol values for deer during the peak of the drought were more consistent with wildlife undergoing acute trauma. For example, values ranged from 23.9 to 114.9 ug/g for musk deer in crowded conditions [40], and deer undergoing high parasite burdens had cortisol values that reached 93–144 ng/g [39] and could result in diminished body condition for deer over time [25]. These published values for deer with ongoing stress are consistent with the values found in this study for St. John deer during drought onset and recovery. However, the highest values found in our study at the peak of the drought (258.5 ng/g) were instead consistent with values that more reflected acute and traumatic events for animals. For example, koalas who had been in vehicle collisions (202 ng/g), were burn victims (200 ng/g), or were in an area during land clearance (669 ng/g) showed cortisol values similar to the deer of St. John during the peak of the drought [41]. High fecal cortisol levels can often be used as a predictor of mortality. Ring-tailed lemurs that died following traumatic events showed average fecal values of 51.1 ng/g [42]. The deer of St. John both endured sustained cortisol levels above that value for the duration of the drought (2015–2016), and also at values five times that amount at peak drought. This highlights the noteworthy resiliency of this population to continue to live with this ongoing stressor present on island for over 2 years.

Stress hormones are also important regulators of energy balance. When food availability diminishes to the point of starvation, cortisol remains chronically elevated, body condition declines, and fat stores are used for energy [43, 44]. Prolonged elevation of glucocorticoids can result in the suppression of reproduction, growth, immune function, and responses to pathogens and parasites [45].

5. Conclusions

This study was able to describe the resilience of an isolated population of white-tailed deer on St. John, US Virgin Islands during a severe drought in 2015–2016. Health observations collected through three successive field seasons representing drought onset, peak, and recovery of the drought enabled novel work documenting physiological changes of deer under the stress of climate change For all six health measures in this study (musculature, coat condition, body condition, the presence of ticks, cortisol, and T-3 levels), there is an association between deer health and time relative to drought (onset, peak, or recovery phase). All measures indicate a rapid return to health following the drought peak.

This study provides a baseline foundation for future research needed to inform the extent to which isolated populations of herbivores cope with increasing climate variability. Further work is needed to explore health differences between males and females during various stages of drought. Future studies would also benefit from using tagging methods for a study design that enables direct observations to individual deer and clarifies how the population maintains resilience in extreme weather. In addition, future work may include a closer analysis of stress hormone levels related to reproduction and survival of individuals. Periodic re-evaluation of the St. John deer population and health will add additional data to population densities that can be supported by the island ecosystem. This work provides a critical baseline to document physiological changes to an isolated Caribbean deer population, and the results of this work can be extended regionally throughout the Caribbean and to analogous species.

Acknowledgements

Thank you to the hard-working field crews who made this work happen, including D. Masters, C. Walls, B. Daughton, D. Duncan, J. DePompolo, B. Wilkins, P. Ewing, M. Johnson, and M. Malone. We thank T. Kelley and A.S. McKinley from the Virgin Islands National Park for their help. C. Anderson and K. Hendricksen contributed to the statistical analysis. Thank you to the Sam Wasser lab for stress hormone analysis. This project was funded by Friends of the Virgin Islands National Park to S. Nelson and the IPHY department at the University of Colorado at Boulder to S. Nelson. This project was completed under permit VIIS-2015-SCI-0027 for 2015 for the Virgin Islands National Park to S. Nelson. The findings and conclusions in this article are those of the author and do not necessarily represent the views of the US Fish and Wildlife Service.

Author details

Suzanne L. Nelson[1*], Nicola Justice[2], Kaitlynn M. Apple[3], Aidan H. Liddiard[4], Madeleine R. Elias[4] and Jon D. Reuter[5]

1 U.S. Fish and Wildlife Service, Portland, Oregon, USA

2 Department of Mathematics, Pacific Lutheran University, Tacoma, Washington, USA

3 Department of Integrative Physiology, University of Colorado Boulder, Boulder, Colorado, USA

4 Department of Mathematics, Pacific Lutheran University, Tacoma, Washington, USA

5 Office of Research Integrity, Department of Psychology and Neuroscience, University of Colorado Boulder, Boulder, Colorado, USA

*Address all correspondence to: suzanne_nelson@fws.gov

References

[1] Rymer TL, Pillay N, Schradin C. Resilience to droughts in mammals: A conceptual framework for estimating vulnerability of a single species. The Quarterly Review of Biology. 2016;**91**(2):133-176

[2] Lee JYJ, Marotzke G. Bala L, Cao S, Corti JP, Dunne F, et al. IPCC. Climate change 2021: The physi-cal science basis. Future Global Climate: Scenario-42 Based Projections and Near-Term Information; Cambridge, UK: Cambridge University Press; 2021: 1-195

[3] Dale VH, Joyce LA, McNulty S, Neilson RP, Ayres MP, Flannigan MD, et al. Climate change and forest disturbances: Climate change can affect forests by altering the frequency, intensity, duration, and timing of fire, drought, introduced species, insect and pathogen outbreaks, hurricanes, windstorms, ice storms, or landslides. Bioscience. 2001;**51**(9):723-734

[4] Angelier F, Wingfield JC. Importance of the glucocorticoid stress response in a changing world: Theory, hypotheses, and perspectives. General and Comparative Endocrinology. 2013;**190**:118-128

[5] Gallana M, Ryser-Degiorgis MP, Wahli T, Segner H. Climate change and infectious disease in wildlife: Altered interactions between pathogens, vectors and hosts. Current Zoology. 2013;**59**(3):427-437

[6] Debow J, Blouin J, Rosenblatt E, Alexander C, Gieder K, Cottrell W, et al. Effects of winter ticks and internal parasites on moose survival in Vermont, USA. The Journal of Wildlife Management. 2021;**85**(7):1423-1439

[7] Beldemenico PM, Telfer S, Gebert S, Lukomski L, Bennett M, Begon M. Poor condition and infection: A vicious circle in natural populations. Proceedings of the Royal Society: B. 2008;**275**:1753-1759

[8] Acevedo-Whitehouse K, Duffus ALJ. Effects of environmental change on wildlife health. Philosophical Transactions of the Royal Society of London. Series B, Biological Sciences. 2009;**364**(1534):3429-3438

[9] Watter K, Baxter G, Brennan M, People A, Murray P. Decline in body condition and high drought mortality limit the spread of while chital deer in north-East Queensland, Australia. The Rangeland Journal. 2019;**41**:293-299

[10] Hing S, Narayan EJ, Thompson RCA, Godfrey SS. The relationship between physiological stress and wildlife disease: Consequences for health and conservation. Wildlife Research. 2016;**43**:51-60

[11] Olivas-Sanchez MP, Vital-Garcia C, Flores-Margez JP, Mora-Covarrubias A, Clemente-Sanchez F. Mule deer forage availability and quality at the Chihuahuan Desert rangelands, Mexico, after a severe 3-year drought. Cogent Biology. 2018;**4**(1):1536315. DOI: 10.1080/23312025.2018.1536315

[12] Lashley MA, Harper CA. The effects of extreme drought on native forage nutritional quality and White-tailed deer diet selection. Southeastern Naturalist. 2012;**11**(4):699-710

[13] Tosa MI, Springer MT, Schauber EM, Nielsen CK. Increased overwinter mortalities of white-tailed deer (*Odocoileus virginianus*) fawns during a drought year. Canadian Journal of Zoology. 2017;**96**(1):55-61. DOI: 10.1139/cjz-2017-0009

[14] Torres-Porras J, Carranza J, Perez-Gonzalez J. Combined effects of drought and density on body and antler size of male Iberian red deer *Cervus Elaphus Hispanicus*: Climate change implications. Wildlife Biology. 2009;**15**:213-221

[15] Reilly AE, Earhart JE, Prance GT. Three sub-tropical secondary forests in the US Virgin Islands: a comparative quantitative ecological inventory. Advances in Economic Botany. 1990;189-198

[16] Ray GJ, Dallmeier F, Comiskey JA. The structure of two subtropical dry forest communities on the island of St John, US Virgin Islands. Forest biodiversity in North, Central and South America, and the Caribbean: research and monitoring. 1998;367-384

[17] Franklin J, Ripplinger J, Freid EH. Regional variation in Caribbean dry forest tree species composition. Plant Ecology. 2015;**216**:873-886

[18] Sanchez DM, Clubbe C, Woodfield-Pascoe N, Bárrios S, Abbott SJ, Heller T, et al. Tropical Important Plant Areas, plant species richness and conservation in the British Virgin Islands. Nature Conservation. 2021:11-39

[19] Carmona EC, Ramírez AV, Cano-Ortiz A. Contribution to the biogeography of the Hispaniola (Dominican Republic, Haiti). Acta Botanica Gallica. 2010;**157**:4, 581-598

[20] Acevedo-Rodriguez P. Flora of St. John, U.S. Virgin Islands. In: Memoirs of the New York Botanical Garden. Vol. 78. Bronx, New York: New York Botanical Gardens; 1996. pp. 1-581

[21] Nelson SL, Taylor SA, Reuter JD. An isolated white-tailed deer (*Odocoileus virginianus*) population on St. John, US Virgin Islands shows low inbreeding and comparable heterozygosity to other larger populations. Ecology and Evolution. 2021;**11**(6):2775-2781

[22] Reuter JD, Nelson SL. Hematologic parameters and viral status for Zika, Chikungunya, Bluetongue, and Epizootic Hemorrhagic Disease in white-tailed deer (*Odocoileus virginianus*) on St. John, US Virgin Islands. Journal of Wildlife Diseases. 2018;**54**(4):843-847

[23] Nelson SL, Durden LA, Reuter JD. *Rhipicephalus microplus* and *Dermacentor nitens* (Acari: Ixodidae) Coparasitize White-Tailed Deer on St. John, U.S. Virgin Islands. Journal of Medical Entomology. 2017;**54**(4):1440-1443

[24] Smiley RA, Rittenhouse CD, Mong TW, Monteith KL. Assessing nutritional condition of mule deer using a photographic index. Wildlife Society Bulletin. 2020;**44**(1):208-213

[25] Potratz EJ, Brown JS, Gallo T, Anchor C, Santymire RM. Effects of demography and urbanization on stress and body condition in urban white-tailed deer. Urban Ecosystem. 2019;**22**:807-816

[26] Audige L, Wilson PR, Morris RS. A body condition score system and its use for farmed red deer hinds. New Zealand Journal of Agricultural Research. 1998;**41**(4):545-553

[27] Millspaugh JJ, Washburn BE. Use of fecal glucocorticoid metabolite measures in conservation biology research: Considerations for application and interpretation. General and Comparative Endocrinology. 2004;**138**:189-199

[28] Washburn BE, Millspaugh JJ. Effects of simulated environmental conditions on glucocorticoid metabolite measurements in white-tailed deer feces. General and Comparative Endocrinology. 2002;**127**:217-222

[29] Ziegler TE, Wittwer DJ. Fecal steroid research in the field and laboratory:

Improved methods for storage, transport, processing, and analysis. American Journal of Primatology. 2005;**67**:159-174

[30] Wasser SK, Hunt KE, Brown JL, Cooper K, Crockett CM, Bechert U, et al. A generalized fecal glucocorticoid assay for use in a diverse array of nondomestic mammalian and avian species. General and Comparative Endocrinology. 2000;**120**:260-275

[31] Nelson SL, Miller MA, Heske EJ, Fahey GC Jr. Nutritional quality of leaves and unripe fruits consumed as famine foods by the flying foxes of Samoa. Pacific Science. 2000;**54**(4):301-311

[32] Hewitt DG. Biology and Management of White-Tailed Deer. Boca Raton, Florida, USA: CRC Press; 2011

[33] McDowell LR. Vitamins in Animal Nutrition: Comparative Aspect Human Nutrition. San Diego, California: Academic Press; 2012

[34] Muchlinski MN, Snodgrass JJ, Terranova J. Muscle mass scaling in primates: An energetic and ecological perspective. American Journal of Primatology. 2012;**74**:395-407

[35] Robbins CT. Wildlife Nutrition and Feeding. San Diego California, USA: Academic Press; 2012

[36] Morales J, Moreno J, Lobato E, Merino S, Tomas G, Martinez de la Puente J, et al. Higher stress protein levels are associated with lower humoral and cell-mediated immune responses in Pied Flycatcher females. Functional Ecology. 2006;**20**:647-655

[37] Musante AR, Pekins PJ, Scarpitti DL. Metabolic impacts of winter tick infestation on calf moose. Alces. 2007;**43**:101-110

[38] Wunschmann A, Armien AG, Butler E, Schrage M, Stromberg B, Bender JB, et al. Necropsy findings in 62 opportunistically collected free-ranging moose (Alces alces) from Minnesota, USA (2003-2013). Journal of Wildlife Diseases. 2015;**51**:157-165

[39] Hu X, Wei Y, Huang S, Liu G, Wang Y, Hu D, et al. Effects of breeding center, age, and parasite burden on fecal triiodothyronine levels in forest musk deer. PLoS One. 2018;**13**(10):e0205080. DOI: 10.1371/journal.pone

[40] He L, Wang WX, Li LH, Bao-Qing L, Gang L, Shu-Qiang L, et al. Effects of crowding and sex on fecal cortisol levels of captive forest musk deer. Biological Research. 2014;**48**:1-6

[41] Narayan E. Physiological stress levels in wild koala sub-populations facing anthropogenic induced environmental trauma and disease. Scientific Reports. 2019;**9**:6031

[42] Pride RE. High faecal glucocorticoid levels predict mortality in ring-tailed lemurs (*Lemur catta*). Biology Letters. 2005;**1**:60-63

[43] MacBeth BJ, Cattet MRL, Obbard ME, Middel K, Janz DM. Evaluation of hair cortisol concentration as a biomarker of long-term stress in free-ranging polar bears. Wildlife Society Bulletin. 2012;**36**:747-758

[44] Bryan HM, Darimont CT, Paquet PC, Wynne-Edwards KE, Smits JEG. Stress and reproductive hormones in grizzly bears reflect nutritional benefits and social consequences of a salmon foraging niche. PloS one. 2013;**8**(11):e80537

[45] Sapolsky RM, Romero LM, Munck AU. How do glucocorticoids influence stress response? Integrating permissive, suppressive, stimulatory, and preparative actions. Endocrinology Review. 2000;**21**:55-89

Chapter 9

Conservation Status of Macro Mammals in the Llanganates-Sangay Ecological Corridor and Anzu River Subbasin

Ruth Arias-Gutierrez, Andrés Tapia and Juan Pablo Reyes Puig

Abstract

With rapid ecological assessments, we diagnose the conservation status of large mammals in the localities Rio Encanto, San Jacinto Las Palmas, Sacha Llanganates, Cabeceras del Anzu, Boayaku, and Flor de Bosque, located within the Llanganates Sangay Ecological Corridor (CELS) and the upper subbasin of the Anzu River, in the eastern foothills of the tropical Andes, sub-Andean and tropical ecosystems in one of the most biologically diverse regions of Ecuador. Camera traps, 5 km linear transect walks, and interviews with local inhabitants recorded 35 species of macro mammals, in 8 Orders and 21 Families, including the Andean bear (*Tremarctos ornatus*), Andean tapir (*Tapirus pinchaque*), Amazonian tapir (*Tapirus terrestris*), puma (*Puma concolor*), and chorongo monkey (*Lagothrix lagotricha*), mammals considered as flagship species, highly endangered. The continuity of natural habitat along the subbasins of the Encanto, Topo, Zuñag, Tigre, and Anzu rivers, as well as the mountain ranges toward the highlands, provide favorable conditions for the flow of individuals, from adjacent areas of the Llanganates and Sangay National Parks to partially protected forest areas, such as the private reserves of Fundation Ecominga, Sumak Kawsay in Situ, community reserves of Boayaku and Flor de Bosque. The protection of the area is a priority.

Keywords: rapid ecological assessment, Ecuadorian flagship species, tropical Andes, threatened species, sub-Andean ecosystems

1. Introduction

Western Amazonia is one of the most biodiverse areas on the planet and the most biologically rich part of the entire Amazon basin [1]. The Llanganates-Sangay Ecological Corridor (CELS) is located between the provinces of Tungurahua, Pastaza, and Morona Santiago, in the central Andean and eastern foothills of the Ecuadorian Andes; it covers about 679 km^2; it is more than 40 km long and between 6 and 20 km wide, between the Llanganates and Sangay National Parks. While the upper subbasin of the Anzu River connects part of the declared corridor and the buffer zone of the

Llanganates National Park, with more tropical areas in the upper Amazon of Pastaza and Napo provinces.

The CELS is one of the areas considered a priority for conservation efforts, due to its strategic location, variety of habitats and ecosystems, and the functions they play in maintaining the planetary climate balance. Likewise, the Anzu River subbasin is located in the conservation hotspot, so the diagnosis of its conservation status is important due to its location in the westernmost portion of the eastern foothills of the central Ecuadorian Andes.

The importance of ecosystems such as the cloud forest and sub-Andean Forest for the regulation of the water cycle is well known, as well as for the alpha diversity and endemism of fauna and flora that these ecosystems contain. On the other hand, both areas are connected with protective forests such as the Abitahua Protected Forest, and are also crossed by cities, population centers, and agricultural activities that exert pressure on this biodiverse area [2].

The corridor was declared a gift of the earth [3] due to the richness of its ecosystems, flora, and fauna. However, the use of this space has not been well analyzed, especially in the area of influence of the Abitahua Protected Forest and the upper subbasin of the Anzu River. There is insufficient information on the fulfillment of its function as a biological corridor. The areas through which the genetic flow of large mammals exists and which may be under pressure from agricultural activities, deforestation, population centers, road traffic, natural barriers, and social-cultural aspects of the population, which influence the ecosystems and species, are unknown.

In this context, the present study sought to diagnose the conservation status in areas of the CELS and the upper subbasin of the Anzu River, through field sampling with camera traps in selected sites in the region between the towns of Rio Negro, in Tungurahua province, and Boayaku, in Pastaza province.

The main results corresponding to a baseline for biological monitoring during 6 months of work focused on the Masto fauna of the upper basin of the Anzu River and other localities belonging to the CELS are presented. The results obtained represent an approach to the full knowledge of the conservation status of this important ecological zone, being nevertheless an important contribution to the continuity of research and conservation efforts in the short and medium term, with a view to safeguarding the great biodiversity of the environment and ecosystems that make up this biodiverse region.

2. Materials and methods

The study area corresponds to the life zones of very humid tropical forest, humid tropical forest, and Andean brow [4], located between 750 and 2100 m above sea level. According to [5], it corresponds to the Pie montane evergreen forest, montane cloud forest, and montane evergreen forest vegetation formation of the Eastern Andes, where annual rainfall of 3000–4000 mm, a temperature ranging between 12 and 24°C and relative humidity of 80–90% is recorded.

Study area is conceived as a macro-thermal rainforest of the eastern region or Amazonian hylea [6], where average temperatures remain within 24°C and rainfall exceeds 6000 mm [6], being within one of the rainiest areas of Ecuador, due to the direct influence of high rainfall areas such as the Abitahua Protected Forest and the CELS.

It is a transition zone between the Amazonian lowlands and the tropical Andes, forming part of the buffer zone of Llanganates National Park and Sangay National Park in the CELS. This geographical feature has given rise to a range of habitat types and microclimates within an altitudinal gradient that ranges from dense forest with sub-canopies below 15 m in the lowlands to open forest with emergent canopies of 20–30 m from 1000 m altitude.

As this is a foothill area of the eastern Andean range, there are buttresses with slopes of 60–70° and abundant bodies of water: springs, streams, and rivers that descend through the foothills of the mountain range. The main subbasins of the area correspond to the Pastaza River, which receives the waters of the Encanto, Topo, Zuñag, Tigre, and Alpayacu rivers, while the Anzu River flows north and empties into the Napo River in the province of the same name.

2.1 Locations and sampling sites

Fieldwork focused on six localities in three provinces, Morona Santiago, Pastaza, and Tungurahua, with emphasis on the subbasin of the Anzu River (**Table 1**), selected based on criteria of connectivity with protected areas, forest quality, as well as previous records or interviews on the presence of endangered species [7].

2.1.1 Rio Encanto

Located in the north-eastern buffer zone of Sangay National Park, Río Encanto protects an important area of evergreen piedmont forest between the provinces of Tungurahua and Morona Santiago. The area is a mosaic of several farm owners, the vast majority of whom have abandoned intensive agricultural activities for more than 40 years, which has kept the forest well preserved. There is a constant connectivity vegetation through the upper lands to Pastaza River's mouth (**Figure 1**).

Location	Province	Municipality	Easting (m)	Northing (m)	Altitude (m) over sea level
Río Encanto (Encanto River)	Tungurahua–Morona Santiago	Baños–Palora	810,024*	9,838,230	1400–1800
San Jacinto las Palmas	Tungurahua	Baños	813,438*	9,850,095	1800–2100
Sacha Llanganates	Tungurahua–Pastaza	Baños-Mera	816,649*	9,847,837	1500–3000
Cabeceras del Anzu (Headwaters of the Anzu River)	Pastaza	Mera	828,840*	9,844,063	1100–1450
Boayaku	Pastaza	Mera	168,411**	9,853,573	819
Flor de Bosque (Forest Flower)	Pastaza	Santa Clara	172,907**	9,857,929	719

**Universal Transverse Mercator, World Geodetic System, zone 17 South.*
***Universal Transverse Mercator, World Geodetic System, zone 18 South.*

Table 1.
Location of study sites. Geographical coordinates collected by the authors in the field work.

Figure 1.
Location of Río Encanto.

Figure 2.
Location of San Jacinto las Palmas.

2.1.2 San Jacinto las Palmas

This place is located toward the high zone of colony Azuay, the southern buffer zone of the Llanganates National Park. It has important remnants of montane cloud forest and includes a mountain range in an easterly direction up to the origins of the Tigre and San Jacinto rivers, forming part of the lands of the San Jacinto las Palmas Association, on its limits with the Zuñag River Basin Ecological Association (**Figure 2**).

2.1.3 Sacha Llanganates

This locality crosses the southern buffer zone of the Llanganates National Park, up to the high zone of the Abitagua mountain range, ascending through the Zuñag river basin to the origins of the Anzu River between the provinces of Tungurahua and Pastaza. It includes part of the conservation area of the Zuñag River Basin Ecological Association and part of the Zuñag River Reserve of the Ecominga Foundation. It conserves a great variety of ecosystems from evergreen foothill montane forest to high montane evergreen forest (**Figure 3**).

2.1.4 Cabeceras del Anzu

This locality constitutes the upper zone of the Anzu River, entering through Mera municipality, in the way to 24th May colony, within the Anzu River

Figure 3.
Location of Sacha Llanganates.

Figure 4.
Location of Cabeceras del Anzu.

Ecological Reserve of the Ecominga Foundation and the Sumak Kawsay in Situ conservation area. The place conserves important remnants of low montane evergreen forest and montane foothills, surrounded by heavily disturbed farms (**Figure 4**).

2.1.5 Boayaku and Flor de Bosque

Those localities are located in the lower portion of Anzu River; correspond to indigenous peasant colonies and communities, made up of one populated center of peasant settlers (Boayaku) and another by the indigenous community (Flor de Bosque). This populated sector forms part of Pastaza municipalities, on the right bank of the Anzu River, and form a part of Mera municipality on the left bank of the same river. Other hydrographic tributaries of the Anzu River, such as the Chontayaku and Challuayaku Rivers, are also found in the area. It is worth noting that a part of the territory of Boayaku (**Figure 5**) is a part of the Abitahua Protected Forest, while almost all of the areas is part of the buffer zone of the Llanganates National Park (**Figure 6**).

Figure 5.
Location of Boayaku.

Figure 6.
Location of Flor de Bosque.

2.2 Methods

2.2.1 Biological and ecological assessments of Masto fauna biodiversity

Field expeditions were carried out covering most of the territory of influence of the populated settlements between the communities of Río Negro (Tungurahua) and

Boayaku (Pastaza), in the area that forms part of CELS and the upper subbasin of the Anzu River. These expeditions are part of the baseline diagnosis that allows, on the one hand, to determine the conservation status of the visited sites, and, on the other hand, to identify potential sampling sites for monitoring with camera traps, taking into account variables such as proximity to population centers, connectivity of habitats and ecosystems, logistical and operational aspects. In each outing, a multidisciplinary team carried out a rapid biological and ecological assessment of the state of conservation of the environment in terms of large fauna (macro mammals).

The information was collected in structured biological and ecological characterization sheets, obtaining records of direct and indirect observations (feces, footprints in tracks, diet, etc.). Interviews were conducted with local guides and members of the communities and the study localities were geo-referenced as well as different points of interest such as areas of faunal abundance and salt licks, among others.

In total, 18 field inspections were conducted, completing 36 effective sampling days from August to March, before COVID-19 pandemic time, at 6 locations with an average of 30 km of trails, with permanent sites for direct and indirect record monitoring, thus determining the best sites for camera trap installation, track recording and field observations.

2.2.2 Macro mammal monitoring with camera traps

Photo-trapping monitoring was carried out using Bushnell Trophy Cam HD camera traps placed along field transects to obtain information on the presence/absence of large mammals such as the Andean tapir, Amazonian tapir, Andean bear, puma, and panthera, among others, considered indicator species of the quality and conservation of the environment in the study area. The cameras were placed at intervals of 20–40 days in the fields in order to identify potential areas of mammal flow and occurrence of indicator species (presence/absence), among other aspects.

2.2.3 Interviews about use and pressure on macro mammals

Semi-structured interviews were conducted with the field guides and communities members with whom we worked during the study. In this way, it was possible to complement the record of species that could not be directly observed during the fieldwork and to learn about their use, conservation status, socio-cultural pressure, and trends in land use and exploitation that generate pressures on the environment and the conservation of large mammals.

3. Results

3.1 Macro mammals' communities in the study area

Analysis of fieldwork records and interviews with local people yielded a richness of 35 species of large and medium-sized mammals for the study area (**Table 2**). The Order Carnivora is the best represented with 11 species, the family Felidae the most numerous with 5 species, followed by the family Mustelidae (n = 3), Procyonidae (n = 2), and Ursidae with only 1 species. Order Rodentia is the second more numerous with 7 species, corresponding to Family Sciuridae (*n* = 3), followed by Dasyproctidae (*n* = 2); Erethizontidae and Cuniculidae have 1 species each. The Order Primates has 5 species: Family Atelidae (*n* = 2), Cebidae, Callitrichidae, and Aotidae (*n* = 1). Order Artiodactyla (2 Families) and Pilosa (4 Families) have 4 species each; Perissodactyl (*n* = 2); Lagomorpha and Cingulate were represented by only 1 species.

In reference to [8], for 2021, according to the threatened categories reported by the International Union for Conservation of Nature (IUCN), there are 2 species in Critically Endangered (CR), 8 species in Endangered (EN), 3 Vulnerable (VU) in the extinction risk. Also, 8 species are in Near Threatened (NT) status and 13 species in the Least Concern (LC); there are Data Deficient (DD) for 1 spicies (**Table 2**).

Condensing data from all sampled localities by the number of records per species, the Andean tapir (*Tapirus pinchaque*) obtained the highest number of records, followed by the tigrillo (*Leopardus tigrinus*), Amazonian cuchucho (*Nasua nasua*), guanta (*Cuniculus paca*), and armadillo (*Dasypus novemcintus*), among the most abundant.

The record of richness by study locality placed the Cabeceras del Anzu as the area with the highest species diversity, followed by Sacha Llanganates and the lower part of the Anzu River corresponding to Boayaku–Flor de Bosque, as detailed below.

3.1.1 Río Encanto

A total of 21 species were registered belonging to 6 orders and 16 families. The most abundant orders being Carnivora (*n* = 7) and Rodentia (*n* = 4), followed by Artyodactyla (*n* = 4), Primates (*n* = 3), Pilosa (*n* = 2), and Perissodactyla (*n* = 1).

3.1.2 San Jacinto las Palmas

In San Jacinto were recorded 15 species of mammals, represented by 6 orders and 10 families. The most abundant order being Carnivora (*n* = 8), followed by Rodentia (*n* = 2), Primates (*n* = 2), and Artyodactyla (*n* = 1), Pilosa (*n* = 1), and Perissodactyla (*n* = 1).

3.1.3 Sacha Llanganates

A total of 19 mammal species were recorded, grouped into 6 orders and 14 families. The order with the highest diversity being Carnivora (*n* = 7), followed by Primates (*n* = 5), Rodentia (*n* = 3), Pilosa (*n* = 2), and Artyodactyla (*n* = 1), and Perissodactyla (*n* = 1).

3.1.4 Cabeceras del Anzu

A total of 30 species of mammals were recorded in this location, belonging to 8 orders and 19 families, the most representative was Carnivora with 10 species, followed by Primates (*n* = 5), Rodentia (*n* = 5), Artyodactyla (*n* = 4), Pilosa (*n* = 2), Perissodactyla (*n* = 2), Lagomorpha (*n* = 1), and Cingulata (*n* = 1).

3.1.5 Boayaku and Flor de Bosque

At these two localities, corresponding to the lower Anzu River, 22 species of mammals were recorded, belonging to 7 orders, and 16 families. The most representative orders

DOI: http://dx.doi.org/10.5772/intechopen.109332

Order	Family	Specy	Local name	Location	Altitude (m) over sea level	Ecuadorian Red List-IUCN	Register type
Carnivora	Felidae	*Panthera onca*	Tigre/Pantera	ANZ	500–1500	EN	I
		Puma concolor	Puma o león	ENC, SJC, SLL, ANZ, BYK, FB	1800–3800	EN	S, F, I
		Herpailurus yagouaroundi	Jaguarundi/Pantera	ANZ, BYK, FB	1100–1300	NT	S
		Leopardus pardalis	Tigrillo/Ocelote	ANZ, BYK, FB	1100–1300	NT	S, F, I
		Leopardus tigrinus	Tigrillo chico	ENC, SJC, SLL, ANZ	1500–2800	VU	S, F, I
	Mustelidae	*Lontra longicaudis*	Nutria o perro de agua	ENC, SJC, SLL, ANZ	1100–1500	VU	S, F, I
		Eira barbara	Cabezamate	ENC, SJC, SLL, ANZ, BYK, FB	1100–3000	LC	S, F, I
		Neogale frenata	Chucuri	ENC, SJC, SLL, ANZ	1500–2000	LC	S, F, I
	Procyonidae	*Nasua nasua*	Cuchucho/Coatí	ENC, SJC, SLL, ANZ, BYK, FB	1100–1800	NT	S, F, I, O
		Nasua dorsalis	Tejón	SJC, SLL	2000–3000	NT	S, I
	Ursidae	*Tremarctos ornatus*	Oso de anteojos	ENC, SJC, ANZ, BYK	1500–3800	EN	S, F, I

Order	Family	Specy	Local name	Location	Altitude (m) over sea level	Ecuadorian Red List-IUCN	Register type
Rodentia	Sciuridae	*Syntheosciurus granatensis*	Ardilla	ENC, SJC, SLL, ANZ, FB, BYK	1500–3000	LC	O
		Hadrosciurus igniventris	Ardilla	BYK, FB	1000	LC	O
		Microsciurus flaviventer	Ardilla chiquita	ANZ	1100–1200	LC	O
	Dasyproctidae	*Myoprocta praatti*	Guatin	BYK, FB	800	LC	O
		Dasyprocta fuliginosa	Guatusa	ENC, SJC, SLL, ANZ, BYK, FB	1100–1800	LC	S, F, I
	Erethizontidae	*Coendou sp*	Erizo	ENC, SLL, ANZ	1800	DD	I
	Cuniculidae	*Cuniculus paca*	Guanta	ENC, ANZ, BYK, FB	1100–1800	NT	S, F, I
Primate	Atelidae	*Alouatta seniculus*	Mono ahullador	SLL, ANZ	1500–2000	VU	I
		Lagotrix lagotricha	Chorongo	ENC, ANZ, SLL, BYK	1500–2100	EN	O, I
	Cebidae	*Cebus yuracus*	Mono cenizo	ENC, SJC, SLL, ANZ, BYK	1100–2000	VU	O, I
	Callitrichidae	*Leontocebus lagonotus*	Chichico	SLL, ANZ, BYK, FB	1100–1800	NT	O, I
	Aotidae	*Aoutus sp*	Mono nocturno	ENC, SJC, SLL, ANZ	1100–2000	EN	O, I
Artiodactyla	Tayassuidae	*Tayassu pecari*	Pecari de labio blanco	ENC, ANZ, BYK, FB	1100–1500	EN	S, F, I
		Dicotyles tajacu	Zajino	ENC, ANZ, BYK, FB	1100–3000	NT	I
	Cervidae	*Mazama americana*	Venado	ENC, ANZ, BYK, FB	1200	NT	S, F, I
		Mazama rufina	Chiva de monte	ENC, SJC, SLL, ANZ	1500–3000	EN	S, F, I, O

Order	Family	Specy	Local name	Location	Altitude (m) over sea level	Ecuadorian Red List-IUCN	Register type
Pilosa	Bradypodidae	*Bradypus variegatus*	Perezoso variegado	BYK, FB	800	LC	I
	Cyclopedidae	*Cyclopes ida*	Flor de Balsa	BYK, FB	1000	LC	I
	Megalonychidae	*Choloepus didactylus*	Perico ligero/ perezoso	ENC, SJC, SLL, ANZ	1100–2000	LC	O, I
	Myrmecophagidae	*Tamandua tetradactyla*	Oso hormiguero	ENC, SLL, ANZ, BYK	1100–2000	LC	O
Perissodactyla	Tapiridae	*Tapirus pinchaque*	Danta o tapir	ENC, SJC, SLL, ANZ	1500–3800	CR	S, F, I
		Tapirus terrestris	Danta o tapir	ANZ, BYK, FB	1100–1200	EN	F, I
Lagomorpha	Leporidae	*Sylvilagus defilippi*	Conejo	ANZ	1100–3600	LC	F, I
Cingulata	Dasypodidae	*Dasypus novemcinctus*	Armadillo	ANZ, BYK, FB	1100–1300	LC	S, F, I, O

Locations: ENC, Encanto River (Río Encanto); SJC, San Jacinto las Palmas; SLL, Sacha Llanganates; ANZ, Headwaters of the Anzu River (Cabeceras del Anzu); BYK, Boayaku; FB, Forest Flower (Flor de Bosque). IUCN register: CR, Critically Endangered; EN, Endangered; VU, Vulnerable; NT, Near Threatened; LC, Least Concern; DD, Data Deficient. Register: S, smears (feces); F, footprints; I, interview; O, observed.

Table 2.
Large mammal's species reported from localities in the CELS and Anzu River subbasin.

being Carnivora with 6 species and Rodentia with 4 species. Order Primate has 3 records, Artyodactyla (n = 3), Pilosa (n = 3), Perissodactyla (n = 1), and Cingulata (n = 1).

3.2 Conservation status of macro mammals in the study area

The six study localities maintain faunal mass communities in a good state of conservation. Taking into account the main groups of herbivores and carnivores expected for the region, representative species were recorded in all sites, demonstrating a relative balance in the food chain and trophic niche of the community. However, there are specific variations in species composition and abundance for each locality, a result of the large altitudinal variation, microclimate, and range of anthropogenic pressures for each sampling site.

Localities in the lower zone are exposed to greater hunting pressure and habitat conversion due to their proximity to communities and population centers, while localities in the upper zone are home to a greater number of records belonging to threatened species due to the difficult topography, vegetation connectivity, proximity to the Llanganates and Sangay National Parks, as well as conservation initiatives developed in the area by different actors.

In terms of threatened categories formulated with IUCN criteria, according to [9], there are 10 globally threatened macro mammal's species in the red list of mammals of Ecuador for 2011 (**Table 3**), corresponding to Endangered (n = 5) and Vulnerable (n = 5).

The IUCN categories applied in the Red Book of Mammals of Ecuador for 2021 [8], increase to 13 species in risk (**Table 3**): Critically Endangered (n = 1), Endangered (n = 8), and Vulnerable (n = 4). From 2011 to 2021, two species reduce risk, evolving from Endangered to Vulnerable, and from Vulnerable to Near Threatened. One species increases the risk from Vulnerable to Endangered and 4 species are considered Near Threatened in 2011 appears increasing the risk to Critically Endangered, Endangered, and Vulnerable.

The most threatened species for 2021 is the mountain tapir *T. pinchaque*, considered Critically Endangered (CR). *T. pinchaque* is one of the species with the highest number of records in the localities of Río Encanto and San Jacinto, these being strategic sites for the conservation of species in the region. Those places are also important sites for the life area of the spectacled bear (*Tremarctos ornatus*), which is considered Endangered (EN).

In the middle and lower zone, the localities of Cabeceras del Anzu, Boayaku, and Flor de bosque present eventual records of the Amazonian tapir (*Tapirus terrestris*) considered Endangered (EN), so it is a threatened species whose long-term viability depends on the plans and programs that can be implemented to ensure the conservation of this and other species that are indicators of healthy ecosystems.

3.3 Ecological aspects of indicator species in the study area

The mammal community in the studied localities is composed of 11 species of carnivores of which 5 species are the main predators. Also, the spectacled bear *T. ornatus* is a typical species of cloud forest and Andean moorlands whose presence in the lower localities may be due to seasonal variations in its distribution, as generally in winter the animals tend to descend to warmer areas in search of food.

During the surveys carried out in the localities of San Jacinto, Rio Encanto, and Sacha Llanganates, several species of palms belonging to the genera *Aiphanes*,

Species	Threat category (2011)	Threat category (2021)
Tapirus pinchaque	Critically Endangered (CR)	Critically Endangered (CR)
Panthera onca	Endangered (EN)	Endangered (EN)
Tremarctos ornatus	Endangered (EN)	Endangered (EN)
Lagotrix lagotricha	Endangered (EN)	Endangered (EN)
Tapirus terrestris	Endangered (EN)	Endangered (EN)
Puma concolor	Vulnerable (VU)	Endangered (EN)
Leopardus tigrinus	Vulnerable (VU)	Vulnerable (VU)
Lontra longicaudis	Endangered (EN)	Vulnerable (VU)
Nasua dorsalis	Vulnerable (VU)	Near Threatened (NT)
Mazama rufina	Vulnerable (VU)	Endangered (EN)
Cebus yuracus	Near Threatened (NT)	Vulnerable (VU)
Aoutus sp	Near Threatened (NT)	Endangered (EN)
Tayassu pecari	Near Threatened (NT)	Endangered (EN)
Alouatta seniculus	Near Threatened (NT)	Vulnerable (VU)

Table 3.
Threatened macro mammals according to the Red Book of Ecuadorian Mammals for 2011 [9], compared with 2021 [8].

Ceroxylum, *Geonoma*, and *Dictyocarium* were observed being consumed by the bear. In addition, some trees of the genus *Vizmia* were observed used exclusively by males for gurgling and possibly to mark territory.

The mountain tapir *T. pinchaque* and its Amazonian counterpart *T. terrestris*, are indicator species of habitat quality, since generally in the areas where *Tapirus* genus was recorded, the general species composition is practically in its original state and the forest is well conserved. Both species are considered gardeners of the forest, as they contribute to seed dispersal, regulation of the growth of some plants, and maintenance of soil fertility. During the sampling, it was possible to record several species of plants consumed by the Andean tapir, among which the genera *Pilea*, *Piper*, *Solanum*, and *Gunnera* stands out, as has been described in previous studies in other localities [10].

4. Discussion

4.1 Flow areas and ecological corridors: importance of the studied localities for the conservation of biodiversity in the area

Large mammals can be good indicators of habitat quality and conservation status. In the present study, sampling areas were distributed along a gradient of human disturbance, from areas close to anthropogenic activities, through secondary forest to pristine forest with little disturbance. The results obtained reflect the presence of 35 species of macro mammals in the study area, 13 of which are at different levels of extinction threat.

Four species of large mammals recorded during the sampling need living areas greater than 500 hectares per individual [9]: *T. pinchaque*, *T. terrestris*, *T. ornatus*, and *Puma concolor*, as they are highly mobile during their daily activities in search of food or avoid overlapping with the territories of other individuals. This implies that the areas sampled constitute biological corridors for these species to and from Llanganates and Sangay National Parks, especially if we take into account that the distance between the Parks and the points sampled varies between 2 and 8 km.

Considering the analyzed sites, just Rio Encanto and San Jacinto show high numbers of records for *T. pinchaque*, while *T. ornatus* has a higher abundance of records in the localities of Rio Encanto, San Jacinto, and Sacha Llanganates. However, the localities of Cabeceras del Anzu and Boayaku could provide seasonal food sources during the winter season for the bear.

In this context, it is important to emphasize the influence of geography and vegetation cover on the availability of habitat for the above-mentioned species. Thus, the rugged geography and absence of access roads for the expansion of human activities have allowed several areas to maintain important extensions of forest along river basins and mountain ranges that descend from the heights of the National Parks; it is precisely through the river basins and mountains ranges that these large mammals descend at different times, due to heavy rains and low temperatures, as well as in search of new sources of food.

The region between the cities of Baños and Puyo constitutes a biogeographic unit demarcated by the upper Pastaza basin, one of the most humid and rainy places in the Amazon basin, together with the complicated geography of the mountains that rise toward the moorlands of the Llanganates and Sangay National Parks, giving rise to an infinity of ecosystems, habitats, and microhabitats. This region conserves a still underestimated biodiversity.

There are important extensions of natural forests that are currently partially protected, thanks to private initiatives, private reserves, and community reserves. Several of the sampling localities have been the subject of research in the field of herpetology, for example, Cabeceras del Anzu and Sacha Llanganates areas have high diversity and endemism of amphibians with more than 60 species including numerous endemic species to Ecuador [10].

4.2 Threats to biodiversity

4.2.1 Hunting pressure

Localities such as Rio Encanto, San Jacinto, and Sacha Llanganates do not currently present strong hunting pressure, due to the conservation initiatives developed in the area, as well as the control carried out by personnel from the Ministry of Environment in the Río Negro sector and its surrounding communities. However, it is known that until a few years ago, people from Amazonian communities came to the forest areas adjacent to the Llanganates and Sangay Parks to hunt.

Subsistence hunting is the main use that local communities have on wildlife, however, due to the characteristic of the Boayaku and Flor de Bosque villagers' land use, hunting is not considered a threat to wildlife; however, a quantitative study is needed during different times of the year and in the long term to determine these impacts in more detail.

Of all existing species, rodents such *Dasyprocta fuliginosa, C. paca,* artiodactyls like *Dicotyles tajacu*, and edentates such *Dasypus novemcinctus* are considered to be the species with the greatest hunting pressure from the communities. It should be noted that there is no evidence of a diversified ethnozoological use of fauna in the communities, as most of the species are destined for food consumption, and very few of them are used for handicrafts, medicinal purposes, and even less for commercialization in the form of meat or breeding animals.

4.2.2 Habitat loss and alteration

Although the large infrastructure and the expansion of roads, provide facilities for the development of communities and population centers, the pressure on natural ecosystems increases exponentially, as does the risk of extinction of endangered species of flora and fauna. This is the case of the localities in the highlands where the Topo Hydroelectric Project is being implemented a few kilometers from the border of the Llanganates National Park. This same project has promoted the extension of a road toward the Leon River near the junction with San Jacinto River, a few kilometers from the present investigation area, which has led to an increase in grasslands and habitat fragmentation. Furthermore, new hydroelectric projects are planned for the future, such as Abitagua, which will form a reservoir on the provincial border between Tungurahua and Pastaza, affecting the mouth of Rio Encanto and an as yet undetermined area in the only probable zone for implementing an effective corridor between the Llanganates and Sangay National Parks.

On the other hand, in the area of Boayaku and Flor de Bosque, are reported latent threats of mining exploitation and oil expansion, as they are located within Block 28 Petroleum concessions in the south-central Amazon, which puts the water sources and territories of the species recorded in this study at high risk.

5. Conclusion

Field sampling in six localities of the Llanganates-Sangay Corridor shows the presence of macro mammals communities in a good state of conservation, representing areas of use and potential corridor toward the Llanganates and Sangay National Parks. Within the study area, the localities of Rio Encanto, San Jacinto, and Sacha Llanganates, constitute a strategic refuge for highly endangered and typically Andean species such as the Andean tapir *T. pinchaque* and spectacled bear *T. ornatus*; while the lowland localities of Cabeceras del Anzu, Boayaku and Flor de Bosque constitute a refuge area for typical Amazonian species such as the ocelot *Leopardus pardalis*, peccary *Dicotyles tajacu*, jaguarundi *Herpailurus yagouaroundi*, and others.

The six localities analyzed are home to globally threatened mammal populations, but there are numerous species of flora and fauna endemic to the region, so conservation initiatives developed by communities, private stakeholders, and other organizations in the region are an important step toward promoting the creation of larger areas of protection. The study area is presented as a potential ecological corridor for the populations of macro mammals monitored, as it allows the flow and connectivity between the high and low zones. The existence of hydrographic subbasins such as the Anzu within the Llanganates-Sangay Ecological Corridor makes it an important refuge for threatened species of the tropical Andes, whose populations have been drastically reduced in the lower zones, close to the colonization frontier.

Finally, it is essential to develop a monitoring program in the six study localities, increasing the sampling effort and the number of camera traps, including visits during different times of the year, with a view to determining activity patterns, habitat use, and population status, with special emphasis on endangered species. It is also essential to involve local actors and communities in the planned initiatives, promoting conservation strategies and sustainable use of resources such as community tourism, forest resource management, and agroecology, as alternatives to traditional exploitation of natural resources.

Acknowledgements

The authors thank Amazonian State University (Universidad Estatal Amazonica) for research funding through two projects: Diagnosis of the state of conservation of the eastern slope of the Central Andes of Ecuador: Llanganates Sangay Ecological Corridor and the Anzu River Subbasin and Eco-cultural Pluralism in Quality Education in the Ecuadorian Amazon. The authors also thank the members of the Kichwa and peasant communities who participated in the research and shared their observations, particularly to Ecominga Foundation's team, Javier Robayo, Recalde family and Environment Minister for the research license N° 08-2015-IC-FLO-FAU-DPAT/MA. The research team members Diana Bermúdez, Karima López de Vargas Machuca, and Luis Sandoval and the assistants are also thanked for their contributions to the fieldwork and insightful discussions and feedback. Special thanks to Dr. Diego Tirira, PhD for pointing out the need to update certain scientific names.

Conflict of interest

The authors declare no conflict of interest.

Author details

Ruth Arias-Gutierrez[1*], Andrés Tapia[2] and Juan Pablo Reyes Puig[3]

1 Amazonian State University, Puyo, Ecuador

2 Zanja Arajuno Ecological Centre, Conservation of the Andean Tapir on the Eastern Slope of the Central Andes of Ecuador (PCTA), Puyo, Ecuador

3 PCTA-Oscar Efren Reyes Foundation, Fundación Ecominga Red de Protección de Bosques Amenazados, Baños, Ecuador

*Address all correspondence to: rarias@uea.edu.ec

References

[1] Finer M, Jenkins CN, Pimm SL, Keane B, Ross C. Oil and Gas Projects in the Western Amazon: Threats to wilderness, biodiversity, and indigenous peoples. PLoS ONE. 2008;**3**(8):e2932. DOI: 10.1371/journal.pone.0002932

[2] Reyes-Puig JP, Yánez-Muñoz MH. Una nueva especie de Pristimantis (Anura: Craugastoridae) del Corredor Ecológico Llanganates-Sangay, Andes de Ecuador. Papéis Avulsos de Zoología. 2012;**52**(6):81-91. DOI: 10.1590/S0031-10492012000600001

[3] Ríos-Alvear G, Reyes-Puig C. Corredor Ecológico Llanganates-Sangay: Un acercamiento hacia su manejo y funcionalidad. Yachana Revista Científica. 2015;**4**(2):11-21 Available from: http://inabio.biodiversidad.gob.ec

[4] Cañadas L. El Mapa Bioclimático y Ecológico del Ecuador. Quito: MAG-PRONAREG; 1983. p. 110

[5] Sierra R, editor. Propuesta preliminar de un Sistema de Clasificación de Vegetación para el Ecuador Continental. Instituto Ecuatoriano Forestal y de Áreas Naturales y Vida Silvestre-INEFAN, Fondo para el Medio Ambiente Global–Banco Internacional de Reconstrucción y Fomento GEF-BIRF, Eco Ciencia: Quito; 1999. p. 194

[6] Acosta-Solís M. Por la conservación de la Amazonia. Revista Geografica. 1992;**115**:77-84. Available from: https://www.jstor.org/stable/40992650?read-now=1&seq=8#page_scan_tab_contents

[7] Tapia A, Nogales F, editors. National Strategy for the Conservation of Tapir in Ecuador (*Tapirus terrestris, Tapirus pinchaque, Tapirus bairdii*). Quito: Tapir Specialist Group, IUCN, Ministerio del Ambiente; 2010. p. 86

[8] Tirira DG, editor. Libro Rojo de los mamíferos del Ecuador. 3th ed, Publicación Especial sobre los mamíferos del Ecuador 13. Quito: Asociación Ecuatoriana de Mastozoología, Fundación Mamíferos y Conservación, Pontificia Universidad Católica del Ecuador y Ministerio del Ambiente, Agua y Transición Ecológica del Ecuador; 2021. p. 87. Available from: https://bioweb.bio/faunaweb/mamiferoslibrorojo

[9] Tirira DG, editor. Libro Rojo de los mamíferos del Ecuador. 2nd ed. Quito: Fundación Mamíferos y Conservación, Pontificia Universidad Católica del Ecuador y Ministerio del Ambiente del Ecuador; 2011. p. 398

[10] Bermudez L, Reyes JP. Diet of the mountain tapir (Tapirus pinchaque) in three localities of the Llanganates-Sangay ecological corridor. Bol. Tec. 10. Serie Zoologia. 2011;**7**:1-13

Chapter 10

An Assessment of the Population Density of Greater One-Horned Rhinoceros in Uttar Pradesh and Their Distribution in India

Richa Kumar and Moharangan Ashokkumar

Abstract

India's alluvial plains once supported the abundance of greater one-horned rhino (*Rhinoceros unicornis*; hereafter rhino). Due to uncontrolled poaching and land degradation due to human intervention, they lost almost 80% of their natural habitat. Around 90% of rhinos are confined to Assam, along the Gangetic plains in northeastern India. Having formerly been a habitat for Indian rhinos, Uttar Pradesh reintroduced them at Dudhwa National Park in 1984. Based on secondary data collected from the study area (1984–2021), we analyzed the rhino population's life-history traits. Seven rhinos were brought from Assam and Nepal to start the seed population, including two males and five females. The rhino population increased steadily in the Dudhwa NP and reached asymptote. Birth rate and mortality rate were 0.17/year and 0.05/year, respectively. Observation of 10 adult females over 47 birth records revealed a higher fertility rate between 8 and 20 years, with a mean inter-birth interval of 4.3 years. According to their large body size, rhinos exhibit similar life-history traits to other large mammals, such as a slow onset of sexual maturity, few young, high inter-birth intervals, and fertility into old age. The current estimated population of Indian rhinoceros is about 35–40 in Uttar Pradesh.

Keywords: *Rhinoceros unicornis*, reintroduction, population estimation, current habitat, threats

1. Introduction

In the Indian megafauna, the greater one-horned rhinoceros (*Rhinoceros unicornis*) [1] is one of the five remaining species of rhinoceros of approximately thirty genera [2]. Rhinos belong to the order Perissodactyla (odd-toed ungulates), which includes horses and tapirs. They are also called the Indian rhino, greater one-horned rhinoceros, or great Indian rhinoceros (hereafter rhino), native to the Indian subcontinent (India and Nepal). Rhino has become extinct from Bhutan and Bangladesh with a decline in the population in their distributional range [3, 4]. The estimated population in India is around 3262 [5] and is 752 in Nepal [6]. It is listed as Vulnerable on the

IntechOpen

IUCN Red List, as populations are fragmented and restricted to distributional areas of less than 20,000 km [2]. This species is listed in CITES Appendix-I and is protected under Schedule I of the Wildlife Protection Act.

Previously, the greater one-horned rhinoceros (*R. unicornis*) was widely distributed from the Hindukush Mountain Range in Pakistan to Myanmar, and it also inhabited the Ganges floodplains [7]. For over 200 years, the Indian rhinoceros has been eradicated from the majority of their former range of distribution due to overhunting, fragmentation of habitat caused by clearing of forests for cultivation, desperate agricultural use, and expansion of tea gardens [5]. It has also been eliminated from most of its former range due to uncontrolled fires and grasslands and swamps being reclaimed by expanding livestock and humans The Indian rhino once ranged throughout the northern part of the Indian subcontinent, along the Indus and Ganges river basins, and as far south as Bangladesh and the southern parts of Nepal and Bhutan. The Terai and Brahmaputra basins are home to a large number of them due to their alluvial grasslands. The Terai grasslands of southern Nepal, northern Uttar Pradesh, northern Bihar, northern West Bengal, and the Brahmaputra Valley of Assam are the only places where it survives as a result of habitat destruction and climatic changes. By the nineteenth century, it was only found in these areas.

We have limited information on the rhino population demographics and life-history traits in India. It is necessary to understand the changes in the population size, mortality, natality pattern, survival rate, and fecundity essential for strategizing conservation measures. We hypothesized that rhino survival, mortality, and natality varied by year, age, and gender during 1984–2021. We attempted to answer the following questions: 1. changes in the population size of rhinos in the Dudhwa National Park, 2. age-specific natality and mortality of rhinos, and 3. reproductive rate, fecundity, and inter-birth interval in the rhinos.

1.1 Current habitat of the greater one-horned rhinoceros in India

There are only three states in India that have rhino populations in different national parks: In Assam [5], there are Kaziranga National Park (2401), Manas Tiger Reserve (32), Pobitora Wildlife Sanctuary (92), and Orang National Park (100). In West Bengal, there are Jaldapara Wildlife Sanctuary (200) and Gorumara National Park (50), and in Uttar Pradesh, where rhinos are found after reintroduction in 1984, there is Dudhwa National Park (35–40). Apart from the population of rhinos in India, Nepal also consists of a good number of rhinos (752); the estimated population in the different protected areas [6] are Shuklaphanta wildlife reserve (17), Bardia wildlife reserve (38), Royal Chitwan National Park (694), and Parsa (3) (**Figure 1**).

1.2 Study area

Dudhwa National Park is located in the Terai region of Uttar Pradesh. The total area of the national park is 490.29 km [2], located along the Indo-Nepal border in Kheri District. Dudhwa was established in 1958 as a wildlife sanctuary for swamp deer and became a tiger reserve in 1979. The Terai ecosystem is one of the most threatened ecosystems in India. Located between the Himalayan foothills and the Gangetic plains, the region extends through Uttarakhand, northern Uttar Pradesh, Bihar, northwestern Bengal, Assam, and southern Nepal. The park falls within the upper Gangetic plain, a vast alluvial plain with an altitude ranging from 150 m in the southeast to 182 m in the north. Sal forests are very dense, and they can be divided into four types: the tropical

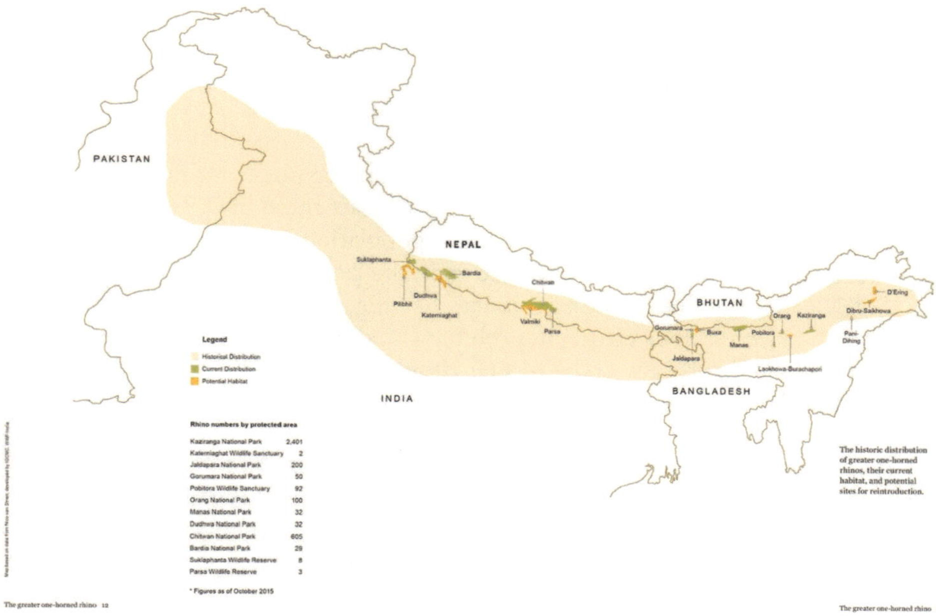

Figure 1.
Existing population of the great one-horned rhinoceros only in India and Nepal (courtesy WWF-India).

semi-evergreen forest, the moist deciduous forest of northern India, the tropical swamp forest, and the tropical dry deciduous forest of northern Guatemala [8]. In total, the park comprises about 19% grassland. In addition to rivers, streams, lakes, and marshes, wetlands are a third major habitat type.

Dudhwa has a subtropical climate that is extremely humid, with a dry winter and three distinct seasons. The summer season begins in mid-March and lasts until mid-June. The weather gets hot with temperatures reaching 40°C. The winter season extends from mid-October to mid-March. The temperature ranges between 20 and 30°C during the day and is cold and foggy at night. Monsoons typically start in mid-July and last until September, resulting in 90% of the 150 cm of rainfall. The temperature during the rainy season ranges from 24 to 30°C.

1.3 Methods

Based on the annual reports and other public sources, we obtained secondary data about the rhino reintroduction program of 1984 and rhino birth records and mortality cases between 1984 and 2021. The annual reports provide detailed information on each case, including the estimated age, gender, cause of death, location, and mortality date. Life-history traits such as natality, age-specific survival, fecundity, mortality, and inter-birth interval were calculated [9, 10]. The rhino population growth model is predicted using a generalized linear model equation (1), specifically logistic non-linear least squares regression [10].

$$y = \beta_1/(1 + \exp(-\beta_0 + \beta_2 * x) \quad (1)$$

Where β_1 is the asymptote value (maximum population size). β_0 is the constant; β_2 is the growth parameter; and x is the year.

2. Result and discussion

2.1 Re-introduction program of the greater one-horned rhinoceros in Dudhwa National Park, that is, the rhino reintroduction program

The great horned rhinos roamed freely in the Terai belt of Uttar Pradesh, including Dudhwa forests, about 160 years ago. The last rhino in U.P. was shot in 1878 in the Pilibhit district near Dudhwa. The rhino population in much of this range was wiped out by hunting and habitat destruction in the last couple of centuries. During the 1979 IUCN Species Survival Commission meeting, the Asian Rhino Specialist Group called for continuous efforts in protecting and monitoring the species and took a step to establish rhinos in the former rhino distribution range. It was concluded that Dudhwa was the most suitable alternate habitat among the various areas considered by the subcommittee. As a result of a visit to Dudhwa NP in 1980–1981, Prof Schenkel, chairman of the SSC Asian Rhino Specialist Group, confirmed the suitability of the park for rhino conservation [11]. Later, it was recommended for translocation of the great one-horned rhinoceros in Dudhwa National Park [12]. Consequently, it was determined to be the best alternate habitat because of its similarities to Kaziranga NP, Assam, the fact that rhinos have been previously recorded there, as well as the adequacy of protection. It was determined by the committee that the Dudhwa NP could provide adequate food for rhinos, so the Botanical Survey of India [13] was commissioned to conduct a study of rhino food plants in the proposed reintroduction area. There were 14 plant species found in Dudhwa NP, including grasses that rhinos consume at Kaziranga in Assam. Furthermore, the area provided a variety of habitats, including grasslands flooded with water, ample shade, water for drinking and wallowing, as well as protection from human activity. Rhinos were relocated from Assam and Nepal to Dudhwa in 1984–1985 as part of efforts to reintroduce rhinos to this area, which has grasslands, swamps, and wooded forests, which are ideal habitats for rhinos.

With the original group of seven growing to approximately forty, the project has been successful. Although the planned soft release did not happen, the population is not released and is kept in fenced enclosures. Two areas in the Dudhwa are well suited for rhinos, namely, Bhadi Tal and Churela Tal, since they provide a combination of grasslands for food, wooded forests for shade, and shallow wetlands where rhinos can forage and wallow.

2.2 Translocation of the greater one-horned rhinoceros from Assam and Nepal to Uttar Pradesh

The first rhino reintroduction program was implemented from 1984 to 1995, and the area where rhinos were to be released is called the Rhino Reintroduction Area. In 1984, five rhinos, that is, two males and three females, from Pabitora Wildlife Sanctuary were translocated to Dudhwa National Park. Before being released into the main area of the Rhino Reintroduction Area (RRA), these animals were kept in specially constructed stockades so that they could receive healthcare and experience electric fencing. A female, out of these five, died of stress abortion before she could be released. On 20.05.1985, animals were released in an area of 27 sq. km in the south Sonaripur range of Dudhwa [14]. The large male was held back until the rest were settled and was released a few days later after being radio-collared. A second female

S. No.	Origin	Name	Sex	Age on arrival
1.	Pobitora WLS, Assam	Bankey	M	7–8 yr
2.	Pobitora WLS, Assam	Raju	M	25 yr
3.	Pobitora WLS, Assam	Pabitri	F	3–4 yr
4.	Royal Chitwan NP, Nepal	Swayambara	F	4–5 yr
5.	Royal Chitwan NP, Nepal	Narayani	F	5 yr
6.	Royal Chitwan NP, Nepal	Himrani	F	4 yr
7.	Royal Chitwan NP, Nepal	Rapti	F	5–6 yr

Table 1.
Released population of rhinos in Dudhwa National Park who started the population in 1984–1995.

died on July 31, 1984, after being tranquilized to treat a wound. Having only one female rhino and two males left, it was necessary to translocate more rhinos, so the government of Nepal agreed to exchange four young adult female rhinos for sixteen domestic elephants. In April 1985, four rhinos were captured near Royal Chitwan National Park and brought to Dudhwa. In this way, these seven rhinoceros, two males and five females, made up the rhinoceros seed population (**Table 1**).

2.3 Establishing a second rhino population within Dudhwa National Park, that is, Phase II

The Rhino Reintroduction Program has been executed by the National Park in two phases, RRA-I, that is, Phase I, and RRA-II, that is, Phase II. A second phase has been established to prevent rhino populations from inbreeding. Over the years, the rhino population in Dudhwa NP has grown steadily and is probably at risk of inbreeding, so to provide better conservation opportunities, it is necessary to move the rhinos to a potential protected area and also to bring in new individuals to make the population genetically healthier. As suggested by the Wildlife Institute of India (WII) [15] and following the management plan of Dudhwa Tiger Reserve, the second RRA of Dudhwa NP has a total area of about 14 sq. km. To create a genetically healthy rhino population, at least one male must be unrelated to Dudhwa's existing population. It was through the collaboration of the Uttar Pradesh Forest Department and WWF-India, with support from the Assam Forest Department and Wildlife Trust of India (WTI), that four rhinos were translocated, a significant milestone in the history of rhino conservation (**Figure 2**).

3. Population ecology of great one-horned rhinoceros in Dudhwa National Park after translocation

When the remains of a newborn calf were found in tall grass after the reintroduction in August 1987, the sign of breeding was detected. The first successful calving occurred in 1989. Three more calves followed this in the same year. According to estimates, 58 rhino calves have been born and 35 rhinos (adults and calves) have died in the past 37 years.

Figure 2.
Rhino mother and calf feeding in the grassland of Dudhwa National Park.

Birth rate: A total of 58 rhino calves were born between 1984 and 2021. The mean number of calves born per year is two calves per year. The population-specific crude birth rate is 0.17 per year for a number of adult females in the reproductive age class. The birth rate varied across the years with a higher birth rate at certain years followed by intermittent lower birth rates. This could be attributed to the number of females giving birth and the inter-calving interval duration (3–4 years) by adult females due to parental care.

Mortality: The mean mortality rate is 0.05/year for the rhino population in Dudhwa National Park. The mortality rate was higher during the initial period of the re-introduction of rhinos (0.4/year), and it was lower (0.2/year) in the subsequent years (**Figure 3**). The major causes of mortality were mal-–male aggression, predation by tigers, and natural causes. Similarly, the major causes of mortality (80%) in rhinos were reported to be self-fight, tiger predation, and poaching in Nepal [16]. Male aggression and fighting among males were more during the breeding time, associated with the territorial behavior of males [17]. Further, male mortalities are significantly higher than females [18]. Retaliatory killing of rhinos due to the human-rhino conflict in Nepal was also reported. An increase in the population and dispersal of rhinos outside the protected area could be the factors attributed to the human-rhino conflict in their distributional range.

Population growth: The rhino population growth data were modeled using the logistic growth model by non-linear least squares. The parameter values (β_1) initial asymptote is 39.68, and the growth parameter value (β_2 is 0.12). The predicted curve is fitted with the actual population data (**Figure 4**), and the model was highly significant. The curve reached asymptote in the year 2010, and the population has been fluctuating near 35–40 individuals for the last 12 years. Further extension of the area or dispersal of individuals could enable the population size to increase.

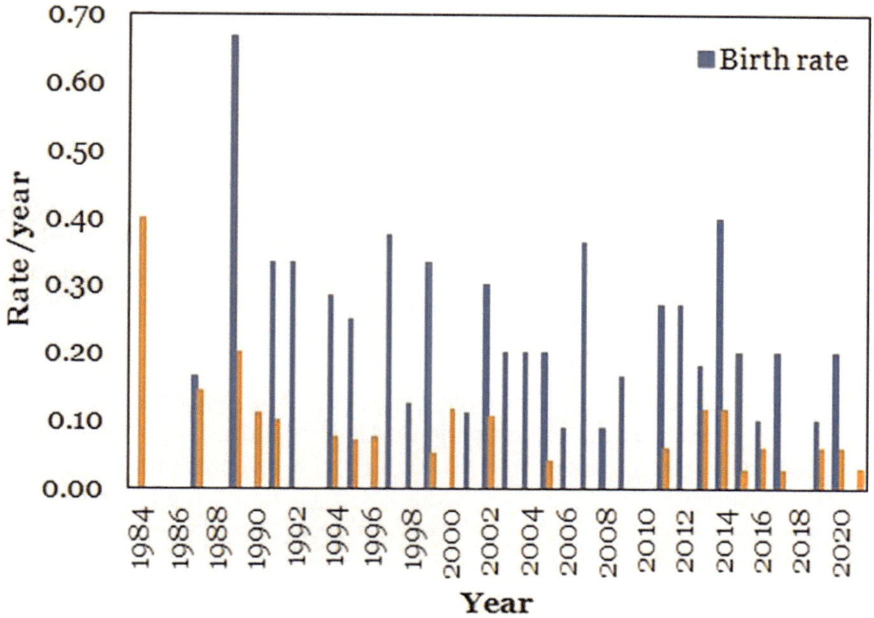

Figure 3.
Birth rate/year/adult female and mortality rate of Indian rhinoceros across the years in Dudhwa National Park from 1984 to 2021.

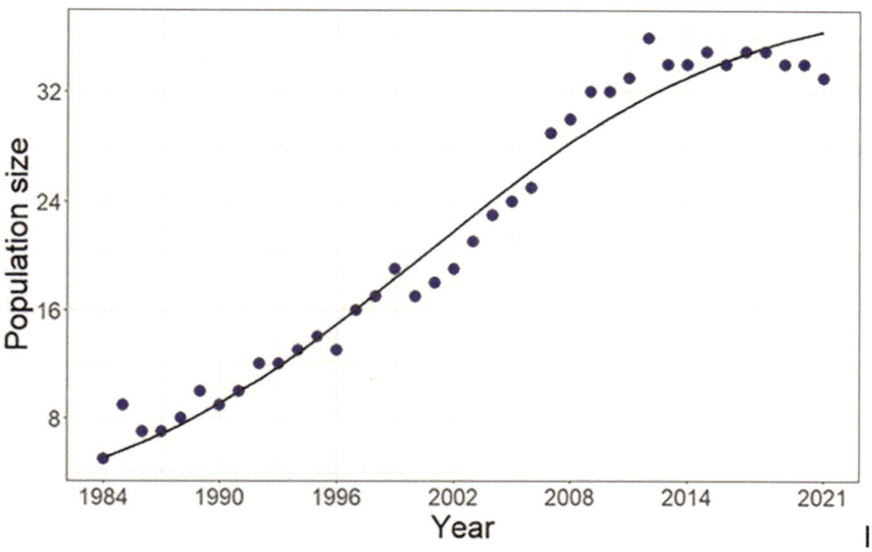

Figure 4.
Actual Indian rhinoceros population and predicted logistic model of population growth in Dudhwa National Park (1984–2021).

3.1 Age-specific survival rate

The proportion of individual rhinos surviving at a particular age class was calculated based on the number of rhinos dying at the specific age class. The survivorship curve is similar to other mammal species with higher mortality at the very young age class followed by moderate mortality in the middle ages and higher mortality at the older age class. Male rhinoceros mortality tends to be higher in the younger age class and female mortality is higher in older individuals. The mortality rate for a very young individual is higher due to predation by tiger (1%). Earlier studies on rhinos reported gender-specific differences in mortality with higher mortality among male rhinos [19, 20]. Male mortality is higher in the dispersal age class. In most of the ungulate species, males die disproportionately due to enhanced growth rates, dispersal behavior, and intra-male aggression [21]. Injuries sustained during the male-male fight may cause mortality directly or indirectly [22]. Male Indian rhinoceros fight each other for territory or mates to establish dominance (**Figure 5**).

Age-specific fecundity rate: The average age at primiparity observed in Indian rhinoceros is five years. The gestation period is 15–16 months. It appears that the female rhinos remain fertile in old age; an adult female gave birth to her sixth calf at the age of 34 years. The fertility rate varied across ages, with a higher fertility rate at the age class between 8 and 20 years, with a mean fecundity rate of 0.24 and a maximum of 0.40. The fertility rate reduces as the female rhinos reach more than 25 years old. The Indian rhinoceros female produces 4–6 calves in their lifetime (**Figure 6**).

Inter-birth interval: The reported inter-calving interval is 3–4 years. The present observation on 10 adult females over 47 birth records indicated a minimum of 1.6–10 years of an inter-birth interval with an average of 4.3 years. Thus, in Indian rhinos, about 61% of births occurred during 3–5 years of interval. The inter-birth interval rate is higher than that of the African buffalo, which is reported to have 1–2 years [23]. Earlier studies in the study area on rhinos indicated that inter-birth interval was associated with the age of the mother [24]. It can be inferred from the earlier graph that the fecundity rate decreases with the age of the female rhinoceros (**Figure 7**).

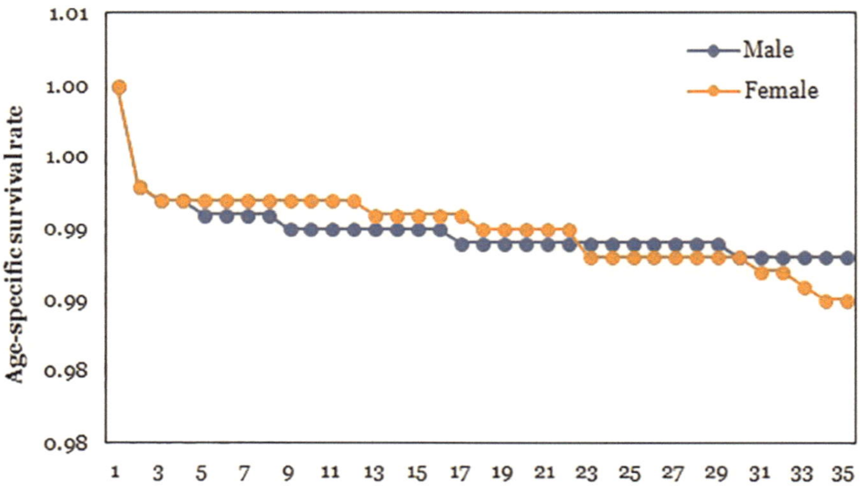

Figure 5.
Age-specific survival rate of Indian rhinoceros in Dudhwa National Park (1984–2021; based on mortality record of male = 12; female = 15).

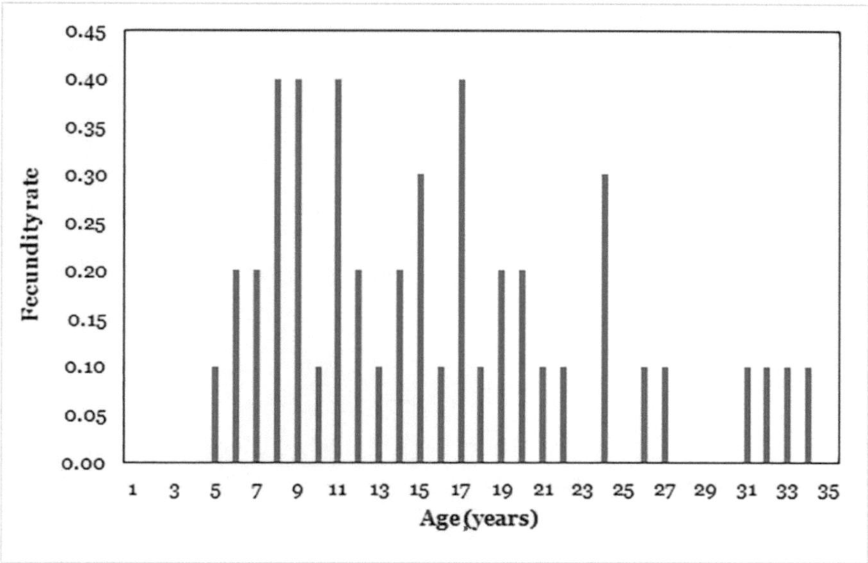

Figure 6.
Age-specific fecundity rate of Indian rhinoceros in Dudhwa National Park (1984–2021; n = 10 adult females; over 47 birth records).

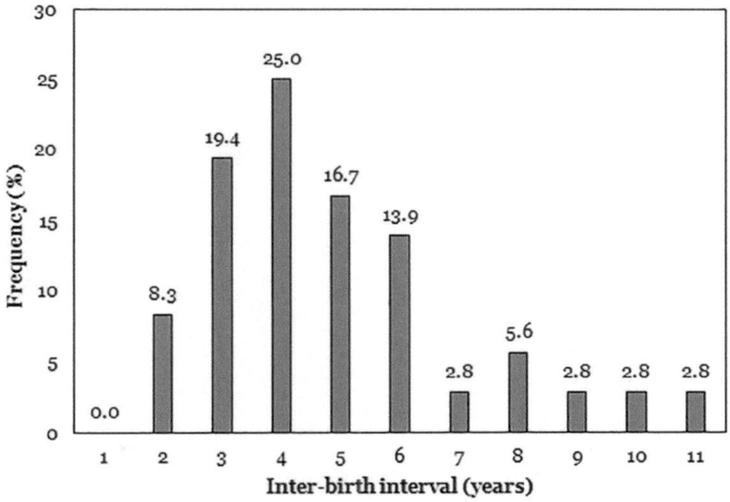

Figure 7.
Inter-birth interval duration of Indian rhinoceros in Dudhwa National Park (n = 10 adult females; over 47 birth records).

4. Conclusions

The Indian rhinoceros population has increased and reached asymptote in the year 2010, and the population has been fluctuating near 35–40 individuals for the past 12 years. Further extension of the area or dispersal of individuals could enable the population size to increase. The mean birth rate and mortality rate of rhinos in

Dudhwa National Park are 0.17/year/adult female and 0.05/year, respectively. Rhinos, in accordance with their large body size, exhibit life-history traits similar to other large mammal species with a slow onset of sexual maturity, fewer number of young ones produced, and higher inter-birth interval and remain fertile in old age. The survival rate of a female is higher than that of a male. The environmental and ecological factors such as carrying capacity, food availability, predation, and environmental stochasticity may influence the life-history traits of rhinos. The natural forest areas adjacent to the protected areas need to be maintained to enable the dispersal of the increasing population in Dudhwa National Park.

Acknowledgments

We would express our sincere thanks to the Principal Chief Conservator of Forest (Wildlife, U.P.) for granting permission for the study. We express our gratitude to Shri. Sanjay Pathak (Field Director) and Shri. Manoj Kumar Sonkar (Deputy Director) of Dudhwa Tiger Reserve for their kind support. We are also thankful to park staff, including range officers, forest guards, and 'Mahouts,' for their help and co-operations from time to time during the study.

Abbreviations

CITES	Convention on International Trade in Endangered Species of Wild Fauna and Flora
DTR	Dudhwa Tiger Reserve
IUCN	International Union for Conservation of Nature
km	kilometer
km^2	kilometer square
n	number
NP	national park
PCCF	Principal Chief Conservator of Forests
RRA	Rhino Reintroduction Area
S. No.	serial number
sq.	square
UP	Uttar Pradesh
WTI	Wildlife Trust of India
WII	Wildlife Institute of India

Author details

Richa Kumar[1*] and Moharangan Ashokkumar[2]

1 Wildlife SOS, Chhattisgarh, India

2 Wildlife SOS, Karnataka, India

*Address all correspondence to: krricha05@gmail.com

References

[1] Laurie WA. The Ecology and Behaviour of Greater One-Horned rhinoceros. Cambridge, United Kingdom: University of Cambridge; 1978. p. 450

[2] Nowak RM, Paradiso JL. Walker's Mammals of the World. Baltimore, MD: Johns Hopkins University Press; 1983

[3] Ellis S, Talukdar B. *Rhinoceros unicornis*. The IUCN Red List of Threatened Species 2019: e. T19496A18494149. DOI: 10.2305/IUCN.UK.2019-3.RLTS.T19496A18494149.en

[4] Jhala HY, Qureshi Q, Jhala YV, Black SA. Feasibility of reintroducing grassland megaherbivores, the greater one-horned rhinoceros, and swamp buffalo within their historic global range. Scientific Reports. 2021;**11**(1):1-15

[5] Asian Rhino Specialist Group (AsRSG) biannual survey report, 2022

[6] National Rhino count 2021, Nepal

[7] Dinerstein E. The Return of the Unicorns: The Natural History and Conservation of the Greater One-Horned Rhinoceros. New York: Columbia University Press; 2003. p. 316

[8] Champion HG, Seth SK. Revised Survey of Forest Types in India. New Delhi: Manager of Publication; 1968. p. 404

[9] Stiling P. Ecology Theories and Applications. 4th ed. New Delhi: Prentice-Hall of India; 2006. p. 403

[10] Rockwood LL. Introduction to Population Ecology. Oxford: Blackwell Publishing; 2006. p. 339

[11] Schenkel R. Report on the Suitability of Dudhwa National Park, UP, as Potential Site for Re-Introduction of the Indian Rhinoceros. Gland, Switzerland: IUCN Species Survival Commission; 1983

[12] Sale JB. Final Recommendations of the Rhino Sub- Committee of the Wild Life Status Evaluation Committee of the IBWL, on the Translocation of Great Indian Rhinoceros. Government of India: New Delhi; 1981

[13] Hajra PK, Shukla V. Dudhwa National Park: Some Botanical Aspects of the Proposed New Habitat for Rhino. Botanical Survey of India: Howrah; 1982

[14] Tripathi AK. Status of Indian One-horned Rhino, Rhinoceros unicornis Linn. 1758 in Dudhwa National Park, Uttar Pradesh, India. Journal of Biological Sciences. 2018:2455-7676

[15] Sinha SP, Sawarkar VB. Management of the Re-introduced Great One horned Rhinoceros (*Rhinoceros unicornis*) in Dudhwa National Park, U. P., India. WII, Dehradun. In: Ryder OA, editor. Rhinoceros Biology and Conservation: Proceedings of an International Conference. San Diego, USA; 1993. pp. 218-227

[16] Bhandari S, Adhikari B, Baral K, Subedi S. Greater one-horned rhino (Rhinoceros unicornis) mortality patterns in Nepal. Global Ecological Conservation. 2022;**38**:e02189

[17] Rachlow JL, Berkeley EV, Berger J. Correlates of male mating strategies in white rhinos (Ceratotherium simum). Journal of Mammalogy. 1998;**79**(4):1317-1324

[18] Subedi N, Jnawali SR, Dhakal M, Pradhan NM, Lamichhane BR, Malla S, et al. Population status, structure and distribution of the greater one-horned rhinoceros Rhinoceros unicornis in Nepal. Oryx. 2013;**47**(3):352-360

[19] Dinerstein E, Price L. Demography and habitat use by greater one-horned rhinoceros in Nepal. Journal of Wildlife Management. 1991:401-411

[20] Subedi N, Lamichhane BR, Amin R, Jnawali SR, Jhala YV. Demography and viability of the largest population of greater one-horned rhinoceros in Nepal. Global Ecological Conservation. 2017;**12**: 241-252

[21] Berger J, Gompper ME. Sex ratios in extant ungulates: Products of contemporary predation or past life histories? Journal of Mammalogy. 1999; **80**:1084-1113

[22] Geist V. New evidence of high frequency of antler wounding in cervids. Canadian Journal of Zoology. 1986;**64**: 380-384

[23] Prins HHT. Ecology and Behaviour of the African Buffalo: Social Inequality and Decision- Making. London, UK: Chapman and Hall; 1996

[24] Pluhacek J, Steck BL, Sinha SP, Houwald FV. Inter birth intervals are associated with age of the mother, but not with infant mortality in Indian rhinoceroses. Current Zoology. 2017; **63**(3):229-235

Chapter 11

Population Density and Age-Sex Composition of Large Mammals in Tropical Forests, Southern India

Mohanarangan Ashokkumar and Rajaratinavelu Nagarajan

Abstract

The population density, structure, and biomass of mammal species were investigated in the tropical forest of Southern India. The population of large mammals was estimated using distance sampling methods at the Mudumalai Tiger Reserve in the dry seasons from 2004 to 2007. The estimated densities of larger ungulate species such as elephant and gaur were 4.8/sqkm and 7.9/sqkm. Medium-sized ungulate species such as sambar, chital, barking deer, and wild pig densities were 7.1, 27.9, 1.1, and 0.4/sqkm, respectively. The estimated density of primate species, *i.e.*, common langur, was 14.9/sqkm. The smaller rodent species, black-naped hare, had a lower density of 1.3/sqkm, which could be attributed to the nocturnal behaviour of hares. Among these mammal species, barking deer and black-naped hare are solitary species, while others are gregarious. The study area supports herbivore biomass density of 15,198 kg/sqkm. Comparison with earlier estimates, the densities of elephant and gaur is increasing in the study area. The population density of large mammals is compared to other areas and discussed.

Keywords: large mammal population, Mudumalai Tiger Reserve, ungulate densities in tropical forest, elephant, gaur, chital

1. Introduction

The assessment of population is the key to understand the demography of animals and for evaluating management effectiveness. Therefore monitoring of population is a crucial component in animal ecology and wildlife conservation. Evaluation of population density and biomass have been used to investigate the complex relationship between a species and its environment [1] and their inter-specific relationship in a community [2]. Management strategies in protected areas must be based on an understanding of the functional relationship between habitat conditions, animal population and the dynamics of these populations. Large herbivores are particularly difficult to conserve, because of their relatively low population densities, unique habitat needs, crop raiding tendencies and their consumption and disturbances due to people [3]. However, herbivores' conservation is crucial because apart from being important economic, nutritional and aesthetic resources, large herbivores directly or indirectly influence forest structure, regeneration and predatory species [4].

They also plays an important role in conservation of large carnivore population, especially tigers. Furthermore, the selective predation of particular prey species, age and size class allowed the coexistence of three carnivores [5–7]. The coexistence of Tigers and Leopard in the tropical forest was due to the specialisation on different sized prey species [7] or based on spatial or temporal partitioning of resources [8]. Therefore it is important to identify the available prey species and their role in shaping the community structure of the carnivores.

Mudumalai Tiger Reserve is part of Nilgiri biosphere reserve with contiguous forest area of about 5520 km^2, supports largest Asian elephant population in India [9]. Study area has heterogeneous habitat, ranging from tropical moist deciduous to thorn forest. These habitats differ in their extent of cover and support diverse ungulate and primate species [10]. This area also supports the highest gaur population in India [11, 12]. The gaur population had recovered significantly since the outbreak of Rinderpest epidemic in the mid 1960s which nearly wiped out the MTR population [13]. The present study provides baseline data on age-sex composition of elephant, gaur, and other mammal species. Our main objective is to obtain base-line estimates of population density and age-sex composition in the area to enable the formulation of viable management strategies.

2. Methods

2.1 Study area

Mudumalai Tiger Reserve is located in the Nilgiris District of Tamil Nadu (11° 32′ and 11° 42′ N and 76° 20′ and 76° 45′ E; **Figure 1**). It extends over an area of 321 km^2 and forms a part of the Nilgiri Biosphere Reserve. The sanctuary is located in the Western Ghats, which is one of the 34 Biodiversity hotspots of the world [14]. The altitude varies from 485 to 1226 m above MSL with a general elevation of about 900 to 1000 m. The annual rainfall is varied from 1001 mm to 1648 mm. Sanctuary receives rain both from southwest (May to August) and northeast (September to December)

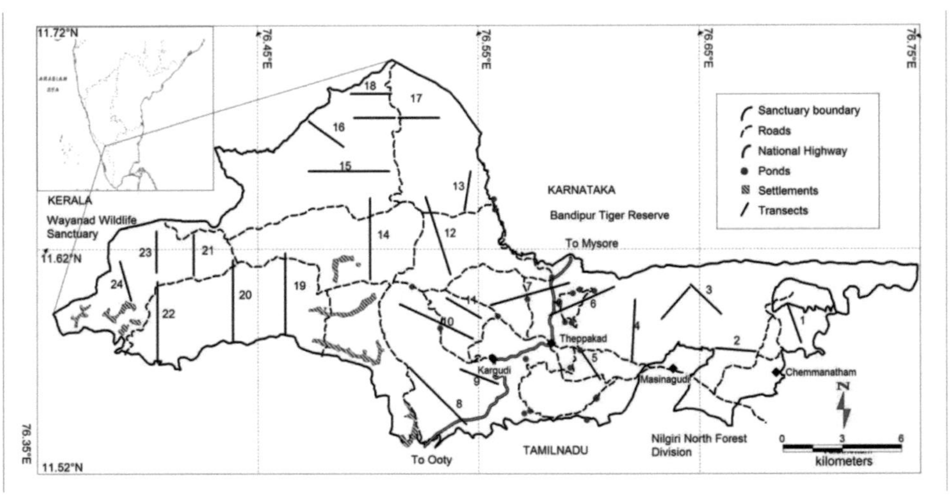

Figure 1.
Map showing Mudumalai tiger reserve and transect locations.

monsoons. The study area has three major forest type namely tropical moist deciduous forest, dry deciduous forest and tropical thorn forest [15]. The dominant tree species associations are "*Acacia-Ziziphus*" association characterising tropical thorn forest, "*Anogeissus – Tectona*" and "*Anogessius – Terminalia – Tectona*" association describing the tropical dry deciduous forest and "*Lagerstoroemia – Terminalia – Tectona*" characterising the tropical moist deciduous forest. The presence of areas of open grassland, numerous permanent water resources and salt licks has resulted in highly productive landscape. The large herbivores include elephant (*Elephus maximus*), three species of cervids, Chital (*Axis axis*), Sambar Deer (*Rusa unicolor*) and Barking Deer (*Muntiacus muntjak*), two species of antelopes, *viz.*, the Four-horned Antelope (*Tetracerus quadricornis*) and the Blackbuck (*Antilope cervicapra*) are the dominant mammals of this area. In addition predators like Tiger (*Panthera tigiris*), Leopard (*Panthera pardus*) and Wild dog (*Cuon alpinus*) can be seen. The Gaur and other herbivores are threatened by habitat degradation from overgrazing and human disturbances.

Densities of herbivores were estimated using distance sampling [16–18]. A total of 24 transects, each 2–4 km in length were monitored within a 321 km^2 survey area encompassing the different habitats from 2004 to 2007. Transects were placed randomly, with stratification by broad forest type (Moist deciduous forest, Dry deciduous forest and Thorn forest; **Figure 1**). Twelve surveys were repeatedly conducted on 24 line transects established throughout the sanctuary from 2004 to 2007 during the dry season in December to March and this provided adequate spatial replicates of the study area (643 km total), covering all vegetation types. On every walk, species, group size, age-sex composition, sighting angle using hand held sighting compass and sighting distance measured by a manual range finder. Population density of large herbivores was estimated using the software Distance v.6 Release 2 [19]. To model detection functions to estimate species density the data for each species per transect was examined for signs of evasive movement and peaking at distances from the line of walk. Following this, the data were either truncated or re-classed so as to ensure a reliable fit of key functions and adjustment terms to the data. The Akaike Information Criterion (AIC) and χ^2 goodness-of-fit test were used to judge the fit of the model. Using the model thus selected, estimates of group density, group size and individual density were derived [19].

In addition to transects, road-strip counts were conducted on a monthly basis in the different vegetation types (1262.5 km total). A total distance of 830, 308 and 124.5 km was sampled in DDF, MDF and TF respectively. In each sighting of gaur, we recorded group size and composition. Only groups where all the individuals counted were used for group size analysis. The sighting data from transect and road-strip count were pooled together to compute season-wise and habitat wise estimate on mean group size and sex ratios and median group size. Mean group sizes obtained from line transect counts actually indicate the size of 'clusters' [16] of animal species being sampled, rather than of social groups. Though these clusters are not always identical to social groups it has been used for population analysis [7]. In addition group size data from transects, counts from road-side counts, foot trails and waterholes were also used to calculate mean group size and group structure. For comparing grouping tendencies of different species the percentage of observed groups in categories such as: solitary animals, family associations (2–3 individuals) consisting of pairs of adult females with their yearlings and young, small groups (4–10 individuals), medium sized groups (11–30 individuals) and large groups of above 30 individuals.

2.2 Population structure

The proportion of different age-sex classes in the population was calculated using transects count supplemented with counts from road-side counts, foot trails and waterholes. The physical characteristics described in the literature [20–22] and observation in captive animals to establish the age-sex categories for classification. Chital, Sambar and gaur were classified as adult male (AM), adult female (AF), Sub-adult male (SAM), Sub-adult female (SAF) and young (YG). Younger elephants (<15 years) were classified by comparing their height to the oldest adult female in the group. Elephants were placed in broad age groups such as calves (<1 year), juveniles (1–5 years), sub-adults (5–15 years) and adults (>15 years) [7, 10].

The mean biomass densities (in kg/km^2) in the study area were derived from multiplying mean ecological density for each species by its average unit weight. Biomass density was estimated from published data on body weights [20, 23, 24] and population structure data from study area.

3. Results

Barking deer and sambar are essentially non-social species with solitary individuals, pairs or family association comprising 80% and 84% groups seen, respectively (**Table 1**). Though solitary individuals accounted for 16% and 25% of elephant and gaur most of these individuals are appeared to be adult males. These two species seems to occupy an intermediate position in terms of sociality with the group size frequencies being relatively evenly distributed among family associations, small groups and medium sized groups. At the other end of the scale, chital and common langur were clearly social groups with solitary individuals comprising less than 5% of all animal classified. The exceptionally large groups of 148 for chital and 32 for gaur were probably temporary agglomerations of several groups.

The study area harboured high ungulate density of 55.4 ± 5.83 animals/km^2. Among the different species chital was the most abundant species with the highest density (29.8/km^2; **Table 2**). The next abundant species was Common langur 11.9 ± 2.31/km^2. Gaur and sambar were the abundant species and their densities were 7.1 ± 1.47/km^2 and 4.2 ± 0.62/km^2 respectively. Barking deer density (0.6 ± 0.20/km^2) was relatively lower than other large herbivores. Among different species chital had the highest group density (3.6 ± 0.55/km^2). Species such as Gaur and sambar had higher density but with lower group density (1.3 ± 0.18/km^2 and 1.7 ± 0.22/km^2). The ecological density varied with group size of different herbivore species. Though there is relatively less variation in the group density among Gaur, sambar and elephant the density estimate varied because of variation in the group size. Among the different species chital and Gaur had higher group size of 8.3 ± 0.85 individuals and 5.59 ± 0.84 individuals respectively. The solitary species such as barking deer, giant squirrel and black naped hare density did not show any variation between the group density and actual density. Overall elephant density was 4.8 ± 0.92/km^2 with mean group size of 4.3 ± 0.40 individuals (**Table 2**).

3.1 Population structure

The adult sex ratios in ungulate species seem to be female biased. The adult male–female sex ratios favour females by 1:33.8 for elephant, 1:4 for gaur and 1:2 for chital in exception equal (1,1) sex ratio was observed in sambar. The composition of adult male

S. No.	Species	n	Range of observed group size	Percentage in each group size class				
				1	2 to 3	4 to 10	11 to 30	30+
1	Barking deer	22	1–2	77.3	22.7	0.0	0.0	0.0
2	Chital	534	1–150	5.1	15.2	36.0	33.1	10.7
3	Sambar	150	1–28	32.7	37.3	22.7	7.3	0.0
4	Gaur	423	1–47	23.4	16.1	34.8	24.0	1.7
5	Elephant	210	1–25	16.2	37.1	41.9	4.8	0.0
6	Four horned antelope	6	1	100.0	0.0	0.0	0.0	0.0
7	Mouse deer	4	1	100.0	0.0	0.0	0.0	0.0
8	Wild pig	29	1–17	44.8	6.9	41.4	6.9	0.0
9	Giant squirrel	20	1–3	70.0	30.0	0.0	0.0	0.0
10	Common langur	117	1–32	0.9	9.4	56.4	32.5	0.9
11	Bonnet monkey	4	4–22	0.0	0.0	50.0	50.0	0.0
12	Black-naped hare	27	1–2	80.0	20.0	0.0	0.0	0.0

Table 1.
Grouping characteristics among large herbivore species: Group size classes, range of observed group size, n = sample size in Mudumalai Tiger Reserve.

S. No.	Species	n	Group density / km²	Mean Group size	Density/ km²	CV (%)	95% confidence interval	
			Dg ± SE	Gs ± SE	D ± SE		lower	upper
1	Gaur	103	1.26 ± 0.18	5.6 ± 0.84	7.1 ± 1.47	20.9	4.69	10.6
2	Chital	212	3.59 ± 0.55	8.3 ± 0.85	29.8 ± 5.50	18.4	20.8	42.8
3	Sambar	117	1.71 ± 0.22	2.4 ± 0.19	4.2 ± 0.62	15.0	3.11	5.60
4	Barking deer	21	0.53 ± 0.18	1.1 ± 0.08	0.6 ± 0.20	34.0	0.31	1.15
5	Elephant	104	1.11 ± 0.19	4.3 ± 0.40	4.8 ± 0.92	19.5	3.25	6.90
6	Wild pig	9	0.12 ± 0.06	1.9 ± 0.59	0.2 ± 0.13	35.0	0.08	0.67
7	Common langur	122	2.42 ± 0.42	4.9 ± 0.41	11.9 ± 2.31	19.4	8.13	17.4
8	Giant squirrel	51	1.33 ± 0.29	1.2 ± 0.06	1.6 ± 0.35	21.9	1.04	2.46
9	Black naped hare	27	1.25 ± 0.51	1.0 ± 0.00	1.3 ± 0.51	41.0	0.57	2.74
10	Mouse deer	2	0.11 ± 0.09	1.0 ± 0.00	0.1 ± 0.09		0.03	0.50

Table 2.
Ecological density (±standard error), group density (±SE) and group size (±SE) of large mammals in Mudumalai Tiger Reserve during the study period.

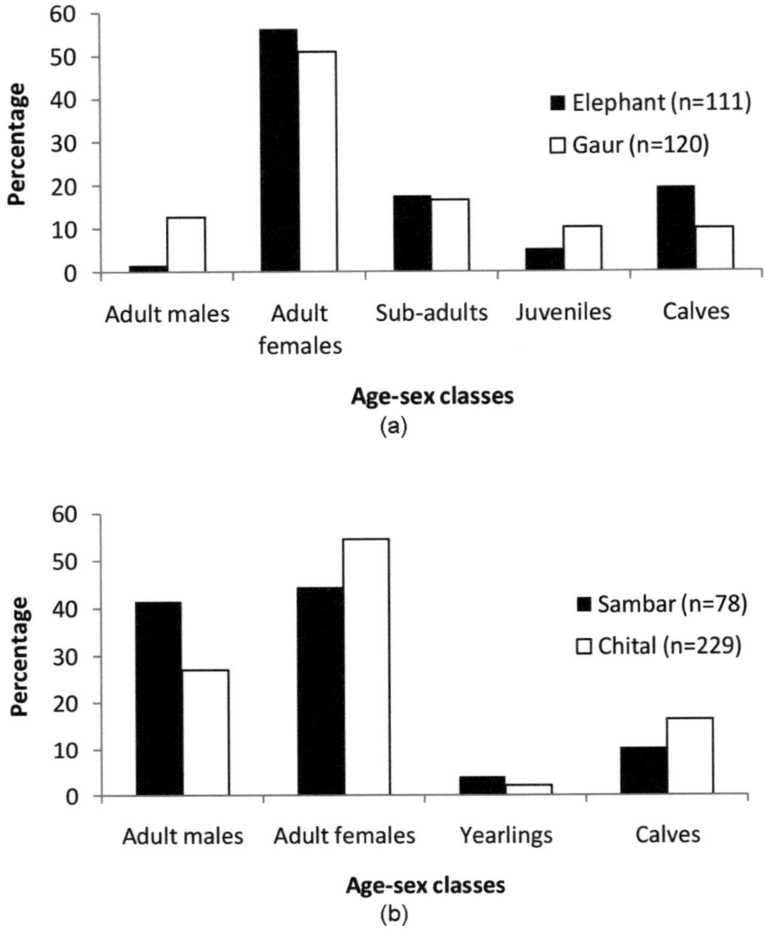

Figure 2.
Percentage of different age-sex classes among four ungulate species (elephant and gaur; sambar and chital) in Mudumalai Tiger Reserve.

was lowest (1.65%) among the elephants (**Figure 2a**). Totally 120 gaurs were sighted during the study period. The composition of adult female (50.79%) was highest in the group. The sub-adult male and female consisted of 16% of the populations. Totally 78 sambar deer were sighted and in which 85% consisted of Adult male and adult female (**Figure 2b**). The calf consisted of 10.34%, the sub-adult male had the lowest composition of 0.49%. While considering age distribution, the proportion of pre-reproductive age classes (sub-adults, juveniles and calves) is 53% in elephant, 36.6% in gaur, 26.4% in chital and 14% in sambar.

4. Discussion

Barking deer and sambar are solitary species. Other than adult males which tend to be solitary, gaur and elephant usually live in small groups. Occasionally they form medium to large groups particularly at favoured feeding sites. Chital is the most gregarious ungulate and commonly form medium to larger groups. The social organisation of elephant and gaur are based around the family unit or the herd consisting of one or more related adult females and their offspring [20, 25–27]. In elephants the most cohesive unit is the family unit and it represents a tightly knit unit both socially and structurally [28]. At puberty, young females remain with the natal group, while the males leave the herd and they tend to form temporary groups with weak social bonds [29]. Little however is known about gaur social structure or about male dispersals.

We looked at the pattern of densities in relation to body size and diet [30] of individual species. Densities of smallest ungulates barking deer is lower than densities of considerably larger species such as chital, sambar and gaur. Because barking deer found in Moist deciduous forest and it selectively feeds on rich but scarce food items such as shoots and fruits [20]. Moist deciduous forest consists of only 33% of the different vegetation types of the study area. Their solitary nature and territorial spacing mechanisms may also contribute to their relatively low densities. Chital which are primarily grazers [22] are most abundant in the study area. Dry deciduous with extensive short grass clearing, Moist deciduous forest with and swamp microhabitat and Open grass lands in thorn forest support high density of chital in Mudumalai Tiger Reserve.

Sambar which is predominantly browsers [20] distributed relatively higher density in Moist deciduous forest in the Western part of the study area. They feed during evening and night hours and rest in large number in open swamp areas in Moist deciduous forest. As could be predicted from body size and diet of sambar are less abundant than chital. However densities of gaur a larger species are higher than sambar. Gaur are mixed feeders and they graze intensively in open swamp areas in Moist deciduous [31] and coarse grass in dry deciduous forest. They also seem to browse on the profuse secondary growth of Bamboo *(Bambusa arundanacia)* in Thorn forest in areas with less human disturbance [10].

Elephant density is higher than earlier estimate (2.4/km^2) [32]. Earlier studies in the year 1980s [25] indicated a much lower density (1/km^2) for the study area. Despite poaching reducing the male numbers and keeping them dispersed breeding is taking place and the population is increasing. The specialist folivore common langur is more abundant than generalist bonnet macaque in the study area. Wild pig that feed selectively on variety of plant and animal foods such as roots, tubers, fruits insects and carrion [33], found in relatively low density (0.4/km^2) in the study area. The factors such as nocturnal feeding habitat and incidence of Anthrax in wild pig in the study area during the study period might reduce wild pig population density. Further monitoring and diseases control measures were taken by the forest department.

The adult sex ratios of ungulate species seem female biased. In ungulate species such as chital, sambar and gaur the males more solitary habits, proneness to injuries from intra-specific aggression, lack of alertness during rut and dispersal behaviour of makes them more vulnerable to predation [20, 24]. Whereas in elephant selective poaching of adult males especially tuskers caused skewed sex ratio (1:34). An earlier study in this area [25] had reported that the adult male to female ratios were far less skewed (1:4.7) than the current ratio of 1:34. This indicates that the adult male population has declined significantly from the mids 1980's. This decline in adult male population in elephants is largely due to poaching of males for their tusks [27].

The biomass of large herbivores is compared with other tropical forests (**Table 3**). MTR supported very high biomass density (15,198 kg/km^2) than similar tropical forests in Southern India. It was almost equal to that of Nagarahole (14,744 kg/km^2 [3]) and Bandipur National park (14,520 kg/km^2 [24]) and more than Kanha (1592 kg/km^2 [20]) and Pench (6013 kg/km^2 [34]) of Central India. Furthermore it was higher than Chitwan, Nepal [35]. Tropical humid forests were generally considered to be poorer habitats than Savannas in terms of supporting high biomass (Eisenberg, 1980) [30].

Region	Area[a]	Habitat type[b]	Biomass density[c] (kg/km^2)	References
Asia	Mudumalai	DDF, MDF and TF	15,198	Present study
	Nagarahole	DDF, MDF and TF	14,744	[3]
	Bandipur	Dry forest-woodland	14,520	[24]
	IGWS & NP	DDF, MDF, TF, EG and GL	14,204	[39]
	Kanha	Moist forest-meadows	1592	[20]
	Pench	Deciduous forest	6013	[34]
	Wilpattu	Dry forest-meadows	766	[21]
	Bardia	Moist forest-grasses	3101	[40]
	Chitwan	Moist forest-grasses	2581	[35]
Africa	Rwenzori	Swamps-savanna	21,373	[36]
	Manyara	Dry savanna	19,259	[36]
	Serengeti	Dry savanna	6840	[36]
	Mara	Dry savanna	19,200	[37]
	Nairobi	Dry savanna	4470	[36]
	Gabon	Evergreen forest	1020	[38]
South America	Masagural	Dry forest-pasture	711	[30]
	Barro Colorado	Evergreen forest	3553	[30]
	Manu	Evergreen forest	1220	[41]
	Pantanal	Moist forest pasture	295	[30]

[a]*IGWS & NP- Indira Gandhi Wildlife Sanctuary and National Park.*
[b]*DDF-Dry deciduous forest; EG- Evergreen; GL-Grassland; MDF-Moist deciduous forest; TF-Thorn forest.*
[c]*The mean body weight of herbivores was used from other published sources [20, 23, 24, 35].*

Table 3.
Comparison of wild biomass density of herbivores in different tropical sites.

But several Savanna and Woodland habitats of Africa seem to support lower biomass densities than MTR [36–38]. For example the biomass estimates from the lowland rain forests of Gabon and Serengeti were considerably lower [38]. Whereas Manyara Lake park and Mara Plains in Kenya and Rwenzori Park in Uganda supported biomass densities higher than that of MTR [36].

5. Conclusions

Mammal population density and social organisation were studied in the tropical forests of Southern India. Among ungulates, the solitary, specialist species barking deer occurred at lower densities. Chital is primarily grazer, highly gregarious and the most abundant species in the study area. Sambar is browser, distributed at higher densities at Moist deciduous forest. Elephant densities are increasing when compared to earlier estimate. But the sex ratio of elephant is highly skewed due to selective poaching of males for their tusks. Mudumalai supports higher biomass densities of mammal species than similar tropical forests in India.

Acknowledgements

We thank Tamil Nadu Forest Department for permission to conduct research within Mudumalai Tiger Reserve. Hill Area Development Program of Ministry of Environment and Forest provided financial assistance. We thank Bombay Natural History Society, Mumbai, Nilgiri Wildlife and Environment Association, Ooty, Indian Institute of Science (Centre for Ecological Studies) and A.V.C. College Department of Zoology and Wildlife Biology for their support.

Conflict of interest

We do not have any conflict of interest to disclose.

Author details

Mohanarangan Ashokkumar[1*] and Rajaratinavelu Nagarajan[2]

1 Field Station, Mudumalai Tiger Reserve, Bombay Natural History Society, Upper Kagudi, India

2 Department of Zoology and Wildlife Biology, A.V.C. College, Mayiladuthurai, India

*Address all correspondence to: vimalashok7@gmail.com

References

[1] Williams BK, Nichols JD, Conroy MJ. Analysis and Management of Animal Populations. New York: Academic Press; 2002

[2] Ahrestani FS, Heitkonig IMA, Prins HHT. Diet and habitat-niche relationships within an assemblage of large herbivores in a seasonal tropical forest. Journal of Tropical Ecology. 2012;**28**(4):385-394. DOI: 10.1017/S0266467412000302

[3] Karanth KU, Sunquist ME. Population structure, density and biomass of larger herbivores in the tropical forests of Nagarahole, India. Journal of Tropical Ecology. 1992;**8**:21-35

[4] Owen-Smith N. Pleistocene examinations: The pivotal role of megaherbivores. Paleobiology. 1987;**13**:351-362

[5] Rosenzweig ML. Community structure in sympatric carnivores. Journal of Mammalogy. 1966;**47**:602-612

[6] Gittleman JL. Carnivore body size: Ecological and taxonomic correlates. Oecologia. 1985;**67**:540-554

[7] Karanth KU, Sunquist ME. Prey selection by tiger, leopard and dhole in tropical forests. Journal of Animal Ecology. 1995;**64**:439-450

[8] Lovari S, Pokheral CP, Jnawali SR, Fusani L, Ferretti F. Coexistence of the tiger and the common leopard in a prey-rich area: The role of prey partitioning. Journal of Zoology. 2014;**295**:122-131. DOI: 10.1111/jzo.12192

[9] Riddle HS, Schulte BA, Desai AA, Meer LVD. Elephants – A conservation overview. Journal of Threatened Taxa. 2010;**2**:653-661

[10] Ashokkumar M, Swaminathan S, Desai AA, Daniel JC. A Study on Ecology and Conservation of Gaur (*Bos frontalis*) in Mudumalai Wildlife Sanctuary and National Park. Ottacamund: Report, Nilgiri Wildlife and Environment Association; 2004. p. 111

[11] Choudhury A. Distribution and conservation of gaur (*Bos gaurus*) in the Indian subcontinent. Mammal Review. 2002;**32**:199-226

[12] Ashokkumar M, Swaminathan S, Nagarajan R, Desai AA. Distribution, ecology and conservation of the gaur (*Bos gaurus* H. Smith, 1827). In: Gupta VK, Verma AK, editors. Animal Diversity, Natural History and Conservation-Vol. 1. New Delhi: Daya Publishing House; 2011. pp. 77-94

[13] Davidar ERC. Cheetal Walk-Living in the Wilderness. Chennai: Oxford University Press; 1997 208p

[14] Mittermeier RA, Myers N, Thomsen JB, Fonseca GAB, Oliveri S. Biodiversity hotspots and major tropical wilderness areas: Approaches to setting conservation priorities. Conservation Biology. 2008;**12**(3):516-520

[15] Champion HG, Seth SK. Revised Survey of Forest Types in India. New Delhi: Manager of Publication; 1968 404p

[16] Burnham KP, Anderson DR, Laake JL. Estimation of density from line transect sampling of biological populations. Wildlife Monograph. 1980;**72**:1-202

[17] Buckland ST, Anderson DR, Burnham KP, Laake JL. Distance Sampling: Estimating Abundance of

Biological Populations, Chapman and Hall, London. Reprinted in 1999 by Research Unit for Wildlife Population Assessment. Scotland: University of St. Andrews; 1993

[18] Buckland ST, Anderson DR, Burnham KP, Laake JL, Borchers DL, Thomas L. Introduction to Distance Sampling. Oxford: Oxford University Press; 2001

[19] Thomas L, Laake JL, Rexstad E, Strindberg S, Marques FFC, Buckland ST, et al. *Distance 6.0*, Research Unit for Wildlife Population Assessment. Scotland: University of St. Andrews; 2009

[20] Schaller GB. The Deer and the Tiger. A Study of Wildlife in India. Chicago: University of Chicago Press; 1967

[21] Eisenberg JF, Lokhart M. An ecological reconnaissance of Wilpattu National Park, Ceylon. Smithsonian Contribution to Zoology. 1972;**101**:47-51

[22] Mishra HR. The Ecology and Behaviour of Chital (Axis axis) in Royal Chitwan National Park, Nepal [Thesis]. Edinburgh: University of Edinburgh; 1982. p. 240

[23] Eisenberg JF, Seidensticker J. Ungulates in southern Asia: A consideration of biomass estimates for selected habitats. Biological Conservation. 1976;**10**:294-308

[24] Johnsingh AJT. Large mammalian prey-predators in Bandipur. Journal of the Bombay Natural History Society. 1983;**80**:1-57

[25] Daniel JC, Desai AA, Sivaganesan N, Rameshkumar S. The Study of some Endangered Species of Wildlife and their Habitats - the Asian Elephant. Bombay: Bombay Natural History Society; 1987

[26] Sukumar R. The Asian Elephant: Ecology and Management. London: Cambridge University Press; 1989 251p

[27] Daniel JC, Desai AA, Mohenraj N, Ashokkumar M, Sakthivel C. Evaluating Population Enumeration Methods and Human Elephant Conflict Mitigation Methods in Mudumalai Tiger Reserve, Tamil Nadu, India, Report. Bombay: Bombay Natural History Society; 2008 70p

[28] Moss C. Elephant Memories. New York: William Morrow & Co; 1988. p. 364

[29] Laws RM, Parker ISC, Johnstone RCB. Elephants and their Habitats: The Ecology of Elephants in North Bunny. Uganda, London: Oxford University Press; 1975

[30] Eisenberg JF. The density and biomass of tropical mammals. In: Soule ME, Wilcox BA, editors. Conservation Biology: An Evolutionary Ecological Perspective. Sunderland: Sinauer Associates; 1980. pp. 34-55

[31] Krishnan M. An ecological survey of the large mammals of peninsular India. Journal of Bombay Natural History Society. 1972;**69**:297-315

[32] Baskaran N, Desai AA. Elephant Census in Mudumalai Wildlife Sanctuary and National Park. Bombay: Bombay Natural History Society; 2000 20p

[33] Prater SH. The Book of Indian Animals. Chennai: Oxford University Press, Bombay Natural History Society; 1980. p. 348

[34] Biswas S, Sankar K. Prey abundance and food habit of tigers (*Panthera tigris tigris*) in Pench National Park, Madhya Pradesh, India. Journal of Zoology. 2002;**256**:411-420

[35] Tamang KM. The Status of the Tiger (*Panthera tigris*) and its Impact on Principal Prey Populations in the Royal Chitwan National Park, Nepal, [Thesis]. East Lansing: Michigan State University; 1982. p. 184

[36] Eltringham SK. The Ecology and Conservation of Large African Mammals. Baltimore: University Park Press; 1979. p. 286

[37] Stelfox JG, Peden DG, Epp H, Hudson RJ, Mbugua SW, Agatsiva JL, et al. Herbivore dynamics in southern Narok, Kenya. Journal of Wildlife Management. 1986;**50**:339-347

[38] Prins HH, Reitsma JM. Mammalian biomass in an African equatorial rain forest. Journal of Animal Ecology. 1989;**58**:851-861

[39] Kumaraguru A. Studies on the Influence of Vegetation Types and Prey Composition on the Feeding Ecology of Tiger, Leopard and Wild Dog in Indira Gandhi Wildlife Sanctuary and National Park, Tamil Nadu, India [Thesis]. Trichirapalli: Bharathidasan University; 2006. p. 203

[40] Dinerstein E. An ecological survey of Royal Karnali-Bardia Wildlife Reserve, Nepal. Biological Conservation. 1980;**18**:5-38

[41] Terborgh J, Emmons LH, Freese C. La fauna Silvestre de la Amazonia: El despilfarro de un Recurso de un removable. Boletin de Lima. 1986;**46**:77-85

Chapter 12

Breeding Ecology and Population Status of Kenyan Amphibians

Alfayo Koskei

Abstract

Although amphibians significantly contribute to energy flow in ecosystem, recent studies show that over 1800 amphibians are at risk of extinction globally and about fifth (235) of species in sub-Saharan Africa are threatened with extinction due to habitat destruction. The results show that coastal biodiversity hotspots including Shimba Hills National Reserve (30 species of anuras) and Taita Hill ecosystem complex (24 species) are leading in amphibian biodiversity and level of endemism. Moreover, the distribution of *Afrixalus slylvaticus* in Shimba Hills Ecosystem was almost equal in both land tenure types (51% in protected and 49% in community land). However, the species was more abundant during the wet season than dry (58 and 42%, respectively) which represented a statistical significant difference (t-value = −3.38, p-value = 0.002, DF = 30). However, the level of endemism declines drastically inland and towards the drier parts of the country. Most amphibian ecosystems in Kenya are facing perturbation from rapid human population growth, expansion of urban and agriculture areas, unplanned infrastructural and unregulated use of riparian zones. This pace of amphibian loss and decline calls for immediate collaborative actions. Reversing the trend requires dire commitment from conservationists and communities.

Keywords: anura, amphibians, ecosystem, biodiversity and endemism, breeding ecology

1. Introduction

This chapter introduces the first annotated status of breeding ecology and conservation status of amphibians in Kenya. The chapter provides a review of recent surveys on the status of habitats and population of amphibians as well as drivers of their population change in areas within the global biodiversity hotspots in Kenya. It also presents the possible course of action for reversing the trend of amphibian decline. The names of most species are given in scientific and where local names are used, explanation is provided. A literature survey and primary data are combined to provide a comprehensive review of the status of Kenyan Amphibians.

2. Research methodology

The population status of amphibians in Kenya was assessed in major biodiversity hotspots that include coastal forests, inland semi-arid areas, central region of Kenya, Western, and lower Eastern. The coastal forests of Kenya are part of the Coastal Forests of Eastern Africa biodiversity hotspot. The distribution of amphibians on the costal forest was assessed in south costal forest in Shimba Hills National Reserve (SHNR) Ecosystem, other central areas of the coast (Rehabilitated quarries near Mombasa) and inland drier forests in Taita Hills Ecosystem. In the leading biodiversity hotspot(s) ecological surveys were conducted for 60 days in the months of dry season and another 60 days wet season.

The sampling was done along six transects distributed equally both in protected and unprotected areas. The unprotected parts were characterized by human activities ranging from agro-ecosystems to settlements and infrastructure. During the fieldwork, visual encounter surveys (VES) [1, 2] were used where all potential microhabitats of the species were searched. The survey was conducted along the river transects (1000 m) but the sampling protocol was adjusted in some areas with terrain limitation or pools of water where line transect was limiting. The number of individuals encountered within 10 m of each side of the transect line was captured, marked, morphometric/biometric and population attributes recorded. Marking was done using toe clipping as adopted by [3]. In other biodivrsity hotspots in Kenya, the status of amphibians were documented using the previous surveys conducted by other scholars and ugmented by the review of their IUCN status online.

3. Results

3.1 Status of wetlands in Kenya

The major cause of decline in wetlands and amphibian habitats in Kenya is human activities particularly land use land cover changes. Other significant drivers include: natural calamities; drought, drop in ground water level, erosion, pollution resulting from industrial as well as domestic waste particularly Lake Nakuru and Agrochemicals from horticulture farms e.g. in Naivasha. Moreover, in other leading watersheds such as Tana River waste from urban centres and agricultural projects along the river channel contribute heavily to its pollution and decline in amphibian biodiversity. In most of these wetlands and lotic systems, characteristics of nutrient and chemical inflow is evidenced by the growth of invasive plant species for example *Salvinia molesta* in Lake Naivasha and Water Hyacinth (*Pontederia crassipes*) in Lake Victoria, algal blooms, prolific growth of macrophytes, frequent death of fishes, and emergence of murky foul water. Another cause of wetland degradation is over-exploitation through water abstraction, drainage to create land for agriculture, burning, timber and other wood species collection as well as fishing. All these activities are common in almost all wetlands in Rift Valley as evidenced by decline in papyrus cover in Loboi and Saiwa swamps, decline in fish harvest in Lake Naivasha, serious fluctuations of eater in Lake Elementeita, Naivasha and Lake Nakuru as well as frequent drying up of many swamps in the region. The major landform features in Kenyan biomes are central plains, coastal strip, the rift valley, ancient block mountains and rift valley lakes. Riverbanks and wetlands provides habitats for many amphibian species and among the notable rivers and wetlands in Kenya are shown in **Figure 1**. The

Ramsar Convention recognizes several wetland areas in Kenya as Ramsar wetlands of international significance such as Tana River Delta, Lake Nakuru, Lake Naivasha, Lake Bogoria, Lake Elmenteita and Lake Baringo. Ramsar defined wetland as "an area covered by fen, marsh, water or peat and can be natural or artificially modified and

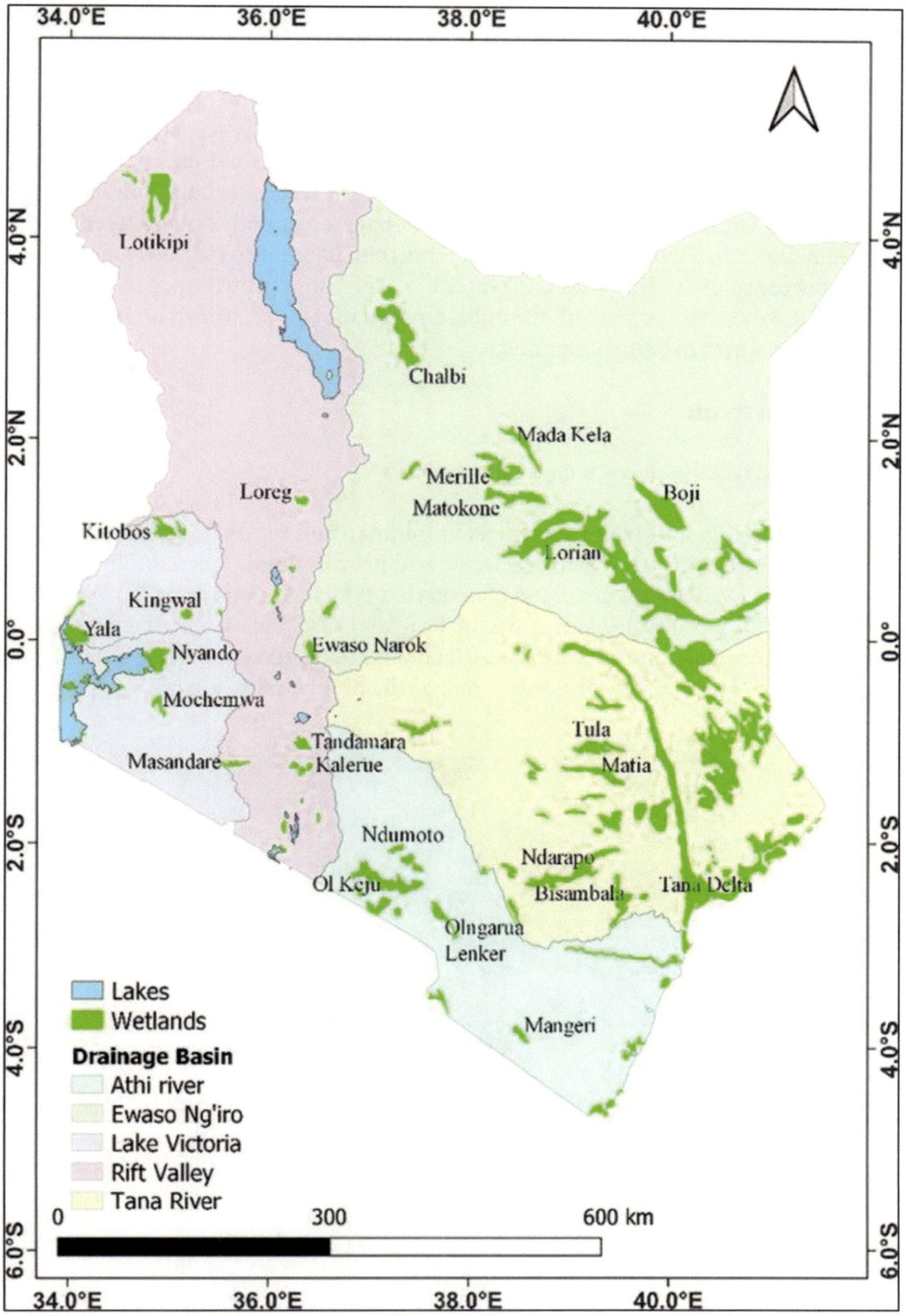

Figure 1.
Map showing major wetlands in Kenya © Munguti, KEMFRI.

can be partly or throughout the year with static or free flowing water that are either salty, blackish and it include areas of marine water whose at low tide the depth does not exceed 6 meters" [4].

It is worth nothing that frogs require moisture of any kind during breeding, for eggs and larvae stage to develop (tadpoles) prior to metamorphosis [5, 6]. However, most species in Africa have evolved astonishing behaviors and lifestyle in face of perturbations from anthropogenic activities to avoid dehydration and manage breeding [7]. As a result, some are no longer reliant on the river banks, lake shores and marshes as those have been degraded and some are now breeding on land especially with little rain and often hibernate underground during dry spells where soil retain some moisture and some in fact developed wrapped cocoons [6]. However, in most undisturbed habitats, amphibians freely float their spawn in water or moist soil or reeds (see **Figure 2**). Unlike reptilian or bird eggs that have a protective outer layer, frog eggs are enclosed in a gelatinous envelope. Eggs then hatch into tadpoles, an aquatic stage that is completely dependent on water to survive and grow. Aquatic ecosystems are therefore key to the success of amphibian populations but, unfortunately, local freshwater systems are being degraded.

3.2 Population status

3.2.1 Status of Afrixalus sylvaticus *in Shimba Hills*

The Shimba Hills stand out as the most amphibian and reptile species rich area in Kenya, predominantly due to the endemic and near-endemic species. The Shimba Hills Ecosystem (SHE) are a dissected plateau that is located between 4°09′–4°21′S and 39°17′–39°30′E in Kwale County on the Kenyan coast. There are currently 30 known amphibians (28 anurans and two caecilians) in this ecosystem, which includes 11 families and 15 genera [8]. The area is one of the biodiversity hotspot with highest

Figure 2.
Spawn of Afrixalus slyvaticus *in Shimba Hills (Source: Author 2016).*

Figure 3.
A & B: Picture of Afrixalus sylvaticus *and* H. rebrovermicalatus.

level of endemism, and face considerable threats relative to the remaining area [9]. For example, two of the endemic and endangered amphibian species in Shimba Hills Ecosystem (SHE) are *A. sylvaticus* (Left) and *Hyperolius rebrovermicalatus* (Right) (See **Figure 3A** and **B**), as well as one snake species *Elapsoidea nigra.* Moreover, the habitat is the only home of sable antelope species in Kenya.

From the surveys of one species (*A. sylvaticus)* we conducted in 2018 in SHE ecosystem, it was evident that many parameters influences the distribution and abundance of amphibians in this ecosystem [10]. One key aspect is the heterogeneous habitats in the SHNR—the area consists of six forest types and woodland and grassland habitats within the reserve [11, 12] which allows for the variety of species from different biogeographic zones. In our survey for example, we conducted both in community land and in protected area and the species distribution was almost equal in both land tenure types (51% in protected and 49% in community land). However, the species was more abundant during the wet season than dry (58 and 42%, respectively) which represented a statistical significant difference (t-value = −3.38, p-value = 0.002, DF = 30). In terms of period and time of the survey, it was evident that more individuals of *A. sylvaticus* were encountered during morning sampling than the other sampling periods though there was no statistically significant difference (F-Value, 0.25 DF = 2, p-value = 0.781). Weather also plays a part in the species sightings, as during the cool and rainy period, we captured the highest number of individuals (47%) as compared to the cool and dry period (8%) where we captured only five individuals [10]. The results are shown in **Table 1**.

3.2.2 Species diversity in Taita Hills ecosystem

Another area within the coastal lowlands rich in amphibians and highest level of endemism is the Taita Hills complex which is located in southern Kenya and it constitutes the northernmost portion of the crystalline block faulted Eastern Arc Mountains (EAM). It belong to the 34 worldwide biodiversity hotspots [13, 14]. According to [15], three (two caecilians and one frog species) out of the 26 species recorded in Taita Hills ecosystem, are endemic and are associated with montane forest. These include *Boulengerula taitana*, *Boulengerula niedeni* and *Callulina dawida* as shown in **Table 2**.

Parameter	Category	N	Mean	SE(±)	StdDev (±)	Min	Max	(%)
Season	Dry	25	1.042	0.0417	0.204	1	2	42
	Wet	34	1.417	0.103	0.504	1	2	58
Land tenure type	Protected area	30	1.364	0.064	0.326	1	2	51
	Community area	29	1.115	0.105	0.492	1	2	49
Time of survey	Evening	18	1.286	0.125	0.469	1	2	31
	Morning	21	1.235	0.106	0.437	1	2	36
	Night	20	1.177	0.095	0.393	1	2	34
Weather	Cool and dry	5	1.667	0.333	0.577	1	2	8
	Cool and rainy	28	1.167	0.078	0.381	1	2	47
	Hot and dry	8	1.000	0.000	0.000	1	1	14
	Warm and dry	18	1.385	0.140	0.506	1	2	31

Table 1.
Shows population size and related parameters of A. sylvaticus.

Species	IUCN status
a. **Taita Hills endemics**	
B. niedeni [16]	Critically Endangered (CR) B1a&b (iii).
B. taitana (Loveridge, 1935)	Vulnerable (VU).
C. dawida [17]	CR B1a&b (ii, iii)
b. **Other montane forest associated species**	
Xenopus borealis (Parker, 1936)	Least Concern (LC)
Arthroleptis xenodactyloides (Hewitt, 1933)	LC
Amietia angolensis (Bocage, 1866)	LC
Mertensophryne taitana (Peters, 1878)	LC
Hyperolius glandicolor (Peters, 1879)	LC
Hildebrandtia macrotympanum (Boulenger, 1912)	LC
Leptopelis concolor Ahl, 1929	LC
c. **Other non-montane species**	
Amietophrynus garmani, A. gutturalis, Amietophrynus xeros, Phrynomantis bifasciatus, Hemisus marmoratus, Ptychadena anchietae, Ptychadena mascareniensis, Ptychadena mossambica, Ptychadena schillukorum, Phrynobatrachus scheffleri, Pyxicephalus adspersus, Tomopterna cryptotis, Hyperolius tuberilinguis, Kassina senegalensis, Chiromantis kelleri, and Chiromantis petersi	LC

Table 2.
Amphibian species diversity in Taita Hills ecosystem.

3.2.3 Species diversity in Bamburi rehabilitated mines

In comparison to these two amphibian rich ecosystems, the survey done in coastal ecosystem that represent a secondary growth after rehabilitation shows low diversity and species richness. In survey conducted in 2013 in rehabilitated mines in Bamburi, central areas of the coast, 12 species were found [18] distributed across nine genera (N = 1994). From the assessment done, there was strong association of two species *Phrynobatrachus acridoides* (46%) and *P. anchietae* (44%) with rehabilitated mines in Bamburi mines near Mombasa as shown by their dominance. Although the alpha diversities were higher among these species in central areas of Mombasa, the species abundances were relatively low (compared to SHNR and Taita Hills) and distribution was uneven due to fluctuations in resource, quality of breeding sites, among other factors [18] as shown in **Table 3**.

3.2.4 Species diversity in Meru National Park

Towards inland ecosystems, as survey conducted in lower eastern ecosystem in Tana River Primate National Reserve, the species diversity was comparatively becoming low. In total 16 amphibians (all anurans) were recorded in 2002 [19]. Moreover, within the drier central parts of the country a survey conducted in Meru national park [20] revealed that species diversity is further declining as only 11 species distributed across seven genera inhabit the area were identified. Although the IUCN status of all species is Least Concern (LC), the population trend of most species remain unknown and low levels of endemism (**Table 4**). In the breeding seasons, the sex ratio is balanced although the males are dominant (as in *H. glandicolor*) which indicted that breeding takes place [3].

3.2.5 Species diversity in Kakamega forest

In another biodiversity hotspot located in the western part of the country, the amount of rainfall increases as compared to central region and the species richness increases. The

Species	n	IUCN Status	Population trend
Phrynobatrachus acridoides (Cope, 1867)	923	LC	Unknown
P. anchietae (Bocage, 1867)	879	LC	Unknown
P. schillukorum (Werner, 1907)	59	LC	Unknown
Amnirana galamensis (Dumeril and Bibron, 1841)	46	LC	Unknown
Kassina maculata (Dumeril, 1853)	23	LC	Unknown
Chiromantis xerampelina (Peters, 1954)	18	LC	Unknown
Xenopus muelleri (Peters, 1844)	14	LC	Unknown
Hyperolius pusillus (Cope, 1862)	11	LC	Unknown
Pyxicephalus edulis (Peters, 1854)	7	LC	Decreasing
Hyperolius argus (Peters, 1854)	7	LC	Unknown
H. tuberilinguis (Smith, 1849)	4	LC	Unknown
Bufo gutturalis (Power, 1927)	3	LC	Increasing

Table 3.
Amphibian population in rehabilitated mines in Bamburi area.

Species (family)	IUCN Status	Population trend
Xenopus laevis	LC	Increasing
Amietophrynus maculatus (*Bufonidae*)	LC	Stable
A. garmani (*Bufonidae*)	LC	Unknown
Amietophrynus gutturalis (*Bufonidae*)	LC	Increasing
P. bifasciatus (*Microhylidae*)	LC	Unknown
H. marmoratus (*Hemisotidae*)	LC	Unknown
H. glandicolor (*Hyperoliidae*)	LC	Unknown
Phrynobatrachus natalensis (*Phrynobatrachidae*)	LC	Stable
Ptychadena porosissima (*Ptychadenidae*)	LC	Unknown
P. mascareniensis (*Ptychadenidae*)	LC	Unknown
P. anchietae (*Ptychadenidae*)	LC	Unknown

Table 4.
Species diversity and IUCN status in Meru National Park.

number of species in Kakamega forest increases to 24 species (anurans only) as altitude increases to 1650 m above sea level as compared to 11 in Meru NP where the altitude of highest point is 850 m asl. More importantly, the level of endemism also increases in Kakamega Forest and its environs [21] as shown in **Table 5** below.

Species	IUCN status	Population trend
Xenopus victorianus Ahl, 1924	LC	Increasing
Bufo kisoloensis Loveridge, 1932	LC	Unknown
Bufo maculatus Hallowell, 1854 *sensu lato*	LC	Stable
Phrynobatrachus graueri (*Nieden, 1911*)	LC	Unknown
Phrynobatrachus. sp. (*aff. Mababiensis FitzSimons*, 1932)	Unknown	Unknown
P. natalensis (*Smith*, 1849) *sensu lato*	LC	Stable
Afrana angolensis (*Bocage*, 1866)	LC	Decreasing
Amnirana albolabris (*Hallowell*, 1856) *sensu lato*	LC	Unknown
P. anchietae (*Bocage*, 1867)	LC	Unknown
Ptychadena porosissima (*Steindachner*, 1867)	LC	Unknown
Ptychadena taenioscelis Laurent, 1954	LC	Unknown
P. mascareniensis (*Duméril & Bibron*, 1841) *sensu lato*	LC	Unknown
P. oxyrhynchus (*Smith*, 1849) *sensu lato*	LC	Stable
Hoplobatrachus occipitalis (*Günther*, 1859)	LC	Stable
Afrixalus osorio (*Ferreira, 1906*) *sensu lato*	LC	Stable
A. quadrivittatus (*Werner*, 1908 "1907") *sensu lato*	LC	Unknown
Hyperolius acuticeps Ahl, 1931	LC	Unknown
H. kivuensis Ahl, 1931	LC	Unknown
H. lateralis Laurent, 1940	LC	Unknown

Species	IUCN status	Population trend
H. viridiflavus (Duméril & Bibron, 1841)	LC	Unknown
Hyperolius cinnamomeoventris Bocage, 1866 *sensu lato*	LC	Unknown
Kassina senegalensis	LC	Unknown
Leptopelis sp.	Unknown	Unknown
Leptopelis bocagii	LC	Unknown

Table 5.
Amphibian diversity in Kakamega forest.

3.3 Major drivers of amphibian habitat degradation

Human activities contribute directly or indirectly to amphibian decline. In Kenya, for instance, unregulated industrial, rapid human population growth, expanding urban and agricultural areas, poor infrastructure and residential waste pose major threats to rivers, lakes and wetlands. This is predicted to decrease the resilience of ecosystem and biodiversity while increasing the prevalence of water borne diseases and parasites. This will result in the decline and extinction of amphibian species across the globe. In some urban areas, for instance, the insufficient basic services such as solid waste removal, untreated sewerage being discharge to water ways often causes water pollution and destruction of amphibian ecosystems.

Notably, human activities have drastically reduced species habitats and degrade their ecosystems. From our survey in Shimba Hills Ecosystem for example, the bush burning (as shown in **Figure 4**), drainage of wetlands, livestock grazing and plantation of exotic tree species (Eucalyptus sp.) has degraded species habitats and limit

Figure 4.
Bush burning as method land clearing in SHE (Source: Author 2017).

areas of occupancy of species such as *Afrixalus sylvaticus* that had been recorded in rice schemes in former habitats in Kaloleni.

4. Discussion

4.1 Status of Wetlands and Amphibian habitats

Notably, the collection of amphibians in East Africa started as early as 1900s, by the colonial officers, explorers and Missionaries from Germany and Britain [7]. During these colonial time, German scientists and explorers created the first account of amphibians in the region [7]. Since then, the growth of molecular analyses using simplified biochemical techniques such as DNA sequencing and next generation sequencing (NGS) of DNA has facilitated the identification of more species and thus phylogenic studies has since flourished [22]. Global biodiversity hotspots are areas with high species richness and endemism but also faces a significant threats vis-à-vis the remaining area [13]. In terms of biodiversity hotspots, at global stage, there are 35 biodiversity hotspots, and we have two in Kenya [23]. One is the Eastern Afromontane hotspot which is represented by the Taita Hills including mountain and highland areas in central region (Aberdare Ranges, Mt. Kenya, and Nyambene Hills) and western parts (Kakamega Forest, Nandi Forests, Cherangani Hills, and Mau Hills). The second is the coastal forests of eastern Africa Kenya represented by the coastal forests of Kenya, namely Shimba Hills and Arabuko-Sokoke Forest.

While Eastern Africa biodiversity hotspot are famed for its high biodiversity and endemism [24, 25], there exists data deficient and strong taxonomic biases in the existing literature and thus there is need for details review of the knowledge gaps through research. The rich biodiversity in the region is attributed to variety of land forms which includes mountains, rift valley, and wetlands [26]. This has fostered the evolution of endemic species of amphibians associated with these habitats. At global stage, wetlands covers an area of about 6 percent of the earth surface and in Kenya it covers 3–4% of the land surface (14,000 km^2) but often extend up to 6 percent during rainy season and 80% of these wetlands are found outside protected areas. Despite many ecological functions that these wetlands including providing habitats for many species of amphibians, they are regarded as wetlands and have been degraded through conversion to agricultural lands, settlements and industrial use. Moreover most wetlands in Kenya have suffered from pollution, catchment destruction and overexploitation. The continued degradation is attributed to lack of awareness and appreciation of the values of these wetlands as well as inconsistent policy implementation. The management of wetlands in Kenya have been influenced by many players that include government agencies such as fisheries, rural development, environment, public works and Agriculture as well as private developers, local fishermen and farmers alongside local and international development organizations. Although wetlands are among the most biologically productive ecosystems globally, local communities often regard it as a wasteland, nuisance, habitat for pests and diseases vectors such as bilharzia, flukes, mosquitoes, ticks, etc. As a consequence, wetlands have been threatened by anthropogenic activities and the cover percent of wetlands areas have declined by about 40 percent and so does the flow rate (discharge) of major rivers by about 30 percent [26] and lakes experienced dramatic fluctuations in water levels with cases of drying out especially in Rift valley lakes.

4.2 Population status

The status of amphibians' population globally shows a serious declining trend, and 41% of 6638 known world's amphibian species are threatened with extinction [27, 28]. More precisely, the recent study revealed that over 1800 amphibians face extinction [29]. The class Amphibia are under the division of vertebrae animals and are classified into three orders namely; Order Caudata (salamanders and newts – not found in sub-Saharan Africa), Order Anura (frogs and toads) and Order Gymnophiona (caecilians). Taxonomically, the amphibians in East Africa are classified into two orders: the order Gymnophiona which includes the caecilians- whose members have annuli that resembles earth worms with moist and shiny skin and do not have legs. They occur globally within the tropics and in east Africa, two families occur. Their habitat is mainly buried in damp soil within forests biome. The second order is the anura whose members are mainly toads and frogs. The order comprises of tailless amphibians mainly frogs. Over 20 families of frogs have been documented globally and 8 families occur in East Africa. Relatively few tadpoles (40%) from East African frogs have been described. About 194 frog species and 9 caecilians have been documented in East Africa [30] and over 100 species in Kenya [7]. Although most of these species occur elsewhere in Africa, about 20 are endemic to Kenya [8]. The largest of the species in Kenya is the huge African Bullfrog, *P. edulis*, who inhabit areas around Tsavo National Parks and can weigh over 1 kg and Snout-Vent Length (SVL) is over 14 cm while the smallest species is the Unguja Puddle Frog, *Phrynobatrachus ungujae* who weigh less than 2 g and its SVL is only 1.6 cm [7].

4.3 Drivers of population decline

Amphibians and reptiles suffer from anthropogenic stress ranging from negative perceptions to impacts of development. Although amphibians are important part of ecosystem many people have negative perception of amphibians and reptiles. They play significant ecosystem functions such as acting as insect and other arthropods pests control, acting as excellent indicators of success in habitats restoration or highlight the degradation of habitats as they are sensitive to changes in the environmental conditions and pollution [14]. As part of ecosystem function, amphibians themselves act as food for secondary consumers such as birds and mammals and thus act as chain for energy transfer from lower tropic level to the higher levels. Toxins from pollutants can easily enter their permeable skins. For example, in an environment where frogs are dying off, this is probably a sign of pollution which have potential effect on humans as well. They also have known and unutilized culinary potential, economic and medicinal values. The skin of some frogs have been discovered to have complex substances such as peptides which have amazingly been used to treat various skin conditions in humans. Going forward, it is predicated that many toxic substances will be found in the skin of most amphibians [28].

Throughout the world, amphibians are suffering an alarming decline due to climate change, increase in ultraviolet radiation and most importantly environmental degradation or shift in land use land cover types [29]. Although most species of frogs have evolved to withstand perturbations, some frogs are becoming less tolerant to factors that was previously in their habitats such as chytrid fungus [31]. Another factor contributing to amphibian habitat destruction is the surface mining as seen in Bamburi area—former mining zones of limestone for cement making. This surface mining not only contribute a direct mechanical threat to amphibians, but is also contribute to water pollution, as well as their habitat. Moreover, some amphibians have been affected by invasive amphibians' species who have considerably pose ecological and socio-economic impact.

4.4 Possible solutions

- Environmental education serves to avert negativity surrounding amphibians.
- Moreover, the support for more amphibian and reptile parks can help draw members of public face to face with amphibians and reptiles and dispel the stereotypes and negativity surrounding amphibians.
- Local engagement in conservation initiatives such as ecotourism ventures, wildlife farming and sustainable practices may provide some sustainable solutions.
- Since the major cause of amphibian decline is the destruction of their habitats, concerted efforts should be focus on wetlands. Since most wetlands occur outside protected areas, some form of extension services should be put in place to create awareness in these areas. While wetlands should be managed as multiple use resource, wetland sanctuaries in areas rich in biodiversity should be established and declare as protected areas or public resource that is governed by public policy.
- Further approaches can include the creation of integrated management authorities through cooperation among all organizations related to wetlands. This should include conservationists, interest holders, government and non-governmental agencies, and ecologists. In order to realize these recommendations, an effective management policies for wetlands should be set up, or reinforced and review the existing policies and where wanting, enact by-laws to protect wetlands from external discharge from industrial effluents, and municipal effluents, commercial and service operations; Repair, servicing of motor vehicle and disposal of industrial chemicals and expired drugs, disposal of waste oils, solid wastes like garbage and plastics, mining activities, damming and irrigation, power generation, agrochemicals such as fertilizers, pesticides, herbicides, livestock grazing and watering.
- Personnel for wetland management should also be trained on management and information disseminate skills on the importance of wetlands and amphibian biodiversity within these wetlands.

5. Conclusion

In the assessment of amphibians' biogeography in major biodiversity hotspots in Kenya, it is evident that their distribution and diversity is influenced by habitat, location of the area, anthropogenic activities in the habitat as well as their level of endemism. Notably, human activities such as land use changes greatly influence the species diversity and richness. For example, in rehabilitated mines in Bamburi, the noted species patterns are possibly due to unfavorable conditions for the tenancy of amphibians that ranges from shelter, the complexity of food web, breeding, hydrological cycle, dispersal corridors in the neighborhood and the edge effect. Moreover, it was evident coastal biodiversity hotspots including Shimba Hills National Reserve and Taita Hill complex ecosystem are leading in amphibian biodiversity and level of endemism. The level of endemism declines drastically inland and towards the drier parts of the country. Anthropogenic activities and climate change is however degrading their habitats and contributing to amphibian decline locally and globally. For example, the notable human activities in most wetland areas in Kenya are, rapid human population growth which has consequently led to unregulated industrial effluents, expanding urban and agricultural areas, poor infrastructural planning and residential waste pose major threats to rivers, lakes and wetlands. Other agents of habitat degradation is the bush burning as a way of land preparation, livestock grazing, drainage of wetlands, and plantation of exotic tree species (Eucalyptus sp.). A concerted efforts are needed to reverse these trends causing amphibian population to decline. Therefore, reversing these trends requires adequate commitment from both government and communities involved.

Author details

Alfayo Koskei
Egerton University, Kenya

*Address all correspondence to: kkalfayo@gmail.com

References

[1] Crump ML. Viual encounter surveys. In: Measuring and Monitoring Biological Diversity for Amphibians. Washington DC: Smithsonian Inst. Press; 1994

[2] Peek R, Yarnell S, Lind A. Visual Encounter Survey Protocol For Rana Boylii in Lotic Environments. Davis, CA: Center for Watershed Sciences, John Muir Institute of the Environment, University of California; 2017. p. 95616

[3] Heyer W et al. Measuring, and monitoring biological diversity, standard methods for amphibians. Argentina: Smithsonian Inst. Press. Editorial Universitaria de la Patagonia; 1994

[4] Howard G. Definition and overview. In: Wetlands of Kenya: Proceedings of a Seminar on Wetlands of Kenya. Gland, Switzerland: IUCN; 1992

[5] da Silva FR, Gibbs JP, Rossa-Feres DdC. Breeding habitat and landscape correlates of frog diversity and abundance in a tropical agricultural landscape. Wetlands. 2011;**31**(6):1079-1087

[6] Laurila A, Aho T. Do female common frogs choose their breeding habitat to avoid predation on tadpoles? Oikos. Apr 1997;**78**(3):585-591 (7 pages)

[7] Spawls S, Drewes RC, Wasonga DV. The Amphibians of Kenya. 2019

[8] Bwong BA et al. Amphibian diversity in Shimba Hills National Reserve, Kenya: A comprehensive list of specimens and species. Journal of East African Natural History. 2017;**106**(1):19-46

[9] Mittermeier RA, Da Fonseca GA, Rylands AB, Brandon K. A brief history of biodiversity conservation in Brazil. Conservation Biology. 2005:601-607

[10] Koskei A et al. Breeding phenology and population dynamics of the endangered Forest Spiny Reed Frog Afrixalus sylvaticus Schiøtz, 1974 in Shimba Hills, Kenya. Journal of Threatened Taxa. 2022;**14**(7):21347-21355

[11] Davis G, Bennum L. Biodiversity overview of Shimba Hills National Reserve, Mkongani Forest Reserve and Mwaluganje Forest Reserve. A report prepared for the Kenya Indigenous Forest Conservation Project (KIFCON). Nairobi: KIFCON; 1993

[12] Luke Q. Annotated checklist of the plants of the Shimba Hills, Kwale District, Kenya. Journal of East African Natural History. 2005;**94**(1):5-120

[13] Norman M. Biodiversity hotspots revisited. Bioscience. 2003;**53**(10):916-917

[14] Mittermeier RA et al. Global biodiversity conservation: The critical role of hotspots. In: Biodiversity Hotspots. Berlin, Heidelberg: Springer; 2011. pp. 3-22

[15] Malonza PK, Lötters S, Measey GJ. The montane forest associated amphibian species of the Taita Hills, Kenya. Journal of East African Natural History. 2010;**99**(1):47-63

[16] Müller H et al. A new species of Boulengerula Tornier (*Amphibia*: *Gymnophiona*: *Caeciliidae*) from an isolated mountain block of the Taita Hills, Kenya. Zootaxa. 2005;**1004**(1):37-50

[17] Loader SP et al. A new brevicipitid species (*Brevicipitidae*: *Callulina*) from the fragmented forests of the Taita Hills, Kenya. Zootaxa. 2009;**2123**:55

[18] Ong'oa DO et al. Diversity and abundance patterns of amphibians in rehabilitated quarries of Bamburi near Mombasa (Kenya). The Wetlands Diversity. 2013;**15**(1):61-72

[19] Malonza PK et al. Diversity and biogeography of herpetofauna of the Tana River Primate National Reserve, Kenya. Journal of East African Natural History. 2006;**95**(2):95-109

[20] Victor Wasonga D et al. Amphibian abundance and diversity in Meru national park, Kenya. African Journal of Ecology. 2007;**45**(1):55-61

[21] Schick S, Veith M, Lötters S. Distribution Patterns of Amphibians from the Kakamega Forest, Kenya. Oxfordshire, United Kingdom: Taylor & Francis and Routledge Imprints; 2005

[22] Ruppert KM, Kline RJ, Rahman MS. Past, present, and future perspectives of environmental DNA (eDNA) metabarcoding: A systematic review in methods, monitoring, and applications of global eDNA. Global Ecology and Conservation. 2019;**17**:e00547

[23] Sloan S et al. Remaining natural vegetation in the global biodiversity hotspots. Biological Conservation. 2014;**177**:12-24

[24] Burgess ND, Clarke GP. Coastal forests of eastern Africa. IUCN-The World Conservation Union: Publications Services Unit; 2000

[25] Myers N et al. Biodiversity hotspots for conservation priorities. Nature. 2000;**403**(6772):853-858

[26] Aron Keche GO, Lekapana P, Macharia G. Status of Wetlands in Kenya and Implications for Sustainable Development. KU Kenya: Ir-Library, Kenyatta University Press; 2007

[27] Walls SC et al. Overcoming challenges to the recovery of declining amphibian populations in the United States. Bioscience. 2017;**67**(2):156-165

[28] Campbell Grant EH, Miller DA, Muths E. A synthesis of evidence of drivers of amphibian declines. Herpetologica. 2020;**76**(2):101-107

[29] Krzikowski M et al. Assessment of the threat status of the amphibians in Vietnam-implementation of the one plan approach. Nature Conservation. 2022;**49**:77-116

[30] Channing A, Howell K. Amphibians of East Africa. Comstock Pub. Associates/ Cornell University Press; 2006

[31] Kielgast J et al. Widespread occurrence of the amphibian chytrid fungus in Kenya. Animal Conservation. 2010;**13**:36-43

Section 4

Ecosystem Services

Chapter 13

Gain and Loss of Forest Cover in Ghana's Forest Reserves in Three Selected National Parks

Kenneth Peprah, Raymond Aabeyir and Bismark Yeboah Boasu

Abstract

Forests play an important role in the ecological, environmental, socio-economic and cultural lives of people. However, human-nature-forest interactions bring imbalances in the state of these resources, hence the need to monitor and manage the forest proactively. This chapter, therefore, assessed the gains and losses of three national parks in Ghana (including Mole, Digya and Kakum) for the period 1986 to 2020. Landsat TM and Landsat8 OLI images were used for the assessment. The Digya National Park has a spatial extent of 4121.89 km^2. In 1986, the extent of the close forest cover was 88.0% of the park and it declined to 82.8% in 2020, a loss of 6.2%. The Kakum National Park covers an area of 463.42 km^2. The close forest cover was 90.15% and 89.52% of the extent of the park in 1986 and 2020, respectively. The Mole National Park covers about 4602.19 km^2. In 1986, 86.73% of the park was open forest, which was reduced to 80.83%, a loss of 5.90%. The study revealed 67% of open forest degradation, 33% loss of close forest and a reported ritual bushfire, wood harvesting and lumbering as the unsustainable practices in the forests. Reclamation of degraded areas is recommended to the Forestry Commission.

Keywords: forest cover, protected areas, forest reserves, gain, loss, degradation

1. Introduction

The forest-human and forest-nature relationships lead to dynamics in the state of many forests across the globe. Forest dynamics thus play a critical role in the services forests provide in support of nature and humankind. The events of climate change and its impacts have increased the need and awareness to monitor the state of forests across the global community.

In spite of the fact that it forms part of the worldwide most valuable resources, the world forests are in a state of fluidity with quickening losses in some regions and gains in others around the world [1]. Forest is a more complex concept ranging from viewpoints of administrative unit, type of land use, and/or type of land cover [2]. The land cover view is however considered for this study. Forest therefore means an ecosystem branded by extensive tree cover, frequently consisting of stands varying in features such as species composition, structure, age class, associated plantations, and wildlife [3]. In each case, forest plays a pivotal role in the development of economies of many countries worldwide [4], especially those within the African continent and particularly the sub-Saharan region where Ghana is located. Forest has been a gainful commodity to the improvement of both humankind and nations and many people have substantial value towards its restoration, conservation and management [5–7]. It provides livelihoods for people, especially the poor [8], serves as attraction for tourists, contributes to the gross domestic product (GDP) and serves as a source of revenue for socio-cultural infrastructure of many nations [9, 10] and Ghana is no exception. Thus, many Ghanaians and the rural folks in particular basically depend on forest productivities for survival and livelihood ventures.

Forest productivity is dependent on the state and nature of the forest [11]. In their study on the importance of forest structure to biodiversity-productivity relationships, for instance, [12] find various relationships (i.e. increasing, constant and decreasing) between species richness and forest productivity for different forest structure classes. This suggests that the structure of any forest including the reserves might determine its diversity. An assessment of the state in terms of gain or loss in extent and quality of forests has therefore become imperative for forest restoration, conservation and management purposes. However, the trend of forest gain and loss indicates a serious global consequence due to the continuous deterioration of forest areas [13, 14]. Hansen et al. [1] conclude that the forest regions have experienced theatrical loss in the last three decades, when they find higher gross forest cover losses compared to the gains in their study on global forest cover based on the analysis of Landsat images across the boreal forests in some selected countries.

Although forests have generally been depleting and many of the reserves are being threatened by both natural [15–17] and anthropogenic factors worldwide [18–21], the menace has been pronounced in Ghana thwarting national commitments to the United Nations Framework Convention on Climate Change (UNFCC) and Intergovernmental Panel on Climate Change (IPCC), national development, making the livelihoods of many forest-dependent populace extremely vulnerable [22] and eventually exposing them to climate change risks in the last three decades [23]. Ghana is a net emitter of CO_2 emissions and contributes to the global imbalance of greenhouse gases and their effects on climate change [24, 25].

The situation of forest degradation even becomes more critical at the beginning of the twentieth century when the pressures posed by anthropologic activities (mainly lumbering, cocoa farming and mining) on the natural forest pushed for the demarcation of portions of the natural forests as forest reserves (protected areas). This was done to protect ecologically sensitive areas, habitat for endangered species and enhance the tourism potential of the country. Extant studies have extensively covered the conservation and management of these forest reserves [5, 26, 27]. However, assessment of the trend of gain and loss of forest reserves and potential implications of the dynamics for forest policy advancement is less studied in Ghana [28], and hence remains a niche in the literature. That is, although some countries in the world have progressed from loss to gain in forest cover [29], Ghana's situation is yet to be ascertained. Therefore, the study was aimed to assess the state of the Ghanaian forest reserves over the last three decades. To accomplish this, we focused on three major forest reserves—Mole, Digya and Kakum National Parks, and employed remote sensing and geographic information systems methods [30–32] to ascertain the trend of the gain and loss (i.e. degree of degradation) of these forest reserves and how the trend potentially influence forest policy development in Ghana and even beyond. By this, we contribute first to other scholars' proposal of integrating the political, socio-economic and methodological aspects to upscale restoration efforts in tropical forest regions around the globe [33]; forest transformation agenda [34]; and the achievements of national and international conservation goals and treaties [35]. The rest of this chapter covers methodology, findings and discussion, and conclusion.

2. The study context and methodology

2.1 Study context

Ghana is located in West Africa and lies between Latitudes 4°44′ N and 11°15′ N and longitudes 3°14′ W and 1°12′ E with an estimated area of 23.85 million ha. It is bordered to the North by Burkina Faso, to the West by Cote d'Ivoire and Burikina Faso, to the East by Togo and to the South by the Gulf of Guinea [25, 36]. The country is endowed with diverse ecosystems with a diversity of plant and animal species. There are 21 protected areas which constitute 5.6% of the country's total surface area that is being managed by the Wildlife Division of the Forestry Commission [37]. The nation's forest resources are under the pressures of mining, agricultural encroachment, legal and illegal logging, woodfuel harvesting, wildfires, and infrastructure development [38]. The deforestation and forest degradation rate as of 2017 was estimated at 3.51% annually. Protected areas are located in all the ecological zones of the country but three of these national parks were selected for the study because of their unique ecological, environmental and socio-economic characteristics. These parks are the Digya, Kakum and Mole National Parks (**Figure 1**).

2.2 Methods

Remote sensing and geographic information techniques were used to assess the gain and loss in the forest cover between 1986 and 2020. These timelines were chosen based on availability and quality of images. Besides, the parks are protected areas and long-term changes are critical for sustainable management decisions. The forest reserves in the country have come under pressure to support the needs of both individuals and the nation as a whole. The same forest reserves are expected to provide ecological and climate change issues. The dynamics of the forests in terms of gain and loss are critical in informing how the current forest reserves are managed for the benefit of the people and nature. Three national parks were selected with one in the northern Savannah area (Mole National Park), one in the forest area (Digya National Park) and one in the coastal area (Kakum National Park) for the assessment of the forest dynamics in Ghana. Landsat images were obtained from the United States Geological Survey (USGS) website and used for the classification. Details of the images are described in **Table 1**. The images were classified in Erdas Imagine, and maps were processed in ArcMap. The quality of the classified images was assessed by comparing random points on the classified images with the same points on high-resolution images from Google Earth [30, 39].

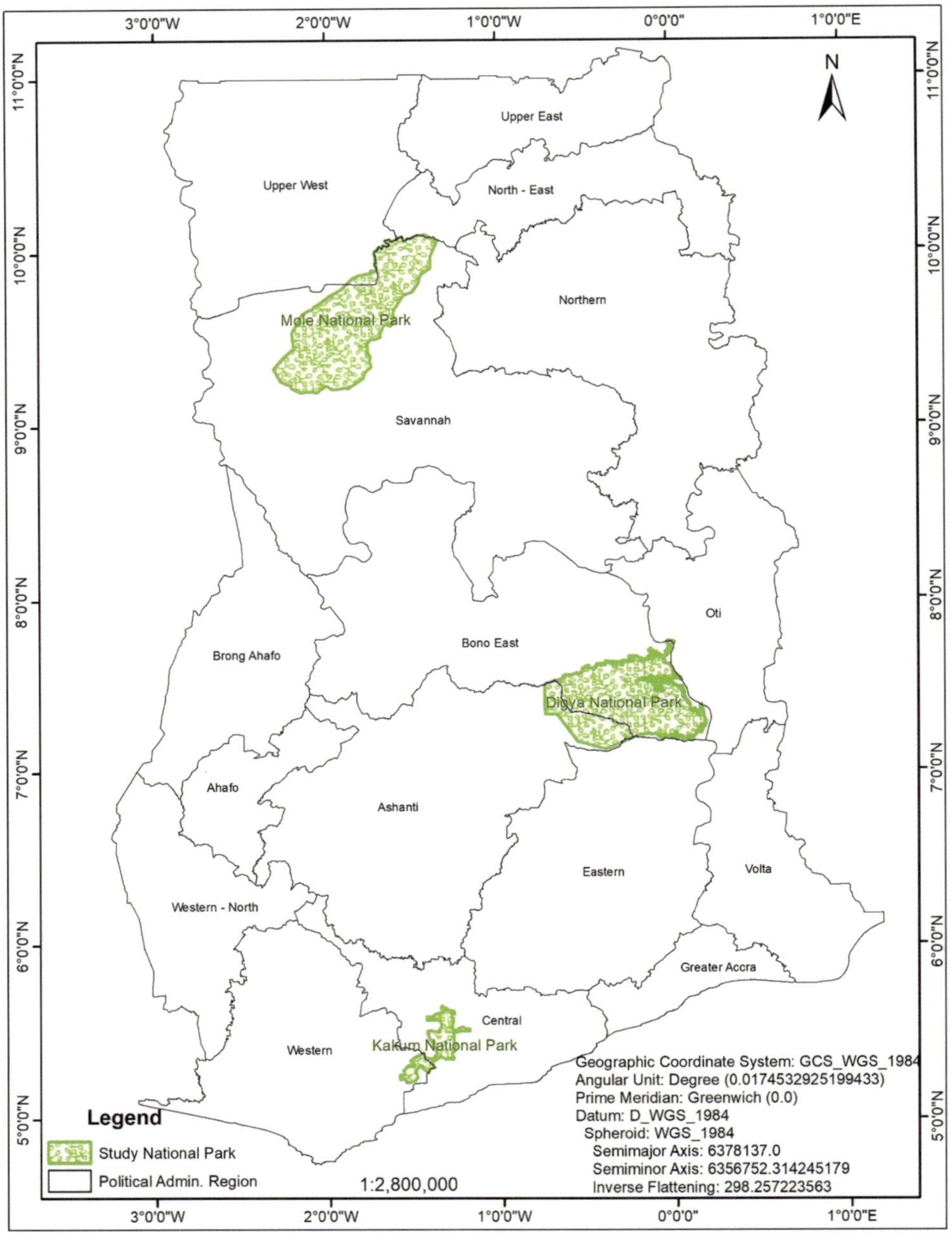

Figure 1.
The selected national parks in the national context. Source: Open-source Map, 2022.

3. Findings

3.1 State of the land cover of the forest reserves

3.1.1 Digya National Park

The Digya National Park (DNP) is the oldest and second-largest national park in Ghana. The park was established in 1909 with a spatial coverage of 650 km^2 [40].

Park	Image	Path	Row	Date	Cloud cover (%)
Digya	Landsat5_TM	193	055	11/01/1986	1.00
		194		05/02/1986	33.00
	Landsat8_OLI	193	055	03/02/2020	0.10
		194		25/01/2020	1.00
Kakum	Landsat5_TM	194	056	29/12/1986	0.00
	Landsat8_OLI			09/01/2020	27.99
Mole	Landsat8_OLI	195	053	18/01/1986	0.00
				01/02/2020	0.13
	Landsat8_OLI	195	054	03/02/1986	1.00
				01/02/2020	0.26

Source: Field survey, 2022.

Table 1.
Characteristics of satellite images used for the study.

When the Volta Lake was created in 1965, DNP was expanded to 3478.3 km^2 to provide protection for parts of the Lake. The Park was gazetted in 1971 as it has become an important habitat for wild animals and provides ecological benefits to Volta Lake. The park contains endangered animal species namely elephants, manatees, and black-and-white colobus. The sustainability of the park is thwarted by poaching of the animals for bush meat and ivory, large-scale grazing by cattle, bush fires initiated for hunting purposes, and logging for wood [41].

The analysis of the satellite images revealed that the park has a spatial extent of 4121.89 km^2. The variation, from what is reported in literature, is attributable to fuzziness in boundary between the park and surrounding vegetation. In 1986 and 2020, the park was dominated by close forest although the close forest has declined in extent while the open forest has increased in extent (**Figures 2** and **3**). The decrease in the extent of the close forest could be as a result of bushfires, and wood logging as observed by Dowsett-Lemaire and Dowsett [41].

Assessment of the extent of the forest cover revealed that 88.0% of the forest was close forest as of 1986 (**Figure 4**). The extent of the close forest declined to 82.8% as of 2020. The decline of 6.2% (**Figure 5**) in the close forest could be attributed to the impacts of wildfires and logging which characterized the park.

3.1.2 Kakum National Park

Kakum National Park (KNP), located in the coastal environs of the Central Region of Ghana, covers an area of 360 km^2 [42]. The Kakum National Park was established in 1931 as a reserve and was gazetted as a national park in 1992 [43]. The park is located about 30 km from Cape Coast, the Central Regional capital. The vegetation of the park is thick evergreen and semi-deciduous forests. Trees found in the forest include Wawa, Odum, Mahogany, bamboo, raffia palms, etc. The fauna includes the forest elephant, bongo, leopard, giant forest hog, duiker and about 200 species of birds [44]. The Park protects the headwaters of River Kakum. It is also a habitat for endangered wildlife species namely African forest elephants, Diana monkeys, yellow-back duikers, giant bongo antelopes, birds and butterflies [45].

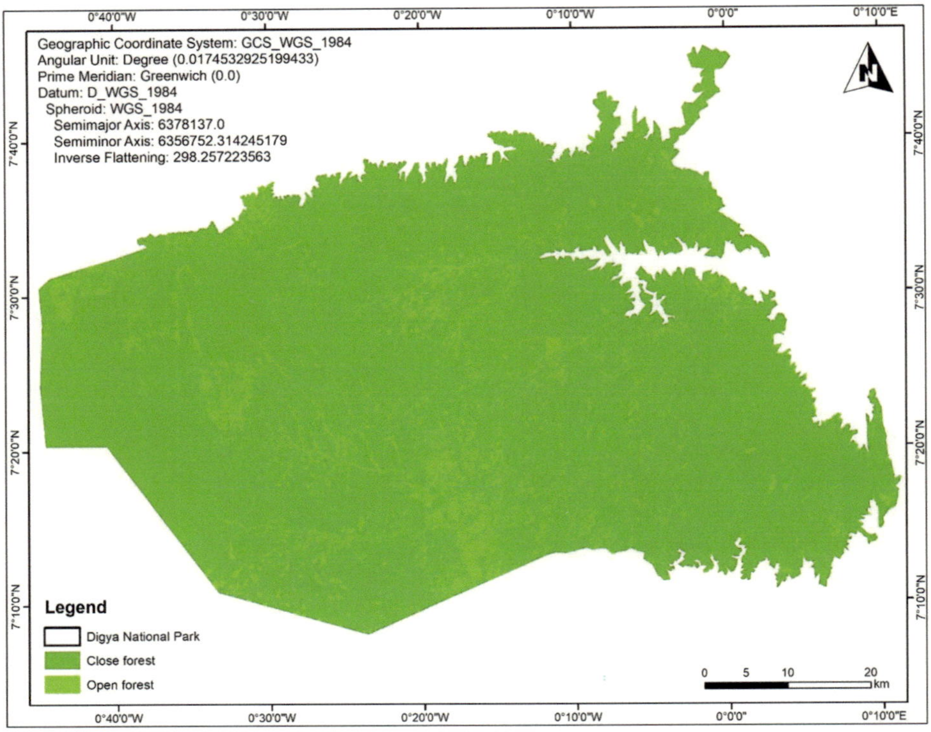

Figure 2.
The land cover state of the Digya National Park as of 1986. Source: USGS, 1986.

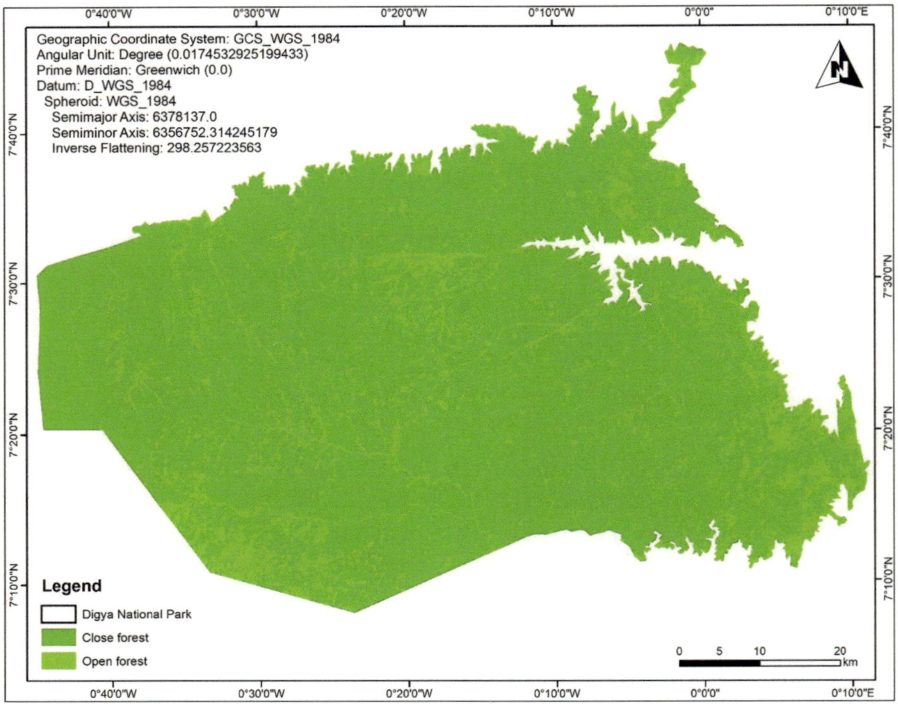

Figure 3.
The land cover state of the Digya National Park as of 2020. Source: USGS, 2022.

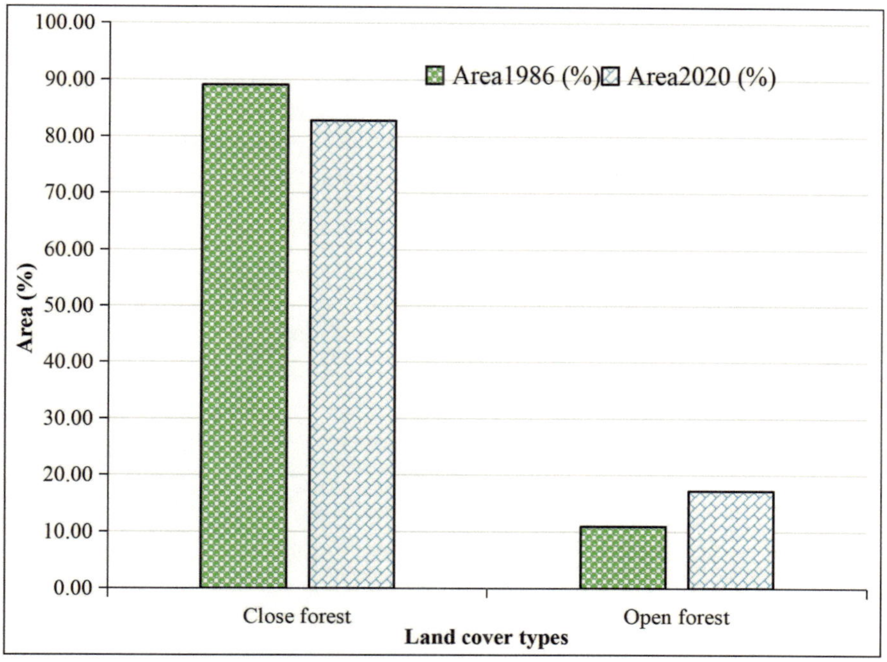

Figure 4.
Comparison of extent of the land cover types of the Digya National Park in 1986 and 2020. Source: Authors' construct, 2022.

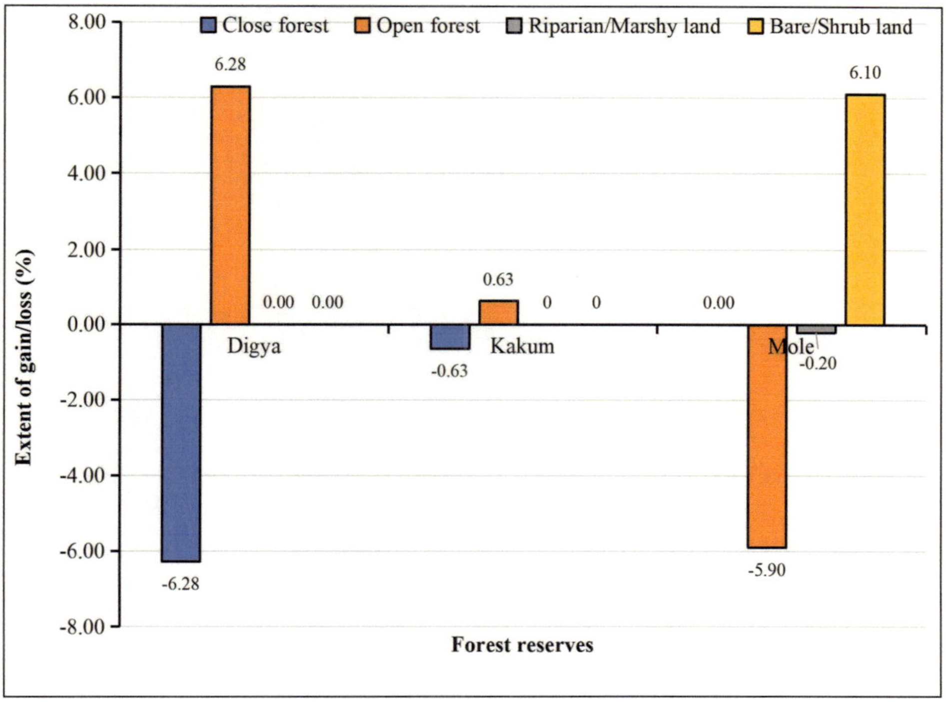

Figure 5.
The gain and loss in the land cover types in Digya, Kakum and Mole National Parks for the period 1986 to 2020. Source: Authors' construct, 2022.

The spatial assessment of the Kakum National Park showed that the park is about 463.424 km^2, which is not so different from the 360km^2 that is reported by Monney and Dakwa [42]. Two main land cover types were identifiable on the raw images of both 1986 and 2020: close and open forests. The close forest covered 327.637 km^2 while the open forest covered 35.787 km^2 (**Figure 6**). The spatial distribution of the land cover of the Kakum National Park in 1986 revealed that

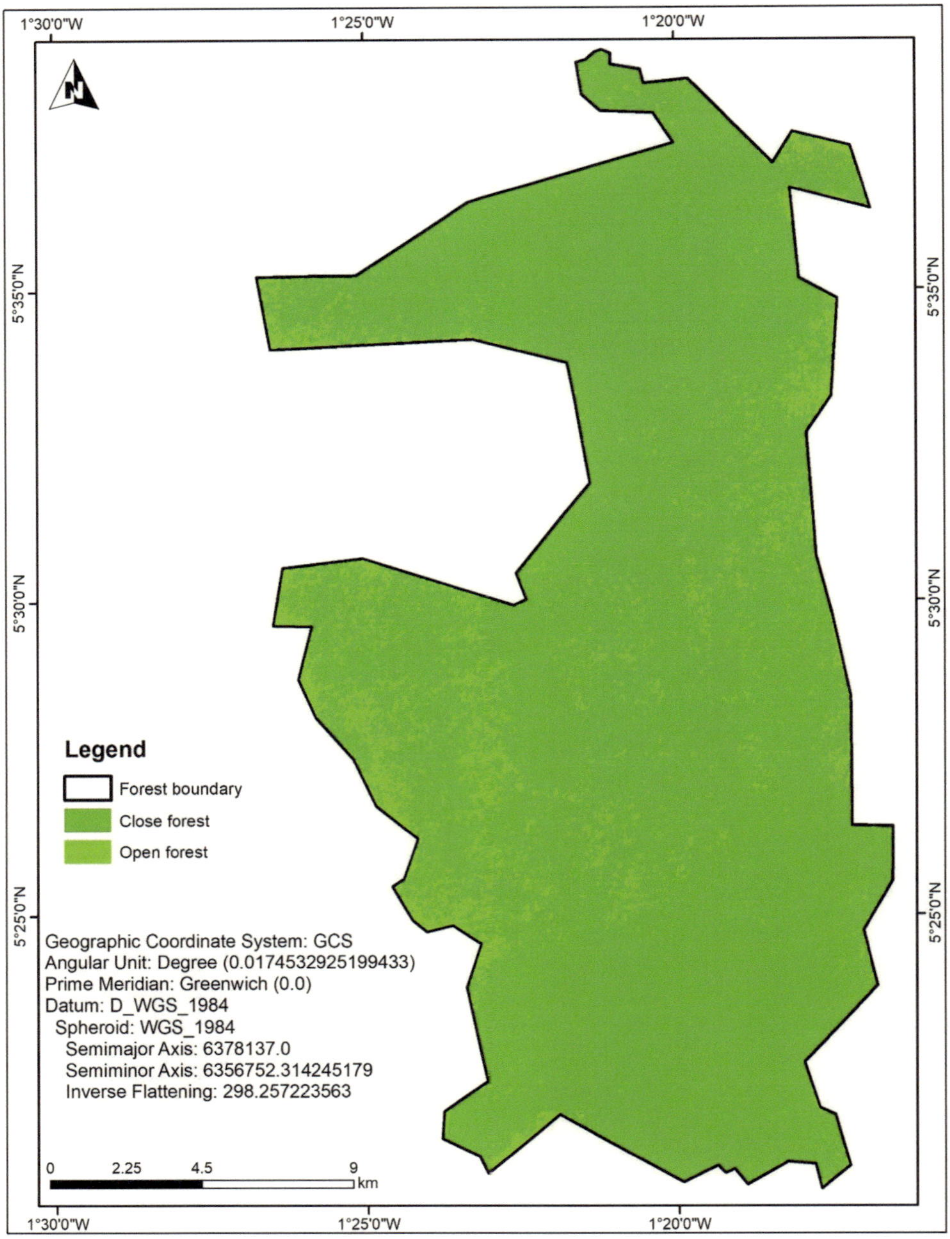

Figure 6.
State of the land cover of the Kakum National Park in 1986. Source: USGS, 1986.

the park is a large close forest with patches of secondary forest dominated in the eastern, middle and western parts of the park. As of 2020, the spatial distribution of the land cover of the park was not too different from that in 1986 (**Figure 7**). However, the dominance of the open forest was in the north-eastern and -western parts of the park.

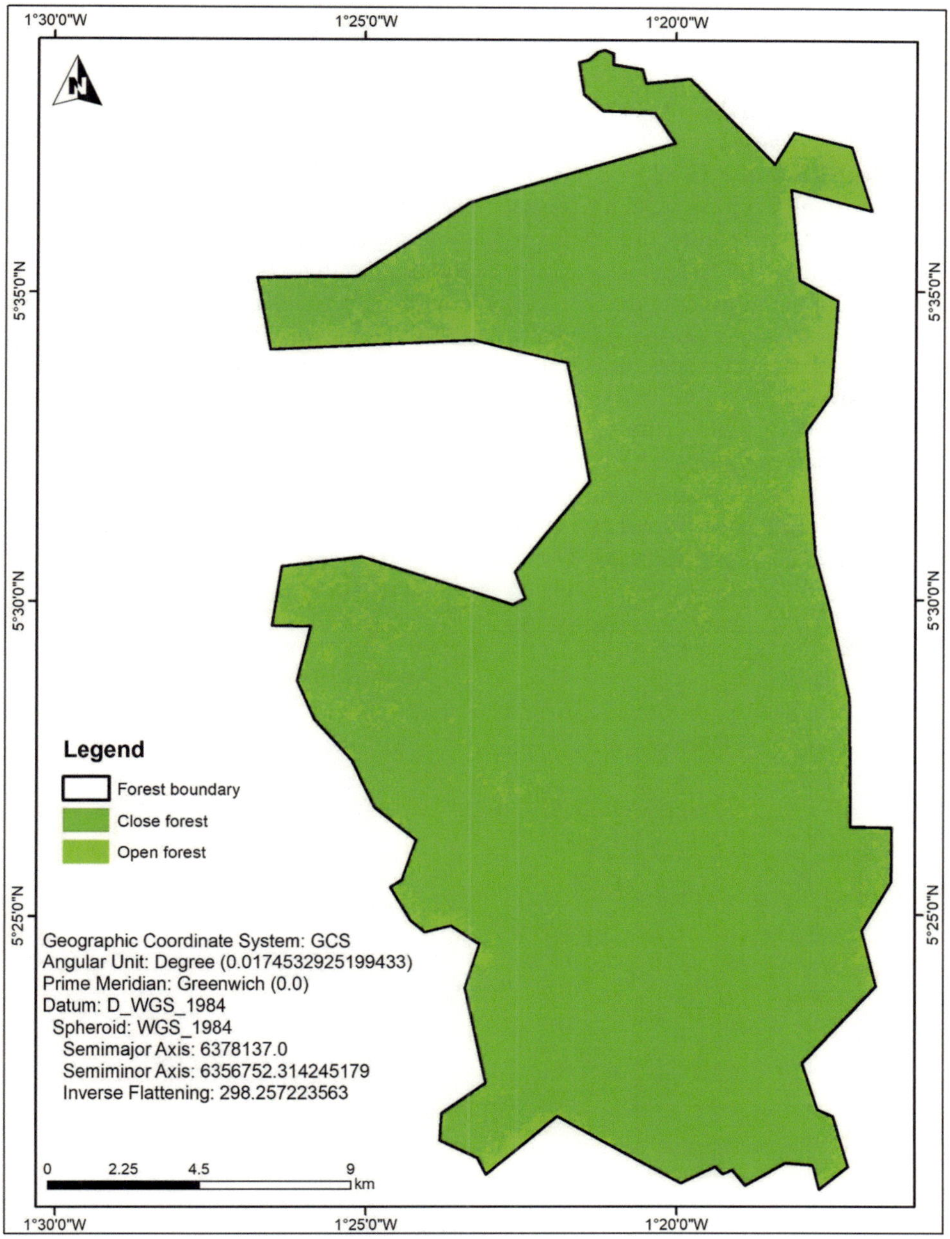

Figure 7.
State of the land cover of the Kakum National Park in 2020. Source: USGS, 2020.

The relatively stable nature of the Kakum National Park in terms of the two land cover types is attributable to the absence of two main anthropogenic pressures that influence the state of vegetation cover in Ghana. These are land conversion and bush-fires. This is reported by the International Union for Conservation of Nature [37]. In their report, these pressures were absent in the park. However, the main pressure in the park is poaching of game.

In **Figure 8**, the close forest cover was 90.153% and 89.523% for 1986 and 2020 respectively. In **Figure 5**, the gain and loss were 0.63% equally.

3.1.3 Mole National Park

The Mole National Park (MNP) is the largest national park in Ghana. The Mole National Park covers about 4577 km^2 and lies between latitudes 9°11′ and 10°10′N and longitudes 1°22′ and 2°13′W [46]. The Park is situated in the West Gonja Municipality of the Savannah Region of Ghana. It stretches to about 20 km north of Damongo, the capital of the Municipality and the Region. It was gazetted as a national park in 1971 for its outstanding wildlife conservation. The park contains a wide range of wildlife species which include elephants, hartebeests, kobs, waterbucks, bushbucks, warthogs, roan antelopes, duikers, oribis, baboons, patas monkeys, vervet monkeys, red-throated bee-eaters, Abyssinia ground hornbills, saddle-billed storks, agama lizards, crocodiles and bush snakes [47]. The Park is also rich in Baobab (*Adansonia Digitata*), Dawadawa (*Parkia biglobosa*), Silk Cotton (*Ceiba pentrandra*), *Burkea Africana, Lannea acida,* Shea tree (*Vitellaria paradoxa*), *Burkea africana* and *Tcrminalia aviccnnioides*. This indicates the ecological and biodiversity importance of the park.

The spatial distribution of the land cover types in 1986 is shown in **Figure 9** and that of 2020 is in **Figure 10**. The total estimated area of the park from the

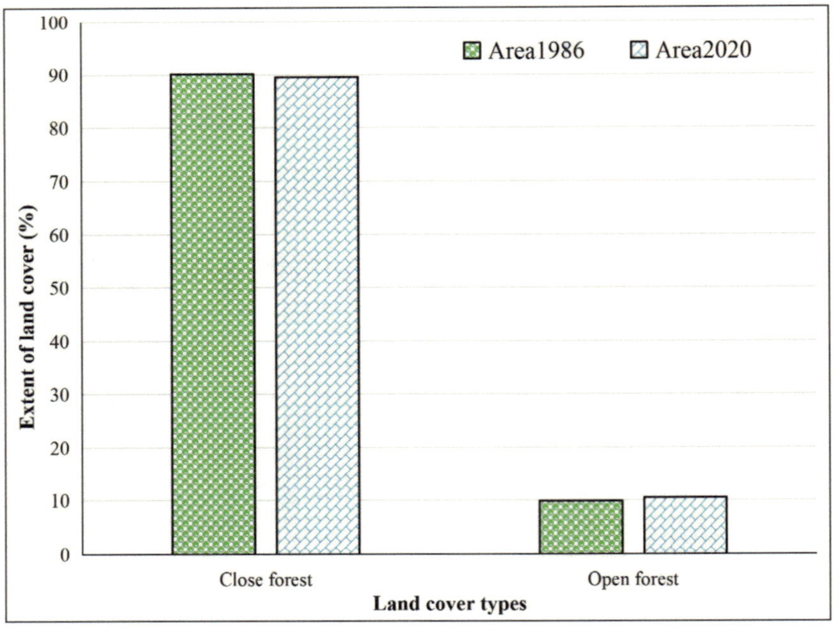

Figure 8.
Comparison of the extent of the land cover types of the Kakum National Park for 1986 and 2020. Source: Authors' construct, 2022.

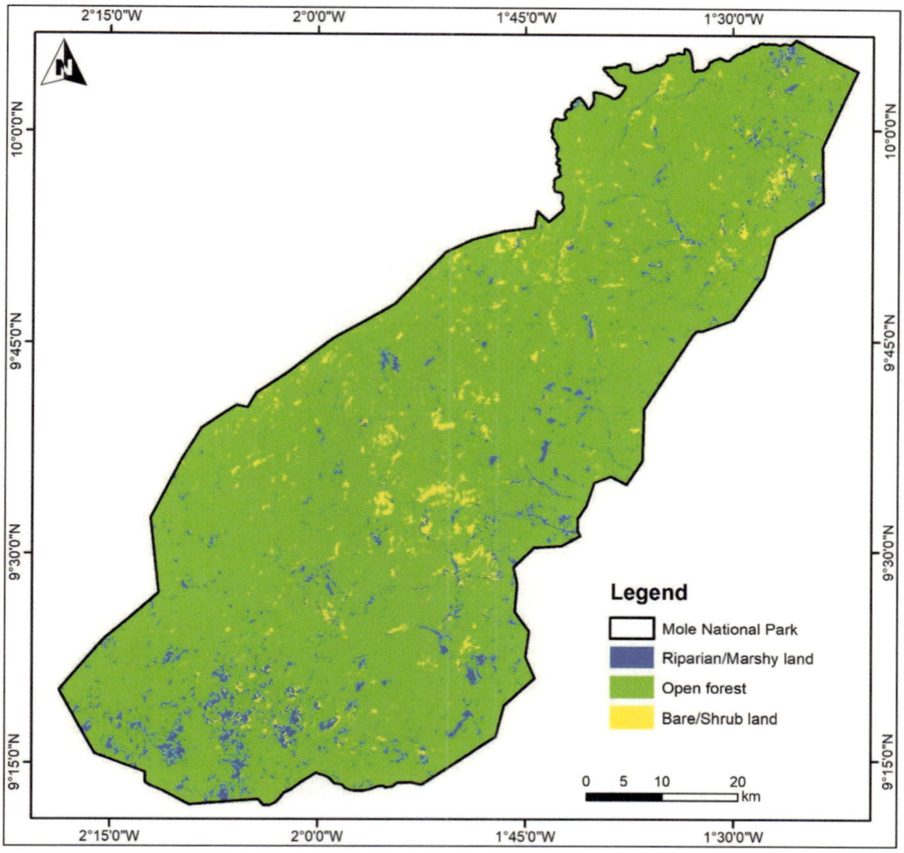

Figure 9.
State of land cover of mole National Park in 1986. Source: USGS, 1986.

image is 4602.185 km^2. The difference in area is attributed to differences in the definition of the boundary of the park. The park is dominated by open forest. The other classes are patches of riparian forest, marshy areas, bare grounds and shrubland.

In the year 1986, 86.73% of the park was open forest with the rest of the extent being riparian/marshy land, bare land and shrubland (**Figure 11**). By 2020, the total forest area constituted about 80.83%, a loss in open forest area of 5.90% (271 km^2) (**Figure 5**). The difference in the extent of the due to the human pressures of woodland conversion, bushfires, and logging for local timber and charcoal production. IUCN [37] rated woodland conversion and annual bushfires high among the pressures that influence the state of the park negatively.

4. Discussion

The study found that about 67% of the national parks use unsustainable conservation practices leading to a high forest cover loss as exemplified in Digya and Mole National Parks. In the case of Mole National Park, located in the interior savannah ecological zone, managers face the problems of annual bushfire, forest conversion to farmland, wood harvesting for charcoal production, and timber exploitation for

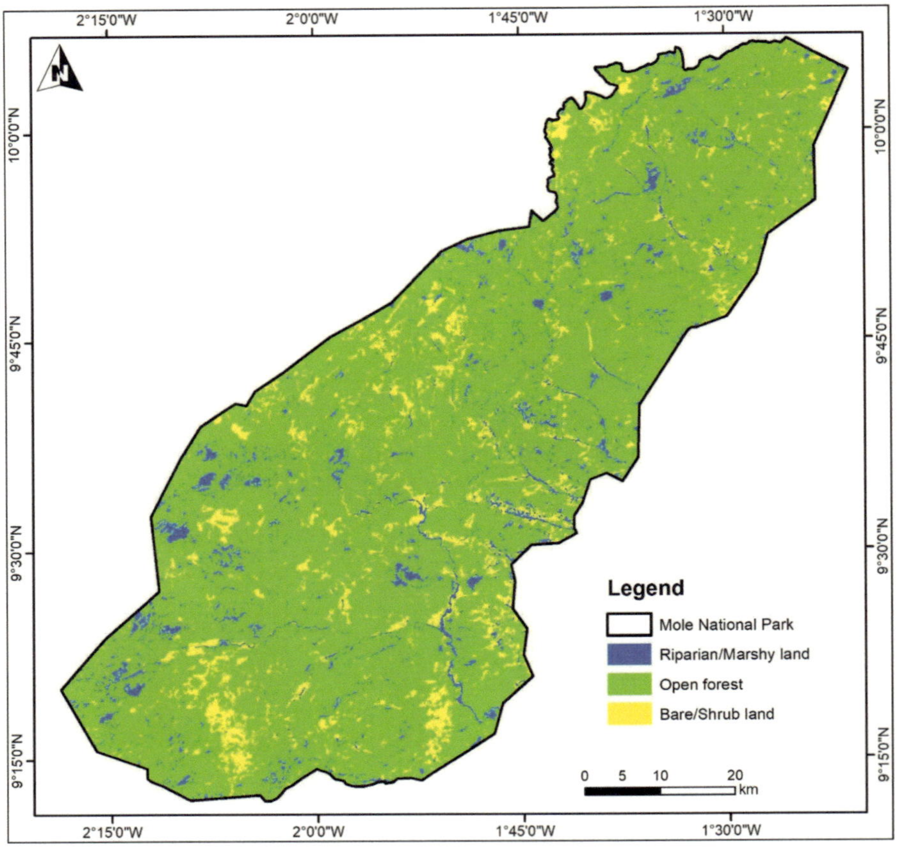

Figure 10.
State of land cover of mole National Park in 2020. Source: USGS, 2020.

local building material and export. Similarly, the Digya National Park, situated in the forest-savannah ecotone, suffers related pressures such as bushfire, logging for fuel wood, and intensive grazing by cattle [37]. The impacts of these pressures are the conversion of close forest to open forest and bare ground, particularly in the Mole National Park. The most affected tree species are rosewood (*Pterocarpus erinaceus)* (lumbering and export), *Burkea africana* (charcoal production) and *Kaya senegalensis* (roofing). There is harvesting of rosewood resulting in a decline in the close/open forest cover in the Digya and Mole National Parks making the rosewood an endangered species [48]. Also, there are 33% unsustainable conservation measures resulting in a minimal close forest cover loss of 0.63% in the Kakum National Park. The reasons for the low forest cover loss are that the communities surrounding the Kakum National Park appear to observe protected area arrangement. There is also strict enforcement of protective regulations due to the importance of the Park to national tourism development. Furthermore, the site of the Kakum National Park close to the rainforest makes natural regeneration and shedding of leaves happen simultaneously. Hence, the forest cover loss is minimal. Another finding is that the forest cover change is not only attributed to anthropogenic factors but climatic and ecological factors as well. For example, the bare grounds in the Mole National Park result from insufficient natural regeneration due to the prolonged dry season (climate-induced).

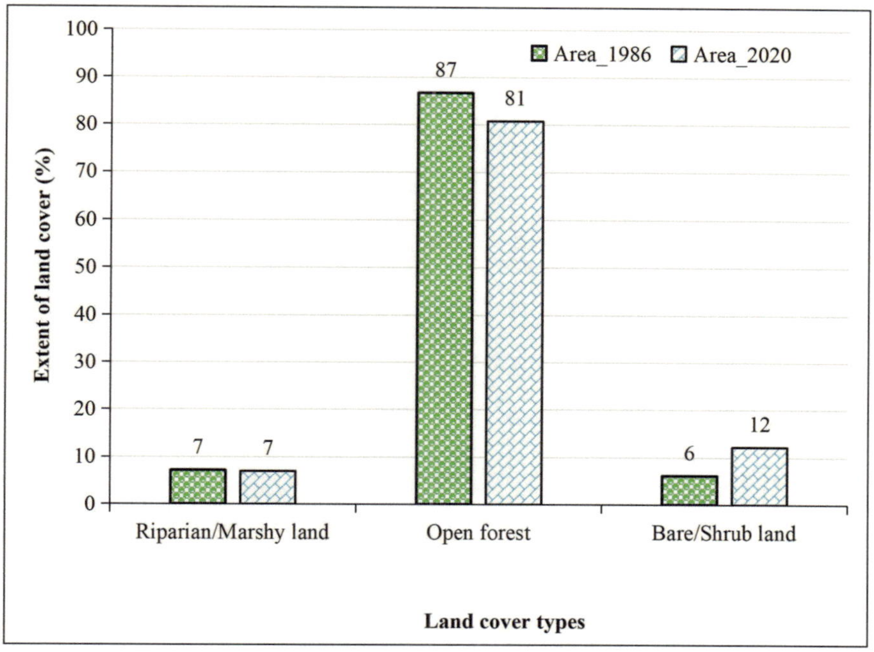

Figure 11.
Extent of the land cover types of Mole National Park for 1986 and 2020. Source: Authors' construct, 2022.

5. Conclusion

This chapter assessed the gain and loss in the forest cover of three selected national parks in Ghana. The Mole National Park was in the northern Savannah zone, the Digya National in the middle zone and the Kakum National Park in the coastal zone. These parks have unique ecological, environmental and socio-economic characteristics. Excessive disturbance in the forest cover state will compromise these characteristics. The analysis of the gain and loss in the forest cover is of relevance for the effective management of these natural resources. As expected, each of the parks exhibited dynamism in terms of gains and losses for the period 1986 to 2020. No detectable grass/bare areas were found in the Digya and the Kakum National Parks. Detectable grass/bare areas were recorded in the Mole National Park.

The Digya National Park experienced the largest extent of gain and loss. The close forest lost 6.28% of its initial cover to open forest. The Kakum National Park experienced the least of changes. The park lost 0.63% of its close forest to open forest for the period.

The Mole National Park experienced the most critical gains and losses in the forest cover. Bare/shrubland gained over open forest. The bare/shrubland gained a little above 6%. The losses forest cover in each of these parks may appear small but they make the parks vulnerable to the pressures of deforestation and degradation. It is therefore recommended that Forestry Commission of Ghana through its divisions should initiate plantation activities in the degraded areas of the park or strengthen the protective measures to allow and maximize natural regeneration in the parks. It is further recommended that similar studies be conducted in the other parks to ascertain the forest cover dynamism for effective management decisions.

Author details

Kenneth Peprah*, Raymond Aabeyir and Bismark Yeboah Boasu
SD Dombo University of Business and Integrated Development Studies, Wa, Ghana

*Address all correspondence to: kpeprah@ubids.edu.gh

References

[1] Hansen MC, Potapov PV, Moore R, Hancher M, Turubanova SA, Tyukavina A, et al. High-resolution global maps of 21st-century forest cover change. Science. 2013;**342**(6160):850-853

[2] Lund HG. When is a forest not a forest? Journal of Forestry. 2002;**100**(8):21-28

[3] Helms JA, editor. The Dictionary of Forestry. Bethesda, MD: Society of American Foresters; 1998

[4] Schweiger AH, Svenning J-C. Analogous losses of large animals and trees, socio-ecological consequences, and an integrative framework for rewilding-based megabiota restoration. People Natural. 2020;**2**:29-41

[5] Abdul-Kadri Y, Zakaria A, Boasu BY. Local actors in the co-management of mole national park and the impacts associated with it. Ghana Journal of Geography. 2021;**13**(3):200-230

[6] Mattsson E, Ostwald M, Nissanka SP. What is good about Sri Lankan homegardens with regards to food security? A synthesis of the current scientific knowledge of a multifunctional land-use system. Agroforestry Systems. 2018;**92**:1469-1484. DOI: 10.1007/s10457-017-0093-6

[7] Nath TK, Jashimuddin M, Kamruzzaman M, Mazumder V, Hasan MK, Das S, et al. Phytosociological characteristics and diversity of trees in a comanaged protected area of Bangladesh: Implications for conservation. Journal of Sustainable Forestry. 2016;**35**:562-577. DOI: 10.1080/10549811.2016.1231615

[8] FAO. Chapter 2: Historical overview of the emergence and evolution of CBF. In: The State of the World's Forests. Forests, biodiversity and people. Rome: Food Agricultural Organization; 2020. DOI: 10.4060/ca8642en

[9] Paradis E. Forest gains and losses in Southeast Asia over 27 years: The slow convergence towards reforestation. Forest Policy and Economics. 2021;**122**:102332

[10] Meyfroidt P, Rudel TK, Lambin EF. Forest transitions, trade, and the 580 global displacement of land use. Proceedings of the National Academy Science USA. 2010;**107**:20917-20922. DOI: 10.1073/pnas.1014773107

[11] Cáceres MEDS, Aptroot A, Lücking R. Lichen fungi in the Atlantic rain forest of Northeast Brazil: The relationship of species richness with habitat diversity and conservation status. Brazilian Journal of Botany. 2017;**40**(1):145-156

[12] Bohn FJ, Huth A. The importance of forest structure to biodiversity–productivity relationships. Royal Society Open Science. 2017;**4**:160521. DOI: 10.1098/rsos.160521

[13] FAO. Global Forest Resources Assessment. Rome: UN Food and Agriculture Organization; 2015

[14] Potapov PV, Dempewolf J, Talero Y, Hansen MC, Stehman SV, Vargas C, ..., Zutta BR. National satellite-based humid tropical forest change assessment in Peru in support of REDD+ implementation. Environmental Research Letters. 2014;9(12):124012

[15] Khan D, Muneer MA, Nisa ZU, Shah S, Amir M, Saeed S, et al. Effect of climatic factors on stem biomass and carbon stock of Larix gmelinii and Betula platyphylla in Daxing'anling mountain

of Inner Mongolia, China. Advanced Meteorological. 2019;**2019**:5692574

[16] Mitchard ETA. The tropical forest carbon cycle and climate change. Nature. 2018;**559**:527-534

[17] Gogoi A, Sahoo UK, Singh SL. Assessment of biomass and total carbon stock in a tropical wet evergreen rainforest of Eastern Himalaya along a disturbance gradient. Plant Biology and Soil Health. 2017;**4**:1-8

[18] Oduro-Appiah J, Agyemang-Duah W. Identifying spatially-explicit land use factors associated with forest patch sizes in a forest reserve in Ghana. Land Use Policy. 2021;**101**:1-10

[19] Sapkota RP, Stahl PD, Norton U. Anthropogenic disturbances shift diameter distribution of woody plant species in Shorea robusta Gaertn. (Sal) mixed forests of Nepal. Journal of Asia Pacific Biodiversity. 2019;**12**:115-128

[20] Agyemang-Dua W, Oduro-Appiah J, Adei D. Protecting the patches from the footprints: Examining the land use factors associated with forest patches in Atewa range forest reserve. BMC Ecological Evolution. 2016;**21**(1):1-13. DOI: 10.1186/s12862- 021-01758-0

[21] Kusimi JM. Characterizing land disturbance in Atewa range forest reserve and buffer zone. Land Policy. 2015;**49**:471-482

[22] Boafo J. The Impact of Deforestation of Forest Livelihoods in Ghana. The Africa Portal Backgrounder. Johannesburg; 2013. p. 49. Available from: https://www.africaportal.org/publications/the-impact-of-deforestation-of-forest-livelihoods-in-ghana/

[23] Van der Werf GR, Morton DC, DeFries RS, Olivier JGJ, Kasibhatla PS, Jackson RB, et al. CO2 emissions from forest loss. Nature Geoscience. 2009;**2**:737

[24] Ghana. Ghana's National Forest Reference Level. Arlington, USA: National REDD+ Secretariat, Forestry Commission; 2017. Available from: www.winrock.org

[25] Forestry Commission The Forestry Commission, Republic of Ghana: Development of Reference Emissions Levels and Measurement, Reporting and Verification System in Ghana. Helsinki, Finland. 2015

[26] Vacik H, Lexer MJ. Past, current and future drivers for the development of decision support Systems in Forest Management. Scandinavian Journal of Forest Research. 2014;**29**(sup1):2-19

[27] Kusimi JM. Assessing land use and land cover change in the Wassa west district of Ghana using remote sensing. GeoJournal. 2008;**71**(4):249-259

[28] Oduro-Appiah J, Agyemang-Duah W, Sobeng AK, Kpienbaareh D. Analysing patterns of forest cover change and related land uses in the Tano-Offin forest reserve in Ghana: Implications for forest policy and land management. Journal of Trees, Forests and People. 2021;**2021**:100105

[29] Pendrill F, Persson UM, Godar J, Kastner T. Deforestation displaced: Trade in forest-risk commodities and the prospects for a global forest transition. Environmental Research Letters. 2019;**14**(5):055003

[30] Galiatsatos N, Donoghue DN, Watt P, Bholanath P, Pickering J, Hansen MC, et al. An assessment of global forest change datasets for national forest monitoring and reporting. Remote Sensing. 2020;**12**(11):1790

[31] Mitchell N, Lung T, Schaab G. Tracing Significant Losses and Limited Gains in Forest Cover for the Kakamega-Nandi Complex in Western Kenya across 90 Years by Use of Satellite Imagery, Aerial Photography and Maps. In Proceedings of the ISPRS (TC7) Mid-Term Symposium Remote Sensing: From Pixels to Processes, Enschede, The Netherlands, 8-11 May 2006. pp. 8-11

[32] Lung T, Schaab G. Change-Detection in Western Kenya: The Documentation of Fragmentation and Disturbance for Kakamega Forest and Associated Forest Areas by Means of Remotely Sensed Imagery. In: ISPRS Archives Vol. XXXV Part B (DVD), Proceedings of the ISPRS XXth Congress. Istanbul, Turkey; 2004

[33] Rodrigues RR, Gandolfi S, Nave AG, Aronson J, Barreto TE, Vidal CY, et al. Large-scale ecological restoration of high-diversity tropical forests in SE Brazil. Forest Ecology and Management. 2011;**261**(10):1605-1613

[34] Sloan S, Meyfroidt P, Rudel TK, Bongers F, Chazdon R. The forest transformation: Planted tree cover and regional dynamics of tree gains and losses. Global Environmental Change. 2019;**59**:101988

[35] Rahman S, Alam A, Salekin S, Belal AH, Saifur R. The COVID-19 pandemic: A threat to forest and wildlife conservation in Bangladesh? Journal of Trees, Forests and People. 2021;**2021**:10019

[36] Ministry of Lands and Natural Resources. The State of the World's Forest Genetic Resources: Ghana Country Report. Rome: Ministry of Lands and Natural Resources; 2012. Available from: https://www.fao.org/3/i3825e/i3825e.pdf

[37] IUCN. Parks and Reserves of Ghana: Management Effectiveness Assessment of Protected Areas. Ouagadougou, BF: UICN/PACO; 2010

[38] Powell RL, Matzke N, de Souza C, Clark M, Numata I, Hess LL, et al. Sources of error in accuracy assessment of thematic land-cover maps in the Brazilian Amazon. Remote Sensing of Environment. 2004;**90**:221-234

[39] Twumasi YA, Coleman TL, Manu A. Biodiversity management using remotely sensed data and GIS technologies: The case of Digya National Park, Ghana. In: Proceedings of the 31st International Symposium on Remote Sensing of Environment. June. Saint Petersburg, Russia Federation; 2005. pp. 20-21

[40] Dowsett-Lemaire F, Dowsett RJ. Exploration of Digya National Park, Ghana (January 2005, March 2008 and March 2009), with special reference to birds. Dowsett-Lemaire Miscellaneous Report. 2009;**57**:24

[41] Monney KA, Dakwa KB. Prospects and potentials of Kakum Conservation Area, Ghana. Journal of Ecology and the Natural Environment. 2014;**6**(5):140-153

[42] Twerefou DK, Adjei-Ababio DK. An economic valuation of the Kakum National Park: An individual travel cost approach. African Journal of Environmental Science and Technology. 2012;**6**(4):199-207

[43] Appiah-Opoku S. Rethinking ecotourism: The case of Kakum National Park in Ghana. African Geographical Review. 2004;**23**(1):49-63. DOI: 10.1080/19376812.2004.9756178

[44] Doe EK. Cocoa Agroforestry Investment in Buffer Zones of a Forest Reserve-the Ghanaian Context. Germany: LAMBERT Academic Publishing; 2013

[45] Doe EK, Aikins BE, Njomaba E, Owusu AB. Land use land cover change within Kakum conservation area in the

Assin South District of Ghana, 1991-2015. West African Journal of Applied Ecology. 2018;**26**(SI):87-99

[46] Mole National Park. Management Plan. Damango: Wildlife Division; 2011

[47] Acquah E, Dearden P, Rollins R. Nature-based tourism in Mole National Park, Ghana. African Geographical Review. 2016;**35**(1):53-69. DOI: 10.1080/19376812.2015.1088389

[48] IUCN. The Red List of Threatened Species. International Union for Conservation Nature and Natural Resources: Washington, DC; 2020

Chapter 14

Ecosystem Services in the Changing Climate: Calling Attention for the Conservation of Tropical and Subtropical Forests

Gul Zareen Ghafoor, Memuna Ghafoor Shahid, Maryam Ali and Naghmana Ghafoor

Abstract

Tropical and subtropical forests have diverse ecological functions but are most disturbed due to climatic changes. Peer-reviewed articles, books and reports were searched and downloaded to analyze the effect of climatic changes on tropical and subtropical forests and highlight the need for conservation efforts to ensure sustainable supply of ecosystem services. These forests store largest share (55%) of global terrestrial carbon pool. But the projected rise in temperature (4°C), CO_2 levels (495 ppm) and changes in precipitation regime by 2100 are expected to cause significant changes in ecosystem productivity and nutrient turnover rate making forests more vulnerable to climate change. High temperature with low rainfall reduces tree growth, increases soil carbon fluxes by accelerating rate of nutrient cycling, restricts the range of pollinators and increases pest infestation (*Phytophthora cinnamomi*) affecting ecosystem health and future food security. Increase in heat waves increase the incidence of wild fires and degrade ecosystem quality. Climate change also reduces the scenic beauty, ecotourism and associated economic and mental health benefits. Proactive measures must be taken to mitigate the likely causes of climate change and efforts should be taken to conserve the existing forest reserves to ensure sustainable supply of the ecosystem services.

Keywords: climate change, tropical forest, subtropical forest, ecosystem services, nutrient turnover, carbon stock

1. Introduction

Forests globally cover 4.06 billion hectares (31%) of land area and provide diverse ecosystem services mainly categorized as regulatory, provisioning, supporting and cultural services [1, 2]. These services have use (tangible) and non-use (in-tangible) values. The use values encompass a diverse array of benefits to the humankind in term of provision of food, timber, shelter and fuel-wood. The non-use benefits include climate regulation, nutrient cycling, flood protection, oxygen production, and prevention of soil erosion, recreation, scenic beauty and spiritual services [3, 4]. All of these services are vital for the mankind and ecological communities to sustain on this planet but are under constant threat from changing climate [5].

The uncontrolled anthropogenic activities related to urban and industrial expansions, fossil fuel burning, deforestation, and land use land cover (LULC) changes have caused rise in atmospheric CO_2 levels. Since the pre-industrial times (back in year 1800), 50% increase (280–412 ppm) in the CO_2 concentration has been observed causing 0.85°C rise in global temperature [6, 7]. The CO_2 concentration has risen with a growth rate of 5.1 GtC/yr during 2011–2020 [7]. The emissions are expected to reach the dangerous level of 495 ppm by the year 2100 and are projected to raise global temperature by 4°C [8–10]. This warming trend has already affected forest ecosystems globally and can further reduce the supply of ecosystem services [11, 12].

Geographically, tropical and subtropical forests are located at 23.5°N and 23.5°S latitude in the regions delimited by 18°C to −3°C isotherms according to the Koeppen Geiger climate classification. These forests are characterized by little variation in overall temperature and comparatively high rainfall. These forests cover parts of Asia, Africa, South America and Australia having distinguished features such as seasonal leaf fall, physiographic position, leaf size and canopy structure [13].

Among all the other forest types, the tropical and subtropical forests host most diverse array of tree species and also contribute significantly in carbon sequestration [14]. Out of 2.4 GtC/yr stored annually by global forests, about 55% of this is stored in tropical and subtropical forests while boreal and temperate forests store only 32% and 14% of it respectively [15]. Along with climate regulation, these ecosystems are also a source of livelihood for the local communities, provide food security and maintain ecological health. Yet these tropical and subtropical forests are at the forefront of changing climate due to their geographical location. The climate change impacts on the ecosystem services of the tropical and subtropical forests are considered a major threat in current century. The projected increase in temperature up to 4°C can cause significant changes and alternations in the ecosystem structure and the functions [8]. The different effects of climate change like increase in temperature, CO_2 concentration and changes in rate of precipitation cause many negative impacts on the ecosystem services of the tropical and subtropical forests [16]. These climatic changes might cause tree species to shift their range towards poles and also affect rates of nutrient turnover [17, 18]. Therefore, there is a dire need to prioritize conservation and restoration efforts to protect the ecological functions of these ecosystems.

2. Methodology

This chapter was aimed at analyzing the effect of climatic changes on ecosystem services provided by the tropical and subtropical forests. The chapter has also been designed to highlight the conservation measures necessary to be taken to ensure sustainable supply of the ecosystem services. Therefore search of literature covered broadly two main issues; the effect of climate change on Ecosystem services of the tropical and subtropical forests and the conservation efforts to save the natural systems. Peer-reviewed journal articles, books and reports were collected from trusted websites and publishers. The key words used during the search were 'ecosystem services, changing climate, conservation efforts, tropical and subtropical forests, precipitation, warming, CO_2 extreme, provisioning, regulatory, supporting and cultural services'. Boolean operators and truncations were used during the search of key words. Once the relevant literature was searched, it was thoroughly reviewed and analyzed and then presented in this chapter.

3. Results

3.1 Provisioning services and climatic extremes

Provisioning services are those services from which society gains benefits in terms of fulfilling their needs. These include provision of food, water, timber, fuel wood of direct domestic consumption. These services also include provision of plant based products such as fibers, medicines and raw materials extracted to prepare clothing materials as depicted in **Figure 1**. These services also render the use of bacteria, algae and fungi for their direct benefits or harvesting resources from these microccopic life forms. Among all of these, provision of food and fodder is the fundamental provisional service for the society [19, 20].

Tropical and subtropical forests have a major influence on the weather patterns, freshwater resources, biodiversity and food. These forests are considered the major sinks of the carbon yet face climatic changes which have caused reduction in their area at an alarming rate [21]. Deforestation and degradation has significantly affected these forests and the services they provide. These forests provide provisioning services like food, wood and timber to the millions of people however; decline of these forests has badly affected livelihood of forest dependent communities. Destruction of

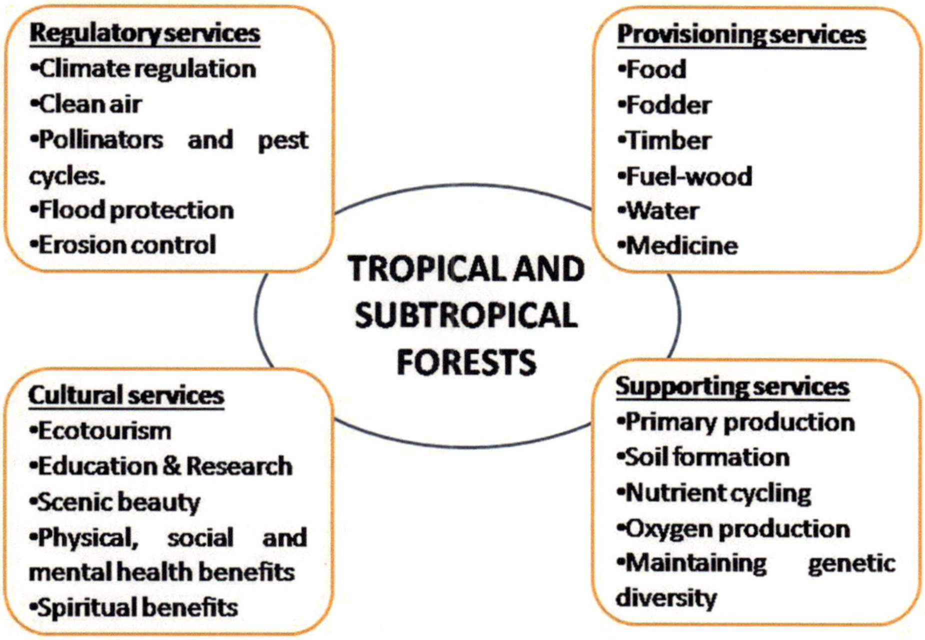

Figure 1.
Diverse array of ecosystem services provided by tropical and subtropical forests.

these forests is also responsible for release of tons of CO_2 in the atmosphere leading to positive feedback to climate change [22].

3.1.1 Provision of water

Tropical and subtropical forests have major role on the rainfall patterns. The destruction of these forests results in the reduction in rate/frequency of precipitation which in turn affect quantity and supply of freshwater to the ecological and human systems [23]. Similarly rising temperature trends coupled with reduction in rainfall frequency and intensity can have profound effects on forested watershed. Water is used for many purposes like household consumption, power generation, and industrial purposes and for the different agricultural purposes etc. [24]. It is also seen that in different parts of the world the water is limited due to climate changes and rising demands of people [25]. Climate change in the future dry years is expected to induce shortage in the availability of surface water and ground water recharge. Climatic changes are also known for triggering wild fires due to heat waves, which may also lead to deposition of sediments and debris in water channels reducing the water holding capacity of reservoirs. During times of high rainfall, the reservoirs may over flow (producing flood events) and can cause damages to human and natural systems. The deposition of debris also affects reservoir water quality increasing treatment costs [26, 27].

3.1.2 Provision of food

Tropical and subtropical forests also link with the global food production both in the direct ways as food (goods) and in the indirect ways in the form of ecological services like pollination and pest control. It has been estimated that about 70% of the principal crops of the world are dependent on the pollinators which is more important for the valuable production of food, fruits and nuts [28]. Climatic changes are known for reducing food production as a result of drought, changes in temperature affecting length of growing season or other extreme events. These climatic changes also affect life cycle of pollinators. When these tropical pollinators decline, the poor farmers are at alarming risk of economic devastation because of their dependence on agriculture for their livelihoods [29, 30]. Besides the deforestation, increased temperature also badly affects the tropical crop production and nutrient quality. The growth and development of the cereal crops is badly affected with the increased temperature and unprecedented rainfall events. So the different regions of the world are facing food security issues due to the climate change [31].

3.1.3 Timber provision

Climatic changes have also been reported for affecting timber production. Climate changes increases the intensity and frequency of wild fire events, floods, storms and prevalence of pathogens affecting plant health. All of these factors contribute to degradation of forest ecosystems and increase in the tree mortality affecting sustainable timber production for the dependent communities [32]. Timber is among one of the important materials required during construction activities and is also a raw material for industries. Low quality timber is also used as a fuel source meeting energy demand for poor people. The decline in the timber production in tropical and subtropical forests has caused poor people to shift to expensive energy alternatives [33].

3.1.4 Medicine provision

Due to minor seasonal changes in temperature (av. 5°C round the year) tropical and subtropical forests host a diverse array of flora and fauna. When the tropical and subtropical forests decline or degrade, the habitat of different species is also affected. These different plant and animal species have the major role in the ecosystem structure and the functions [34]. Among these functions, an important role these species are playing is their medicinal value. The lack of formal health services and synthetic medicines in the developing tropical countries cause people to rely on herbal medications. Many of the present day medicines derive their active ingredients from these tropical/subtropical floral and faunal species. The market of ethno-medicine and traditional health care system is quite large in tropical Asia, Africa and Latin America. However, climate change and subsequent habitat loss is significantly affecting the medicinal value of these ecosystems [35, 36].

3.2 Supporting services and climatic changes

Supporting services are those services which help in supporting and sustaining life on this planet. These services provide the indirect benefits to the natural systems itself and to the society. These services include photosynthesis (primary production), soil formation, nutrient cycling, oxygen production, water cycling and maintaining genetic diversity as shown in **Figure 1** [37]. Tropical and subtropical forests provide these supporting services playing an important role in sustaining life on this planet.

3.2.1 Nutrient cycling and soil formation

Nutrient cycling and soil formation are among the major functions of the ecosystem. Nutrient cycling is a step by step process in which the nutrients are mineralized, absorbed and stored/immobilized in the biomass, enter into the food chain via trophic levels and finally enter into the soil when organisms die or shed their waste products. Once in the soil, the matter decomposes and mineralized to become part of the cycle again. Phosphorus, nitrogen, sulfur, carbon and oxygen are the basic elements that are used in the biomass production [38]. Climate change affects the rate of nutrient cycling and the soil formation of tropical and subtropical forests in different ways [39]. Climatic warming is known for accelerating the pace of decomposition thus affecting rates of nutrient cycling and soil microbial community composition in the tropical and subtropical ecosystems [40]. A fast pace of nutrient cycling results in positive feedback to climate change due to increased soil respiration and decomposition emitting more CO_2 and N_2O (by product of nitrogen cycle). Climate warming has also been known for increasing the rate of production of root nodules accelerating nitrogen uptake, making nitrogen rich leaf litter and ultimately more N_2O emissions from decomposing litter [41].

Climate warming and increase in precipitation is known for causing reduction in litter residence time in soil due to accelerated decomposition rates. This means that warmer and wetter future will make soils deficient in its organic carbon fraction. Soil organic matter is an important indicator of soil health. The organic fractions help in the conditioning of soil, making soil aggregates to improve its structure for effective root penetration, infiltration and aeration. On the contrary, elevated CO_2 levels are reported for suppressing the rate of litter decomposition n the tropical and subtropical environment due to accumulation of recalcitrant fractions and low leaf nitrogen level necessary for microbial decomposition processes [18]. The geographic location of tropical and subtropical forests make these areas to be naturally warm, while the projected rise in temperature and precipitation can have devastating effects on soil quality, nutrient dynamics and overall ecosystem productivity [40, 42].

3.2.2 Primary production

Biomass production (primary production) is the one of the fundamental function of the tropical and subtropical forests and is linked to other services such as erosion control (roots binding soil) and evapotranspiration rate affecting water cycle [43]. The ecosystem primary production is linked with the rate of photosynthesis. Climate change affects the rate of biomass production and the services they provide [44]. It affects the rate of the photosynthesis in different ways (either positive for some species or negative for others) so the biomass production or net primary productivity of these forests is altered [45]. Long term warming is estimated to cause increase in the rate of photosynthesis by 7–29% for six subtropical tree species [46]. While short term warming affects the rate of stomatal conductance and stomatal density, rate of photosynthesis and carbon accumulation (biomass production) in *Schima superba* in subtropical environment [12]. The stored biomass can be used for the different purposes like the production of food, fodder, biofuel and timber.

3.3 Regulating services and climate change

Regulatory ecosystem services are the benefits obtained from regulating ecological processes. Ecosystems regulate the natural environment in various ways such as by regulating climate or maintaining air quality through carbon sequestration (removing excess CO_2 from atmosphere), water purification, erosion control, flood protection, pollinator and pest cycles and regulating intensity and frequency of natural disasters as depicted in **Figure 1**. Regulatory services are linked to other ecosystem services like they affect the supply of provisioning and supporting services as well [47].

3.3.1 Climate regulation

This is a fundamental regulatory ecosystem service important for maintaining the quality of air and regulating local and global climate patterns. This service refers to the removal of harmful greenhouse gases (CO_2) and sequestering it in plant biomass through the process of photosynthesis. The global forests has fundamental role in maintaining climate whereas, the tropical and subtropical forests present their largest share (55%) in the supply of this service [15]. The role of forests in carbon storage and sequestration is becoming of fundamental importance as the pace of climate change is accelerating. Countries that are signatory to REDD+ (Reducing Emissions from Deforestation, Forest Degradation, conservation, enhancement and sustainable forest management) get monetary incentives by enhancing their forest carbon stock [48]. However, the rate of carbon sequestration is mainly defined by climatic factors which if disturbed, affect global climatic patterns along with delivery of linked ecosystem services. It has been estimated that the rising atmospheric CO_2 concentrations can have fertilizing effect on plant growth in the tropics which might enhance rate of carbon sequestration. But elevated CO_2 levels have been estimated for accelerating the metabolic activities of the soil microbial flora. This will accelerate the rate of decomposition of organic matter and thus will speed up the rate of carbon turnover. A high soil CO_2 efflux means positive feedback to climate change [49, 50]. The rising CO_2 levels have been reported for altering leaf litter chemistry in terms of increased lignin concentration (recalcitrant fraction), high C/N ratio but with low leaf nitrogen level in the tropical and subtropical ecosystems [49].

Other climatic factors such as precipitation can also affect this service. Tropical and subtropical species have been reported for experiencing reduction in growth rate and production of roots with low biomass in response to drought stress and increase in soil carbon loss in areas with high rainfall [51, 52]. Similarly warming coupled with elevated CO_2 levels is also observed for increasing the rate of soil CO_2 efflux and evapotranspiration (affecting hydrological cycle) with overall reduction in plant growth rate and potential for carbon sequestration [53]. Inter-specific differences have also been observed for tropical and subtropical forests in response to climate warming. The subtropical evergreen broadleaved species have been documented for showing increase in overall growth and biomass carbon accumulation in response to projected climatic warming while the subtropical deciduous species have been reported for reduction in their regulatory ecosystem service [41].

3.3.2 Pollination, disease and pest regulation

The pollination, disease and pest regulation service has direct effect on ecological and human health and well being. This regulatory service also affects other provisioning services like food and fodder production important for human and livestock survival. Climate change can greatly influence the pollinator species. Reduction in the range and distribution of Colombian stingless bees (important for honey production) has been documented under various climate change scenarios. This means that reduction in pollinators and pollination due to climate change would result in food security issues in the tropics in future affecting agriculture productivity and rural livelihoods [54]. Similarly, increase in rainfall intensity and rising temperature in the subtropical regions has caused increase in the incidence of infestation of *Phytophthora cinnamomi* in the woody plants resulting in root rot, cankering and death of plant [55]. Warm temperature is also reported for increasing the incidence of *Cylindrocladium quinqueseptatum* infection causing leaf blight in tropical and subtropical forests in Southeast Asia and Australia [56].

3.4 Cultural services and climatic changes

Cultural ecosystem services are the non-material benefits that people obtain from ecosystems such as education, recreation, spiritual experiences, social relations and esthetic appreciation [57]. The array of cultural ecosystem services is depicted in **Figure 1**. These services contribute to the overall well being of the dependent communities in terms of provision of mental, physical and social health benefits and improve the cognitive ability of a person [58]. Climatic extremes have been documented for reducing the physical and mental health benefits derived from these forest ecosystems. Ecotourism/recreation is linked with economic growth of a country however, rising temperature is also responsible for reducing this cultural ecosystem service due to increased onset of climatic extreme events [43]. The increase in the frequency and intensity of climatic disasters such as floods, heat waves, droughts results in degradation of natural areas thereby reducing their scenic beauty and associated ecotourism and health benefits. These climatic extremes also lead to loss of job/rural livelihood affecting community's physical, mental and socio-economic health [59].

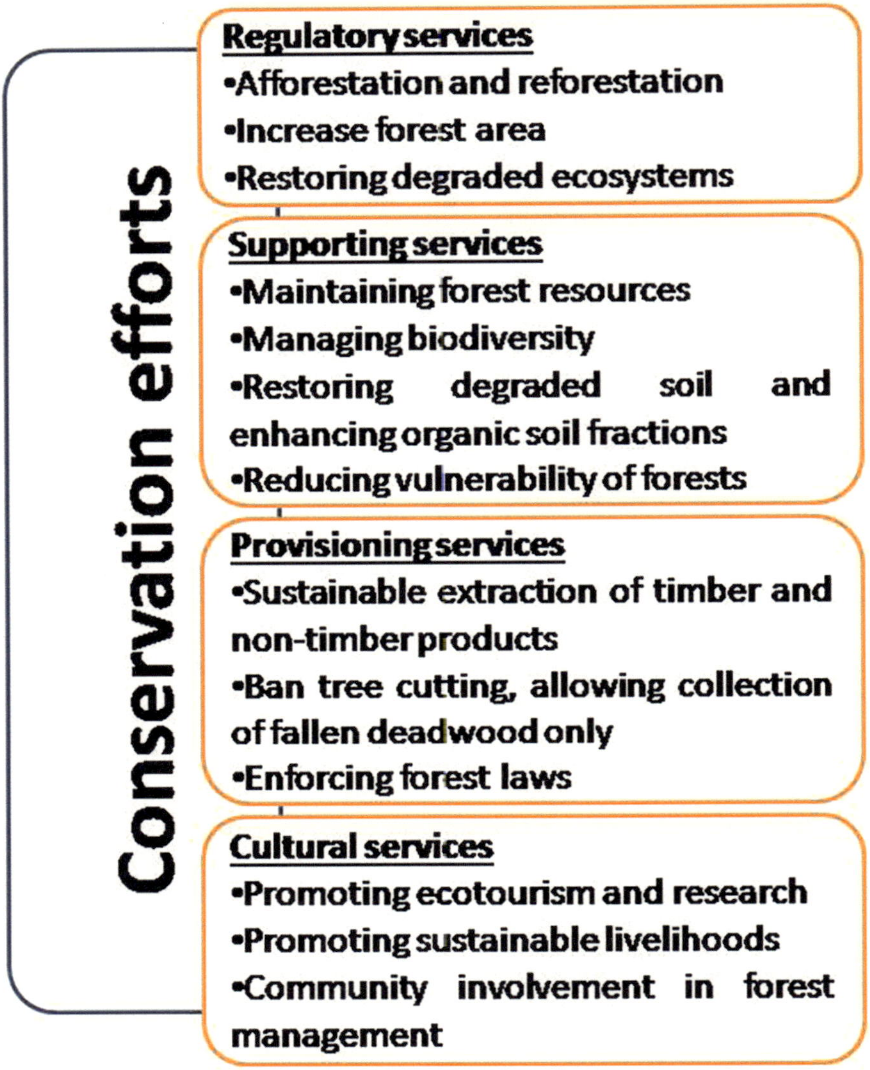

Figure 2.
Conservation efforts aiming at protection of ecosystem service.

4. Conservation of natural systems

Considering the wide ranging effects of climate change on regulatory, provisioning, supporting and cultural ecosystem services, there is a dire need to take proactive conservation measures to increase the resilience and resource base of the tropical and subtropical forests. There exist research gap highlighting the effect of individual climatic factors on each of the ecosystem services. Significant number of studies has been conducted on the effect of climatic changes on provisioning and regulatory services, nutrient cycling and primary production of the tropical and subtropical forests, but to the best of knowledge little is known for cultural services. Climate change is a gradual process and is linked with anthropogenic activities. The rising demand for energy supplies (coming mainly from fossil fuel resources), unintended land use land cover changes, increase in the air pollution and un-sustainable harvesting of forest resources are all linked with rising global atmospheric CO_2 concentration and temperature. In the tropic and subtropics, parts of the region might experience drought while other might experience floods in future. Therefore, the forests are expected to behave in different manner when dryness or wetness couples with warm temperature and elevated CO_2. Restricting the use of fossil energy resources and shifting to sustainable and renewable energy resources might solve the problem to some extent. However, conservation efforts focusing on reforestation and afforestation of the areas combined with restoration of degraded ecosystems can help in increasing the area of the forests and thus supply of the services as depicted in **Figure 2**. Other measures such as promoting sustainable harvest and maintaining biodiversity and genetic resources might reduce vulnerability of the forests to changing climate. The communities should be banned from removing wood and permission should be granted for the collection of fallen deadwood only to meet the fuel wood need of the rural poor. There is generally a lack of forest conservation regulations and community involvement in the management of tropical and subtropical resources. Effective implementation of forest management practices, stakeholder involvement along with enforcement of laws prohibiting un-sustainable extraction of forest resources might help conserving existing resources as shown in **Figure 2**.

5. Conclusion

Tropical and subtropical forests host a diverse array of species and play a significant role in delivery of ecosystem services. These forests are essential for ecological health and rural well being. The geographic location of these forests makes them more vulnerable to the impacts of changing climate. The projected rise in global temperature, changes in precipitation regime and elevated levels of CO_2 will affect the supply of ecosystem services which might cause devastating effects on ecological and human systems. However, forest conservation and restoration efforts combined with reduction in greenhouse gas emission sources, community involvement and strengthening the relevant local institutions can protect the natural resource base of tropical and subtropical regions. Further research is need of the hour to investigate role of each of the climatic variable on individual ecosystem services to fill data gaps.

Conflict of interest

The authors declare no conflict of interest.

Author details

Gul Zareen Ghafoor[1*], Memuna Ghafoor Shahid[2], Maryam Ali[1]
and Naghmana Ghafoor[3]

1 Sustainable Development Study Centre, Government College University Lahore, Pakistan

2 Department of Botany, Government College University Lahore, Pakistan

3 Department of Economics, Lahore College for Women University Lahore, Pakistan

*Address all correspondence to: zareen.sdsc@gcu.edu.pk

References

[1] Baskent EZ. A framework for characterizing and regulating ecosystem services in a management planning context. Forests. 2020;**11**(1):102

[2] FAO. Global Forest Resource Assessment 2020: Key Findings. Rome: Food and Agriculture Organization; 2020. DOI: 10.4060/ca8753en

[3] Mengist W, Soromessa T. Assessment of forest ecosystem service research trends and methodological approaches at global level: A meta-analysis. Environmental Systems Research. 2019;**8**(1):1-18

[4] Cheng X, Van Damme S, Li L, Uyttenhove P. Evaluation of cultural ecosystem services: A review of methods. Ecosystem Services. 2019;**37**:100925

[5] Bellard C, Bertelsmeier C, Leadley P, Thuiller W, Courchamp F. Impacts of climate change on the future of biodiversity. Ecology Letters. 2012;**15**:365-377

[6] Prentice IC, Farquhar GD, Fasham MJR, Goulden ML, Heimann M, Jaramillo VJ, et al. The Carbon Cycle and Atmospheric Carbon Dioxide. Switzerland: IPCC; 2001

[7] Friedlingstein P, Jones MW, O'Sullivan M, Andrew RM, Bakker DC, Hauck J, et al. Global carbon budget 2021. Earth System Science Data. 2022;**14**(4):1917-2005

[8] Szulejko JE, Kumar P, Deep A, Kim KH. Global warming projections to 2100 using simple CO2 greenhouse gas modeling and comments on CO2 climate sensitivity factor. Atmospheric Pollution Research. 2017;**8**(1):136-140

[9] World Meteorological Organization. WMO Statement on the State of the Global Climate in 2019. Switzerland: World Meteorological Organization (WMO); 2019

[10] Olivier JGJ, Peters JAHW. Trends in Global CO2 and Total Greenhouse Gas Emissions. Netherlands: PBL Netherlands Environmental Assessment Agency; 2020

[11] Chuckran PF, Reibold R, Throop HL, Reed SC. Multiple mechanisms determine the effect of warming on plant litter decomposition in a dryland. Soil Biology and Biochemistry. 2020;**145**:107799

[12] Wu G, Liu H, Hua L, Luo Q, Lin Y, He P, et al. Differential responses of stomata and photosynthesis to elevated temperature in two co-occurring subtropical forest tree species. Frontiers in Plant Science. 2018;**9**:467

[13] Pearcy RW, Robichaux RH. Tropical and subtropical forests. In: Chabot BF, Mooney HA, editors. Physiological Ecology of North American Plant Communities. Dordrecht: Springer; 1985. DOI: 10.1007/978-94-009-4830-3_13

[14] Slik JWF, Arroyo-Rodríguez V, Aiba S-I, Alvarez-Loayza P, Alves LF, Ashton P, et al. An estimate of the number of tropical tree species. Proceedings of the National Academy of Sciences. 2015;**112**:7472-7477

[15] Pan Y, Birdsey RA, Fang J, Houghton R, Kauppi PE, Kurtz WA, et al. A large and persistent carbon sink in the world's forests. Science. 2011;**333**:988-993

[16] Dawson TP, Jackson ST, House JI, Prentice IC, Mace GM. Beyond predictions: Biodiversity conservation in a changing climate. Science. 2011;**332**(6025):53-58

[17] Fei S, Desprez JM, Potter KM, Jo I, Knott JA, Oswalt CM. Divergence of species responses to climate change. Science Advances. 2017;**3**(5):e1603055

[18] Rai A, Singh AK, Singh N, Ghoshal N. Effect of elevated CO2 on litter functional traits, mass loss and nutrient release of two subtropical species in free air carbon enrichment facility. Environmental and Experimental Botany. 2020;**172**:103994

[19] Urry J. Climate change and society. In: Why the Social Sciences Matter. London: Palgrave Macmillan; 2015. pp. 45-59

[20] Haines-Young R, Potschin MB. Common International Classification of Ecosystem Services (CICES) V5.1 and Guidance on the Application of the Revised Structure. Nottingham, UK. 2018. Available from: www.cices.eu

[21] La Vina A, de Leon AM. Two Global Challenges, One Solution: International Cooperation to Combat Climate Change and Tropical Deforestation. Washington, DC, USA: Center for Global Development; 2014

[22] Bazzaz FA. Tropical forests in a future climate: Changes in biological diversity and impact on the global carbon cycle. In: Potential Impacts of Climate Change on Tropical Forest Ecosystems. Dordrecht: Springer; 1998. pp. 177-196

[23] Spangenberg JH, Görg C, Truong DT, Tekken V, Bustamante JV, Settele J. Provision of ecosystem services is determined by human agency, not ecosystem functions. Four case studies. International Journal of Biodiversity Science, Ecosystem Services & Management. 2014;**10**(1):40-53

[24] Brockerhoff EG, Barbaro L, Castagneyrol B, Forrester DI, Gardiner B, González-Olabarria JR, et al. Forest biodiversity, ecosystem functioning and the provision of ecosystem services. Biodiversity and Conservation. 2017;**26**(13):3005-3035

[25] Jewitt G. Can integrated water resources management sustain the provision of ecosystem goods and services? Physics and Chemistry of the Earth, Parts A/B/C. 2002;**27**(11-22):887-895

[26] Luce C, Morgan P, Dwire K, Isaak D, Holden Z, Rieman B. Climate Change, Forests, Fire, Water, and Fish: Building Resilient Landscapes, Streams, and Managers. USDA For. Serv. - Gen. Tech. Rep. RMRS-GTR 1-155. Colorado: U.S. Department of Agriculture, Forest Service, Rocky Mountain Research Station; 2012

[27] Warziniack TW, Elmer MJ, Miller CJ, Dante-Wood SK, Woodall CW, Nichols MC, et al. Effects of climate change on ecosystem services. In: Halofsky JE, Peterson DL, Ho J, Little NJ, Joyce LA, editors. Climate Change Vulnerability and Adaptation in the Intermountain Region [Part 2]. Rocky Mountain Research Station, Fort Collins, CO: U.S. Department of Agriculture, Forest Service; 2018. pp. 376-403

[28] Cardinale BJ, Matulich KL, Hooper DU, Byrnes JE, Duffy E, Gamfeldt L, et al. The functional role of producer diversity in ecosystems. American Journal of Botany. 2011;*98*(3):572-592

[29] Vitousek PM, Hooper DU. Biological diversity and terrestrial ecosystem biogeochemistry. In: Biodiversity and Ecosystem Function. Berlin, Heidelberg: Springer; 1994. pp. 3-14

[30] Gowda P, Steiner JL, Olson C, Boggess M, Farrigan T, Grusak MA. Agriculture and rural communities. In: Reidmiller DR, Avery CW, Easterling DR, Kunkel KE, Lewis KLM, Maycock TK, Stewart BC, editors.

Impacts, Risks, and Adaptation in the United States: Fourth National Climate Assessment II. Washington, D.C., USA: U.S. Global Change Research Program; 2018. pp. 391-437

[31] Challinor AJ, Ewert F, Arnold S, Simelton E, Fraser E. Crops and climate change: Progress, trends, and challenges in simulating impacts and informing adaptation. Journal of Experimental Botany. 2009;**60**(10):2775-2789

[32] Buongiorno J, Zhu S, Zhang D, Turner J, Tomberlin D. The Global Forest Products Model: Structure, Estimation, and Applications. California, USA: Elsevier; 2003

[33] Hodder KH, Newton AC, Cantarello E, Perrella L. Does landscape-scale conservation management enhance the provision of ecosystem services? International Journal of Biodiversity Science, Ecosystem Services & Management. 2014;**10**(1):71-83

[34] Speak AF, Mizgajski A, Borysiak J. Allotment gardens and parks: Provision of ecosystem services with an emphasis on biodiversity. Urban Forestry & Urban Greening. 2015;**14**(4):772-781

[35] Zhang W, Ricketts TH, Kremen C, Carney K, Swinton SM. Ecosystem services and dis-services to agriculture. Ecological Economics. 2007;**64**(2):253-260

[36] Colfer C, Sheil D, Kishi M. Forests and Human Health: Assessing the Evidence. Indonesia: CIFOR; 2006

[37] Carpenter SR, Mooney HA, Agard J, Capistrano D, DeFries RS, Díaz S, et al. Science for managing ecosystem services: Beyond the millennium ecosystem assessment. Proceedings of the National Academy of Sciences. 2009;**106**(5):1305-1312

[38] Richter DDB, Oh NH, Fimmen R, Jackson J. The rhizosphere and soil formation. In: The Rhizosphere. USA: Academic Press; 2007. p. 179

[39] Li Y, Xu Y, Li Y, Wu T, Zhou G, Liu S, et al. Warming effects on morphological and physiological performances of four subtropical montane tree species. Annals of Forest Science. 2020;**77**(1):2

[40] Ghafoor GZ, Sharif F, Hayyat MU, Shahid MG, Shahzad L. Effect of climatic factors on leaf litter decomposition dynamics of a subtropical scrub forest under field and microcosm conditions. All Life. 2022;**15**(1):1175-1187

[41] Ghafoor GZ, Sharif F, Khan AUH, Shahid MG, Siddiq Z, Shahzad L. Effect of climate warming on seedling growth and biomass accumulation of Acacia modesta and Olea ferruginea in a subtropical scrub forest of Pakistan. Écoscience. 2021;**2021**:1-14

[42] Bothwell LD, Selmants PC, Giardina CP, Litton CM. Leaf litter decomposition rates increase with rising mean annual temperature in Hawaiian tropical montane wet forests. PeerJ. 2014;**2**:1-12

[43] Malhi Y, Franklin J, Seddon N, Solan M, Turner MG, Field CB, et al. Climate change and ecosystems: Threats, opportunities and solutions Philos. Transactions on Royal Society B Biological Science. 2020;**375**:1-8

[44] Reyers B, Biggs R, Cumming GS, Elmqvist T, Hejnowicz AP, Polasky S. Getting the measure of ecosystem services: A social–ecological approach. Frontiers in Ecology and the Environment. 2013;**11**(5):268-273

[45] Colvin SA, Sullivan SMP, Shirey PD, Colvin RW, Winemiller KO, Hughes RM, et al. Headwater streams and wetlands

are critical for sustaining fish, fisheries, and ecosystem services. Fisheries. 2019;**44**(2):73-91

[46] Li Y, Liu J, Zhou G, Huang W, Duan H. Warming effects on photosynthesis of subtropical tree species: A translocation experiment along an altitudinal gradient. Scientific Reports. 2016;**6**(1):1-14

[47] Sutherland IJ, Villamagna AM, Dallaire CO, et al. Undervalued and under pressure: A plea for greater attention toward regulating ecosystem services. Ecological Indicators. 2018;**94**:23-32

[48] Bayrak MM, Marafa LM. Ten years of REDD+: A critical review of the impact of REDD+ on forest-dependent communities. Sustainability. 2016;**8**(7):1-22

[49] Cha S, Chae HM, Lee SH, Shim JK. Effect of elevated atmospheric CO2 concentration on growth and leaf litter decomposition of Quercus acutissima and Fraxinus rhynchophylla. PLoS One. 2017;**12**(2):1-19

[50] Chen S, Zhang Y, Wu Q, Liu S, Song C, Xiao J, et al. Vegetation structural change and CO2 fertilization more than offset gross primary production decline caused by reduced solar radiation in China. Agricultural and Forest Meteorology. 2021;**296**:108207

[51] Gray SB, Brady SM. Plant developmental responses to climate change. Developmental Biology. 2016;**419**(1):64-77

[52] Dannenberg MP, Wise EK, Smith WK. Reduced tree growth in the semiarid United States due to asymmetric responses to intensifying precipitation extremes. Science Advances. 2019;**5**(10):eaaw0667

[53] Wu Y, Wang D, Qiao X, Jiang M, Li Q, Gu Z, et al. Forest dynamics and carbon storage under climate change in a subtropical mountainous region in Central China. Ecosphere. 2020;**11**(3):e03072

[54] Gonzalez VH, Cobos ME, Jaramillo J, Ospina R. Climate change will reduce the potential distribution ranges of Colombia's most valuable pollinators. Perspectives in Ecology and Conservation. 2021;**19**(2):195-206

[55] Jung T. Beech decline in Central Europe driven by the interaction between *Phytophthora* infections and climatic extremes. Forest Pathology. 2009;**39**:73-94

[56] Booth TH, Jovanovic T, Old KM, Dudzinski MJ. Climatic mapping to identify high-risk areas for *Cylindrocladium quinqueseptatum* leaf blight on eucalypts in mainland South East Asia and around the world. Environmental Pollution. 2000;**108**:365-372

[57] Csurgó B, Smith MK. The value of cultural ecosystem services in a rural landscape context. Journal of Rural Studies. 2021;**86**:76-86

[58] Rall E, Bieling C, Zytynska S, Haase D. Exploring city-wide patterns of cultural ecosystem service perceptions and use. Ecological Indicators. 2017;**77**:80-95

[59] Sandifer PA, Sutton-Grier AE, Ward BP. Exploring connections among nature, biodiversity, ecosystem services, and human health and well-being: Opportunities to enhance health and biodiversity conservation. Ecosystem Services. 2015;**12**:1-15. DOI: 10.1016/j.ecoser.2014.12.007

Chapter 15

High Equatorial Andean Forests and Their Socioecological Problems

Alfonso Avellaneda-Cusaría

Abstract

The so-called Tropical Forests, in the case of Colombia, which is located biogeographically in the Intertropical Convergence Zone, that is, in the Equatorial Zone, are really equatorial forests. The high Andean forests that correspond to those above 2500 m above sea level are ecosystems of high ecological importance, due to their location on the border of temperate zones, known as coffee climates that are between 1200 and 2200 m above sea level and the cold zones corresponding to the Andean subpáramos and moors above 3200 m above sea level. In these ecosystems, there is great biodiversity that configures them as ecotones of great importance for the survival and conservation of species that have been adapting to difficult exploitation conditions, which have them at high risk of extinction, due to the expansion of the agricultural frontier and climatic factors such as torrential rains and extreme droughts exacerbated in recent decades. Due to the climate variability that currently characterizes global warming processes. In recent years, peasant resistance movements have emerged for the defense of the territory, which configures it as a scenario of active and growing socio-ecological conflicts.

Keywords: high Andean forests, socioecology, environmental conflicts, equatorial forests, peasant population

1. Introduction

This chapter develops a vision of the current situation of the equatorial high Andean forests of Colombia and the socio-ecological conflicts that define their current situation. It begins with a proposal to approach its biogeographic context, the processes of occupation that have been transforming the natural landscape during the last 70 years, the processes that have defined these changes until its current situation of profound transformations that lead it to define it as an ecosystem in danger of extinction since only less than 5% of the original coverage remains.

Faced with this situation, environmental planning and management approach to date, as well as public policies, have been inefficient in guaranteeing their conservation and sustainability. As a result of the above, the last 10 years have been emerging processes of peasant resistance from mobilizations, strikes, and social organizations such as committees for the defense of water and forests, community action boards in towns and villages, and increasingly strong community aqueduct boards that demand participation in the decision on delimitation and conservation of areas of ecological importance such as the high Andean forests. Wetlands and water recharge areas. On the situation and balances of these social processes, an exhaustive investigation has been carried out that has been based on the participation of the author either as a direct author or as a public official, as a social leader, or as a researcher from the university academy.

In response to this situation and in the search to propose a new analytical approach to address environmental conflicts in these territories, a geo-ecosystem approach is proposed as an option that has environmental governance as a dynamic axis based on the recognition of the role for this purpose, must be fulfilled by rural communities that have historically been playing the main role in knowledge processes, conservation, management and adaptive strategies in these ecosystems.

2. Methodology

The research presented on the current situation of the high Andean forests summarizes the research products carried out by the author in several periods on the equatorial high Andean ecosystems.

First period (1986–1995): As a public official of the National Institute of Renewable Natural Resources (INDERENA) I toured a large part of the Andean mountains attending aspects of conservation, management, and management of environmental conflicts in the Eastern and Central mountain ranges and in the Sierra Nevada de Santa Marta [1].

Second period (2002–2006). As an advisor to Non-Governmental Organizations and the Ministry of Environment, Housing and Territorial Development for the evaluation of the implementation of the National Policy on Ecosystems of the High Andean Mountain [2].

Third period (2006–2022). As an advisor and member of social organizations for the defense of water and moors against energy mining extractivism.

Throughout this investigative process following the parameters of the methodological approach of research, Acción Participativa (IAP), method proposed, by the sociologist Orlando Fals Borda, was recognized nationally and internationally as an appropriate method for this type of socioecological work [3]. This method basically consists of a dialog with the communities and local actors of a territory, known as a dialog of knowledge.

3. Definition and biogeographic contexts of high Andean forests in Colombia and their sociocultural aspects

3.1 Biogeographical aspects

It could be said that the high Andean forests are located in a strip that goes from 2400 m above sea level to 3800 m above sea level. They are called high Andean forests or high montane forest, which are:

> *Forests included in the strip between 2900 to 3800 meters above sea level that are characterized as a stratum of trees and shrubs between 3 and 8 m high, with a predominance of compounds. (Col) are representative of this category oak groves and cloud forests, the vast majority located in relictual areas in the Sinú-Caribe, Caquetá,*

> *Meta, Patía, Catatumbo River, Alto and Medio Magdalena, Medio Cauca, Río Atrato and Sabana de Bogotá basins. [4]*

However, it should be noted that the Colombian high Andean zone, which bioclimatically would be between 2400 and 3200 m above sea level, is characterized as cold climate. The following factors would explain this.

There are three large mountain ranges that have their own biogeographical characteristics and with it, the location of these natural forest masses varies in the altitudinal strip from one mountain range to another.

The diversity of the floristic composition according to scholars of this is due to the great variety of factors that have acted in its genesis and that is particularly associated since the geological origin of the Andes, which have been forming an extraordinary variety of geological formations that continues throughout the millennia and that constitutes the basis of the origin of the diverse and complex floristic and faunal diversities of the ecosystems of the high mountain equatorial. During the last great period of the Quaternary, due to climatic changes, the strip of forests throughout the Andean corridor has been changing and with its composition and structure due to migrations of some species from temperate zones to cold zones, this diversity in floristic composition in some especially wetter areas, Known as cloud forests have led to define these as areas of megabiodiversity or recognized as one of the main centers of diversity and speciation in the world [5–7].

In the northern Andean zone of the Cordillera Oriental, the most characteristic species on its eastern slope are the enennillo (Wenmania tomentosa) and on the western slope the oak (Quercus humboldtii).

There are relatively few in-depth studies of the forests of the equatorial high mountain despite the fact that these ecosystems have been under great pressure, especially due to the expansion of the agricultural frontier.

> *"Several estimates suggest that less than 10% of the original Andean forests remain in Colombia" (Henderson et al. 1991) and, probably, less than 5% of the high Andean forests. [8]*

Recent studies have found that higher altitude forests are more heterogeneous, therefore not finding a dominant species [9].

There is also the Sierra Nevada de Santa Marta as a set of ecosystems independent of the Andean mountain ranges and older than these, where the presence of equatorial forests at the heights mentioned for the high Andean forest, depends on its biogeographic characteristics of high complexity. Both the forests of the Andean areas and those of the Sierra Nevada de Santa Marta above 2400 m above sea level can be grouped as equatorial forests of the high mountains.

3.2 Occupation of the spaces of the high Andean forests and sociocultural aspects

3.2.1 Construction of mestizo peasant culture

The mestizo peasant culture of the high mountain or cold climate, which is characterized by the use of the ruana, the scarf or the blankets, as one of the main founding peoples of the Colombian nation of origin especially Muisca, Paez, and Pasto is located in these territories and for centuries has contributed to constitute the current territories in the high Andean mountains.

"The territory of the páramo" has been forged in the amalgam between settlers and high Andean nature, is the genesis of the crossing of their realities. Recognize that these ecosystems have been undergoing transformation processes mediated by human presence, to the extent that lagoons and their sacred environments since time immemorial had been treated as sacred sites by the Native Americans of the Chibcha civilization and others earlier and for 500 years by the Hispanic presence that displaced the indigenous cultures of the valleys and fertile slopes, The poor peasant, product of miscegenation and the indigenous peoples survivors of the humanitarian catastrophe of extermination took refuge in these natural scenarios and configured since then new environmental systems. Thus, peasant peoples or peasant indigenous peoples emerged who became the authentic environmental subjects of these territories to the extent that they chose as an action of resistance the defense of the land as a unifying element, not seen as circumscribed to private property, but closely linked to the concept of territory, nature, common good and the sacredness of it, while it is the giver of the benefits in food, water, work and rest. The mythical imaginary of the Chibcha ancestor, as the spirit of the current environmental subject, survives in the mestizo culture of the peasant of the region of the Cundiboyacense highlands, and in a similar way we can say that, in the Nariño and Cauca mountains, in the Serranía del Cocuy, in Perijá, the Central mountain range and the Sierra Nevada de Santa Marta [10, p. 199].

The high mountain ecosystems that until the mid-twentieth century was preserved almost intact, have been suffering a process of deterioration due to social and economic processes, related to the displacement or migration of populations from the mountain slopes or the high valleys and Andean plateaus, to higher areas, due to the deterioration of the soils, to the intensification of potato cultivation, propitiated by the green revolution, which advised the use of the soils of these ecosystems and those of the high Andean forest as the propitious ones for this crop. In this way, the high mountains became the last refuge of peasant and indigenous inhabitants who were also dragged there by the various political violence that occurred throughout the twentieth century. The peasant or the indigenous who had kept sacredness in front of these territories and their lagoons were surprised by the aforementioned social processes and through strategies of adaptation to the difficult climatic conditions of these lands, settled there hunting, fishing, and cultivating tubers and roots. Throughout this period of ancestral recognition, as of the occupation, the relationship of the inhabitants with nature has not differentiated the páramo itself, as a non-forested ecosystem, from the high forests that were established as limits between the areas of the high mountain and the lowlands of the mountain slopes in the cold land, traditionally occupied by peasants from the highlands and mountain ranges in Boyacá, Cundinamarca, Santanderes, Tolima, Cauca, Nariño, Valle, Antiguo Caldas, and Sierra Nevada de Santa Marta. This differentiation did not occur because they constituted a set of habitats where the cultures of the cold lands developed and the highest areas were considered common properties, which could not be transformed into the regime of private property that characterized the other Andean territories during the Colony and the Republic.

3.2.2 Occupation of spaces in the Colombian high Andean mountains and the use of forests, consolidation of a peasant economy

The areas of the greatest presence of the forests of the high equatorial mountains correspond to the Cauca-Nariño mountain massif, better known as the Colombian Massif, where several of the main rivers of Colombia are born, such as the Cauca and the Magdalena, which in turn constitute the Magdalena-Cauca hydrological system, the most important for the development of the Colombian territory; there is also born the Patía River that drains toward the slope of the Choco Biogeográfico; and the Caquetá and Putumayo rivers that go to the Amazon region.

These spaces were occupied for millennia by peoples who possibly, for the most part, came from the Caribbean, who would constitute the ancestral indigenous peoples and in the most recent centuries by the Afro-descendant peoples who arrived in Colombian territory as a result of slavery during the sixteenth to nineteenth centuries.

With the Hispanic occupation, a new indigenous mestizo-Hispanic peasant was consolidated in the areas of the high equatorial mountain, which with different degrees of intensity has populated the high mountains of the former departments of Caldas, Antioquia, Valle del Cauca, in the Western and Central mountain ranges and the highlands of Cundiboyacense and Santander in the Eastern mountain range of Colombia. It should be noted that, during the second half of the nineteenth century, German migration arrived in Colombia, which was located in part of the mountains of Santander and Boyacá and Arab immigration in part of the Antioquia mountain, contributing to enrich the variety of peasant miscegenations.

This process of occupation was accompanied by the use of countless species for the manufacture of utensils such as baskets, backpacks, crafts, sieves, and other objects to facilitate the daily work of the people, who collected the indigenous heritage and perfected it for centuries with the skill of the artisans, who in each subregion were and still are elements of cultural identity in Antioquia. the Coffee Axis, Boyacá, Santander, Tolima, Cauca and Nariño. Species such as the enennillo (Wenmania tomentosa) was used until well into the 1960s to extract from its bark substances for tanning hides and countless vines and pastures for crafts.

3.2.3 From peasant agriculture to wheat-barley monocultures fourteenth and twentieth centuries

Prior to the arrival of the Hispanic occupation, the most characteristic indigenous peoples of the high equatorial mountains were the various peoples of the Cauca River, the Patía and the Putumayo, the Quimbaya peoples of the Central Mountain Range, the Pijao people in Tolima, the great Muisca people in the Cundiboyacense highlands and the Arawak peoples. Tairona, kogi, arsarios, and others in the Sierra Nevada de Santa Marta.

From these peoples and their miscegenation with the Hispanic element, a mestizo peasant people emerged that today characterizes the majority of the Colombian population in the highlands.

From the sixteenth century with the Hispanic occupation, cereal crops such as wheat and barley and deciduous fruit trees were introduced, which displaced, either by intensity or by express prohibition of the Spanish crown to indigenous crops. Cattle and sheep also entered, which became an element of high importance for peasant life, due to its milk, meat and wool products, the latter replacing cotton in blankets throughout the highlands of Cundiboyacense and Santander mainly. In this way what brought the Hispanic invasion to the Colombian Andean zone was not its culture but its Mediterranean nature forged by long centuries of Roman, Mongolic and Arab occupation.

Throughout the period from the sixteenth century until the first half of the twentieth century, the occupation of these peoples was made in the areas of the equatorial forests of high mountain. During the Colony, it was these forests that provided wood inputs for the nascent and flourishing peoples during this period and the first phase of the birth of the Republic between the nineteenth and twentieth centuries. The occupation of species such as oak encenillo, tuno, cucharo el cedar, and guayacán stands out.

Peasant agriculture in these high mountain areas was characterized by crop rotation between corn, potatoes, quinoa, and legumes typical of the area and others brought from the old continent such as barley, wheat, and oats.

In particular, these last two crops formed the basis of peasant economies that provided for a long period of inputs for bread and beverages in urban areas of towns and cities. Whole wheat bread, soups of various varieties of wheat and barley were staples of families in the cold lands and especially in cities such as Bogotá, Tunja, and Popayán. The peasant arepa based on corn and its multiple forms of preparation is today one of the staple foods of the Colombian population.

3.3 Industrialization of the countryside and extractivism

3.3.1 Industrial potato monoculture and its impact on high Andean ecosystems

Since pre-Hispanic times, the peoples of the highlands of the Andes and the Sierra Nevada de Santa Marta cultivated potatoes as one of their most precious products along with corn and arracacha. Until the early twentieth century in Boyacá, various varieties of wild potatoes were collected, which were born spontaneously in the clearings of the high Andean forests. Due to the population growth of urban areas after the 1950s, the demand for this tuber grew and with it research on new varieties. Prior to the Green Revolution that emerged strongly in the 1960s, potato crops were fertilized with organic materials resulting from the composting of livestock and poultry manure, as well as ash from the burning of firewood as fuel in peasant homes. At this time, potato crops were made in the area of Andean forests up to 3000 m above sea level. The introduction of new varieties of potato dependent on agrochemicals for fertilization and control of pests and diseases, generated during the Green Revolution, was propitiating the occupation of moors and areas of high Andean forest, encouraging the monopolization of land in fewer and fewer smallholder owners or tenants of peasant plots, which altered the mosaic of high Andean mountain ecosystems where the main victim was the forest. In this way, the large forest masses have been disappearing at the expense of the agricultural frontier for potato planting with rotation with semi-extensive livestock that today characterizes much of the territory. During the 1990s, especially in the Central mountain range, poppy crops proliferated in these areas. One of the great socio-environmental problems of today is the uncontrollable growth of industrial potato crops, with high consumption of biocides that affect the recharge areas and birth of water resources of thousands of streams and hundreds of rivers. During this process, the peasant went from small smallholder with his plot of multi-stratum crops to day laborer in potato crops and cattle farms in these mountainous regions.

3.3.2 Second half of the twentieth century: coal mining, iron, other mining: ecosystem fragmentation, paralysis and destruction of peasant territories, and planting of eucalyptus and pines

During the 1950s, a process of industrialization in Colombia began, which led to the expansion in the exploitation of coal that had been used since the nineteenth century to feed domestic stoves and was now required to feed the hydroelectric plants of Paipa, Zipaquirá in the highlands of Cundiboyacense and others in Valle del Cauca. In the same way, the development of the steel industry in Paz del Río Boyacá required large tons of this energy mineral. Toward the 1980s and up to the present, this mineral nourishes cement and other steel companies. Due to this, the exploitation of sinkhole coal in these departments increased during the following decades until today in areas of the high Andean mountains, seriously affecting aquifers and water currents, until they led to their contamination or disappearance. To stabilize the mine sinkholes, during the decade of 1960–1970, the secretaries of agriculture increased the planting of eucalyptus forests (*Eucalyptus globulus*) as a timber input for this exploitation work and with them were replaced from native forests by forests planted with this exotic species. Today, the landscape of the Cundiboyacense highlands at the forest level is mostly covered by this type of forest community. The damage to plant and animal biodiversity is considerable, expressed in the disappearance of several species of birds such as eagles and other raptors and mammals such as spectacled bears and deer.

4. Current situation: environmental conflicts and an ecosystem in danger of extinction

The dynamics of the expansion in the occupation of spaces during the last 80 years in the equatorial high mountain areas of Colombia have brought as a consequence and dramatic decrease of forests in these regions and taking into account the figure of less than 5% of existence of the original Andean forests we could affirm that this ecosystem vital for the regulation of biogeochemical cycles and ecological corridors of fauna and the flora that it lends to the whole of the Main Ecological Structure of the nation, is today in danger of extinction [8].

The following environmental problems, which have become real socio-ecological conflicts, which are at the root of this worrying situation are:

i. Expansion of agricultural frontiers to affirm the post-Thispánicos settlements in the mountain ranges, which were constituted during the time of La Colonia (1530–1819) as a basis for the creation of the Republic of Colombia from the War of Independence, which although it brought contributions from European Modernity and gave birth to a mestizo people with the peasantry, It was carried ahead with its demographic catastrophe product of violence and diseases and until the near extinction of indigenous cultures such as the Muisca, the Tairona, and others cornered them in the depths of the jungles and mountains. During this period and for the construction of nascent cities such as the cities of Popayán, Santafé de Antioquia, Cali, Popayán, Bogotá, Tunja and Pamplona, Villa de Leyva, and Guaduas among the main colonial settlements of the Andean mountain, the wood of the equatorial high Andean forests was used and use of the fauna was made that was intense until the second half of the twentieth century and still continues hunting and Illicit trade in the fauna and flora of these ecosystems. The Conquest and the Colony with its disastrous traces of destruction of nature and ancestral peoples and the peasant people, the main victim of all the violence of the nineteenth and twentieth centuries, after the birth of the Republic, has not yet ended in these territories. However, despite the destruction of important forest masses during this period, until the late 1950s when the Agustín Codazzi Geographic Institute generated the cartography of Colombia to support Law 2 of 1959 that created the Zones of Forest Reserve of the Nation, large areas of high Andean forests existed throughout the country.

ii. With the establishment of Hispanic colonial capitals such as Villa de Leyva in Boyacá for the planting of wheat and barley, two clean crops, as they are called from modern agricultural systems, forests were destroyed and land was exhausted in the highlands of Cundiboyacense to generate a deep ecological crisis that has been narrated by Alexander von Humboldt in one of his visits to Colombia for the first decade of the nineteenth century noted that it was necessary to suspend the use of firewood as fuel due to the depletion of forests including those of the Eastern Hills of Bogotá and recommended replacing the source of domestic energy based on firewood with coal from Nemocón With this, in the following years the energy substitution would begin in Colombia, which would be the replacement of renewable energy sources by non-renewable ones, that is, by fossil fuels such as coal [11, 12].

iii. This destruction of the Andean landscapes since the 1950s has continued in these Boyaca and Cundinamarca territories of the high Andean mountains with the proliferation of technified crops of flowers, tomatoes, legumes, and fruits.

iv. With the depletion of native wood, either by the use or by expansion of the agricultural frontier throughout the twentieth century, particularly in the period between 1950 and 2010, pine (*Pinus patula*) and eucalyptus (*Eucalyptus globulus*) plantations have increased in the high equatorial mountains of departments of Boyacá, Santanderes, Cundinamarca, Antioquia, Caldas, Risaralda, Quindío, Tolima, Valle, Cauca, and Nariño, a situation that has brought a serious socioecological problem due to the depletion of water resources and the destruction of habitats of wild fauna and flora, introducing profound changes in the landscape that today barely hides in small refuges the native species still existing but threatened.

v. Potrerization, that is, the alteration of peasant and indigenous production systems by large pastures, is the main indicator in these ecosystems of the crisis of the agricultural sector and the violence of the periods of 1946–1958 and 1982 to the present. Large areas, which can be estimated at some 2 million hectares of peasant and indigenous productive systems, have been being made potrerized or technified with systems highly dependent on agrochemicals and biocides such as potatoes, onions, vegetables, deciduous fruits, and highly technical floriculture with greenhouses, affecting the subsistence economies of these peoples and introducing modernizing patterns of agriculture that are threatening the existence of Peasant peoples and small indigenous settlements, which still survive throughout the Colombian high equatorial mountains.

vi. During the decades of 1980–1990, there were oil bonanzas in the Eastern Plains and piedmont areas of Arauca and Casanare and coal bonanzas in the departments of Guajira and Cesar. All of the above on warm climate ecosystems. From 1990 to the present, the exploration and exploitation of coal and oil have been located in the high Andean mountains on ecosystems of the high Andean forests of the mountain massifs of the moors of Pisba, Guerrero, Guacheneque [13] for coal extraction due to the high prices of this energy source in the international market and oil in the areas of high Andean forests of the Sogamoso valley in the municipalities to the northeast of this as are corrales, Beteitiva, Paz del Rio, and others. This has led to strong and dynamic socioenvironmental conflicts between the peasantry and foreign oil and coal companies favored by the prevailing extractivist policies [14, 15].

vii. At present, new threats, due to the growth of large cities and the megalopolization of cities such as Bogotá D.C., Medellín, Cali, Bucaramanga, Pereira, Manizales, Pasto, and Popayán, loom over these ecosystems: courbanization, tourism, and the installation of solar energy parks. Foreign immigrants and urban dwellers with good economic resources have begun a process of buying peasant land at low costs, to establish country houses, generating a process of dispossession of their lands from poor peasants who normally end up being employed as semi-schooled servitude by their new employers and begins a strong trend of so-called ecotourism that although it constitutes a flow of economic resources that can favor the income of Peasant families also loom as a threat to the high Andean relict forest and páramos by altering the scarce habitats of wild species during bird watching and camping and altering peasant life and the extraction of wild species for urban gardens. In addition to this, the urban growth in the indicated cities and the corridor of the Sogamoso valley or Chicamocha river valley in the department of Boyacá between Tunja and Paz del Río, has been destroying the tutelary hills of urban areas, affecting the native species of the high Andean forest and with it the water resources by conversion of the streams into rainwater channels and waste deposit.

viii. In recent years, in the department of Boyacá, favored by public policies to introduce solar energy as an energy source, projects for large solar energy parks have been proposed in areas of high Andean forests and páramos that threaten the supply and quality of water resources and the existence of native forest ecosystems. This has already manifested itself as an environmental conflict generating important mobilizations of peasants in the municipalities of Paipa and Sotaquirá against these projects, the fragility of public policies and the corruption of government entities in the face of initiatives of new and juicy businesses promoted by foreign investments in these projects has been evident [16].

ix. This currently constitutes the basis of socio-environmental conflicts over water and in defense of forests and moors in the department of Boyacá which is one of the departments with the greatest presence of high Andean forests in Colombia, along with the departments of Cundinamarca, Cauca, Tolima, Santander, and the Coffee Axis.

4.1 Protected areas: overvaluation of the páramos and disregard of the high equatorial Andean forest

During the two decades of this century, the governments of Colombia have been implementing a public policy for the protection of the ecosystems of the high Andean mountains called "Delimitation of the páramos", generating great environmental conflicts due to the way in which it has been built from a technical and scientific perspective that has ignored the territorial realities in the regions where these ecosystems exist. The strategy of delimitation of the páramos advanced by the Alexander von Humboldt Institute, entity in charge of the protection and conservation of biodiversity in Colombia, between 2012 and 2018 has ignored that the high Andean ecosystems, which include the páramos and the high Andean forests are complex sets of deeply dependent ecosystem mosaics, where, In addition, due to the processes of destruction of the high Andean forest in the last 80 years, a process of paramization has been propitiated, which consists of the fact that old areas of high Andean forest are now areas of páramos. In the expression of biologist Germán Márquez, what has happened is that many of the current páramos are degraded high Andean forests. A serious mistake is made, ignoring in the delimitation of high Andean ecosystems the importance of forests, which continue to be pressured by livestock processes and potato monocultures in large areas, without actions to protect these by environmental authorities. This is today the cause of one of the main socio-ecological conflicts in these regions. The equatorial high Andean forest is the main ecosystem barrier that favors water recharge zones and allows the regulation of wind currents and the regulation of water and soil resources, necessary for the existence with quality and quantity of peasant production systems, as well as the habitat of large mammal species such as the spectacled bear, deer and birds such as the eagle, Las Pavas de Monte, and Los Gavilanes.

5. Results: perspectives for its recovery from the Geoecosystems-territory-environment (GTA) model and environmental governance

Although since the 1990s, a policy was developed for the management and protection of high Andean ecosystems and páramos whose application was evaluated by the author of this writing as part of a team of researchers from the Ministry of the Environment. The actions developed for the protection of these areas, especially during the governments of President Santos (2010–2018) focused on the delimitation of the páramos, showing a contempt for the importance of the high Andean forests and their population. Aspects such as its importance in the regulation of water resources and as a bioclimatic barrier between temperate zones and high mountains were not taken into account despite the diagnoses about their critical situation in terms of depletion of the areas that characterize them. During the 20 consecutive years of this century, the exploitation of wood of species such as oak and cedar continues without making effective the application of laws that prohibit this practice and important areas of natural forests continue to be cut down for the planting of pine (*P. patula*) and eucalyptus (*E. globulus*) or more critically for the expansion of the livestock frontier and potato monoculture [2, 17].

The precarious situation of the peasants, due to the abandonment of part of the State, has favored the continuation of the destruction of these ecosystems and the increase of livestock breeding, an agricultural system of very low productivity and generation of employment. This socioeconomic factor is unknown or weakly addressed in government policies and programs for the conservation of high mountain ecosystems and has been the main cause for the emergence of socio-ecological conflicts in the last 10 years within which the movements for the defense of water against gold mining in the Santurban páramo in northeastern Colombia are mentioned. For the defense of water and territory against coal mining in the Pisba páramo in Boyacá and for the defense of peasant territories and micro-watersheds against the mining of construction materials and the construction of solar parks in the wetlands of the Chontales páramo in the villages of Sativa, The Stock Exchange and the Carrizal in the municipalities of Paipa and Sotaquirá.

According to "For example, when the formulation of the territorial planning plans of the municipalities at the national level was carried out in the 2000s, environmental sustainability became merely an instrument requirement to access economic resources, behaving as a polysemic concept and an empty signifier formulated (2000–2010 generally); situation that has not been corrected, since most of these plans to date have not been reformulated or updated" [18].

Due to the above and seeking to solve the crisis in the predominant environmental planning and management, collecting the proposals of the extensive research, arising mainly from the workshops with communities carried out in the processes of the evaluation of the Program of the research carried out by the Water, Health and Environment group on ecosystem services and water governance and the requests made by the communities in conflict against mining in the Paramo of Pisba (Bibliography) and in the villages of Sativa and La Bolsa of the municipality of Paipa has come within the framework of strengthening environmental governance by proposing the Geoecosystem, Territory, Environment, GTA approach, fed from geoecology and landscape theory. This model is based on considering that there are five major geoecosystems that manifest themselves in all territories: water, soil, biodiversity, climate, and social system [2, 15, 19].

Within this approach, the territory as a biogeographic and socioeconomic cultural space, in permanent transformation and construction by anthropic activities and the dynamics of nature, is the main object-subject of environmental analysis.

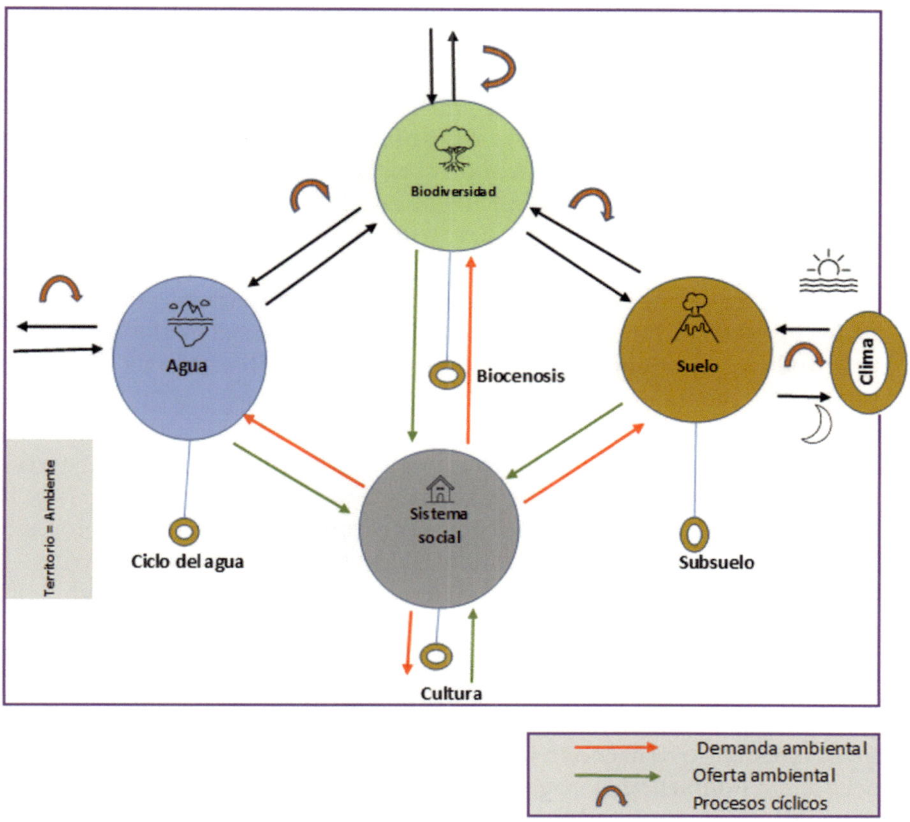

Figure 1.
Geoecosystems-territory-environment model, GTA Fountain: [19].

This object-subject connotation refers to the fact that in this scenario the four great physical-biotic geoecosystems are interacting: water, soil, biodiversity, climate, and the social system that by its conscious nature serves as a transforming subject of the territorial environmental complex (**Figure 1**).

Each of the four geoecosystems: water, soil, biodiversity, and social system nest in the climate geoecosystem, the most important of all and which is the ultimate determiner of the possibility of existence and/or sustainability of the other systems in the environment that constitutes the territory where they are located. Water, soil, and biodiversity systems in natural conditions, that is, not mediated by human action, have cyclical thermoecological exchanges that allow them to be sustainable and maintain their resilience. The social system differs from the previous ones to the extent that it generates more and more environmental demands, which are characterized by thermo-ecological-economic, non-cyclical exchanges, which, given the intensity of the magnitude of the spaces occupied and the intensity of the exploitation of natural resources, are affecting the environmental supply or ecosystem services in the territory, And the human ecological footprint grows continuously, due to an extractivist and consumerist development model that characterizes the current capitalist production system [19].

6. Final conclusions and reflections

i. The equatorial forest ecosystems of the high Andean mountains and the Sierra Nevada de Santa Marta in Colombia are in a serious environmental crisis due to the devastation of recent decades that has reduced them to 5% of their original cover.

ii. The centralist environmental management and ignorance of the environmental systems and the socio-ecological realities of the territories of the high equatorial mountains of Colombia has allowed great extractivist and predatory forces of these ecosystems to continue deepening the ecological crisis and the threat of extinction of the equatorial high Andean forests in Colombia looms.

iii. In recent years, countless and growing environmental conflicts have arisen against extractivism and for the defense of territories based on the defense of water and forests as regulators of climate patterns, wildlife habitats, and water suppliers for thousands of inhabitants.

iv. Corresponding to this socio-ecological problem, an increasingly strong movement has been emerging from civil society, especially from the peasant population that demands the implementation of environmental governance processes that recognize peasants as a population subject to rights to a healthy environment, water, and dignified life on their lands. A law recognizing the rights of the peasant population is currently being drafted in the Colombian parliament, an important step to facilitate the protection, recovery, and conservation of the equatorial high Andean forests in Colombia.

v. Environmental governance actions must be complemented by the change of focus of environmental management that is ignorant of socio-ecological realities toward management based on environmental governance and the geo-ecosystem approach proposed in this article, an approach that has been emerging and acquiring more purposeful force from social movements and academia.

7. Recognitions

I express my gratitude to the communities and social organizations of peasants and community aqueducts of the municipalities of Paipa and Sotaquirá in the departments of Boyacá, Cundinamarca, Santander, and Tolima, who have taught me many of their expressions, knowledge, and customs about life in the mountains, to my parents who educated me recognizing the peasant wisdom to the following universities: Universidad El Bosque, Universidad del Tolima, Universidad P edagógica y Tecnológica de Colombia, and Universidad de Cundinamarca, for having given me the opportunity to investigate with the research groups and with the students the environmental realities of the population of the high Andean equatorial mountains and Alegría Fonseca, Margarita Marino de Botero and Julio Carrizosa Umaña for opening spaces for me to investigate, publicize and publish the results of several investigations on the high Andean mountains in the last 30 years. To INDERENA, for having been the cradle for 10 years of my concerns as an environmental researcher.

Conflict of interest

The author declares that there is no conflict of interest.

Author details

Alfonso Avellaneda-Cusaría
Universidad El Bosque, Faculty of Engineering, Research Group: Water, Health and Environment, Bogotá D.C., Colombia

*Address all correspondence to: avellaneda.alfonso@gmail.com

References

[1] Barbosa C, Rodríguez W, Avellaneda A. Estudios Ambientales En La Sierra Nevada De Santa Marta Afectados Por Cultivos De Marihuana Y Fumigación Con Glyfosato - 1986. Santa Marta: Fundación pro Sierra Nevada de Santa Marta; 1986

[2] Marino de Botero M, Avellaneda-Cusaría A, Rubio F, Rudas G. Programa para el manejo sostenible y restauración de ecosistemas de alta montaña andina colombiana: Páramos (PNP). Bogotá: Ministerio de Ambiente, Vivienda y Desarrollo Territorial; 2008

[3] de Oliveira Figueiredo G. Investigación Acción Participativa: una alternativa para la epistemología social en Latinoamérica. Revista de Investigación. 2015;**39**(86):271-290

[4] Avellaneda-Cusaría A. Ministry of Environment, Housing and Territorial Development. Environmental Thesaurus. Bogotá D.C; 2005

[5] Galindo-T R, Betancur J, Cadena-M JJ. Estructura y composición florística de cuatro bosques andinos del "Santuario de Flora y Fauna Guanentá-Alto Río Fonce", Andes del Este de Colombia. Caldasia. 2003;**25**(2):313-335

[6] Van del Hammen T. History and paleoecology of neotropical Andean montane forests. In: Brown MKAD, editor. Santo Domingo de Heredia. 2001. pp. 63-84

[7] Churchill SP. Biodiversity and Conservation of Neotropical Montane Forest. New York: The New York Botanical Garden; 1995

[8] Carrizosa-U J. La selva andina. Bogotá: l ello Editorial; 1990. pp. 151-184

[9] Galindo-T R, Betancur J, Galindo R. Estructura y composición florística de un bosque humid tropica del Parque Nacional Natural catatumbo-Barí. Revista Colombia Forestal. 2007;**10**(20)

[10] Avellaneda-Cusaría A. Gestión Ambiental y Planificación del Desarrollo-El Sujeto Ambiental y los Conflictos Ecológicos Distributivos. Bogotá D.C.: ECOE Ediciones; 2007. p. 199

[11] Molano-Barrero J. Villa de Leyva: Ensayo de interpretación social de una catastrophe ecológica. Bogotá: Fondo FEN de Colombia; 1990

[12] De la Pedraja Tomán R. Historia de la energía en Colombia. Bogotá: El Ancora Editores; 1985. p. 173

[13] Amaya-Franco S, Durán-Caro JS. Analysis of the Mining-Environmental Conflict in the Páramo De Pisba (in the Jurisdiction of Tasco-Boyacá) from the Perspective of Local Communities. Bogotá D.C.: Universidad Libre; 2017

[14] Basto Torrado SP. El conflicto socioambiental del humedal de Santurbán.Un análisis bioético con enfoque en ecología política. Revista Colombiana de Bioética. 2017;**12**(1):8-24

[15] Magazine S. Inhabitants of Paipa, in Boyacá, denounce environmental destruction by mining projects. Semana Magazine. 2021:7-10

[16] Radio C, Ospina A. Sotaquirá and Paipa Reject Solar Panel Project in Paramo Area, Tunja. 2021

[17] Ministerio de Ambiente. Política nacional para el manejo sostenible y restauración de ecosistemas de la alta montaña colombiana: Páramos o. Bogotá: Imprenta Nacional; 1998

[18] García-Cabana AC, Robayo-Triviño DA. Reflections on the perspectives of environmental and institutional sustainability for the territorial planning of the socioecological systems of Páramo and high Andean forest in Colombia. Investigacion y Desarrollo. 2021;**29**(2):8. ISSN 2011-7574 (online)

[19] Avellaneda-Cusaría A, Fuentes-Cotes M, García-Delgadillo JM, Quintero-Reyes MP. Ecosystem Services and Water Governance, Quebrada El Cedro, Locality of usaquén Dogota D.C. Bogotá D.C.: Universidad El Bosque; 2021. pp. 128-130

Chapter 16

Bertholletia excelsa: Key Species for Sustainable Livelihoods and Forest Conservation

Philippe Waldhoff and Saulo Eduardo Xavier Franco de Souza

Abstract

Brazil nut (*Bertholletia excelsa* Bonpl.) is vital for livelihoods in the Amazon region as a source of income and food, and since it is collected mainly from old-growth forests, it is also a cornerstone for forest conservation. This chapter presents socioeconomic and cultural characteristics of Brazil nut collectors, depicting their production systems and identifying individual perceptions about the effects of this activity on their livelihoods and on forest conservation. We interviewed 119 collectors, organized in associations, cooperatives, or autonomous informal organizations. Collectors live from agriculture, fishing, and forest extractivism, and reported the of use more than 30 forest products, highlighting their refined knowledge on forests. On the other hand, they have little access to formal education and information technology. The collection of Brazil nuts proved to positively impacts livelihoods' capitals, being a key species for forest conservation and maintenance of Amazonian livelihoods. Forests are plentiful in Brazil nut trees and collectors are much more likely to remain in forests, incentivizing their conservation. Moreover, the organization into associations or cooperatives is especially positive to broaden sales options and achieve better market prices, breaking historical cycles where middlemen played a preponderant role in Brazil nut commercialization processes across the Amazon region.

Keywords: non-timber forest products, Brazilian Amazon, traditional communities, sustainable livelihoods, Brazil nut tree

1. Introduction

Bertholletia excelsa Bonpl. (Brazil nut tree) is one of the most prominent species of the Amazon forests. It has called the attention to human populations since early occupation by the first Amerindians to the present day [1], since scientific evidence suggests that some dense *B. excelsa* stands have anthropogenic origins [2]. Its seeds are widely consumed and highly valued in regional, national and international markets. It is of great economic importance, especially across Brazil, Bolivia, and Peru [3, 4], being a source of income for thousands of local communities, as indigenous, riverine, and "*quilombola*" communities [5]. Brazil nuts are the only internationally marketed seed product collected exclusively from old-growth tropical forests [6, 7].

Management of Brazil nut groves is seen as an important strategy for forest conservation and local development [3, 6, 8, 9].

Brazil nut economic importance gained greater dimensions with the crisis in the natural rubber economy, starting from the 1930s. Commercialization of Brazil nuts was structured on the same basis as the commercialization of natural rubber, following the logic of the "*aviamento*," where the commercial relationship is based on the exchange of products, without the flow of money. In this system, extractivists provide products such as rubber and Brazil nuts to the "bosses," who, in return, provide basic subsistence products. Historically, in this exchange relationship, extractivists ended up indebted, characterizing themselves as social groups devoid of economic strength and political insertion, and presenting themselves as fragilely organized and distant from decision-making centers [10].

Most extractive forest products undergo economic cycles marked by three phases: growth in volumes extracted due to the awakening of economic interest; the limit of supply capacity in view of available stocks; and the decline in extraction with the start of commercial substitute plantations [11]. The commercial collection of nuts, through extractivism, has lasted for more than a century, making it an exception to this typical cycle of forest extractives. Between 1998 and 2017, Brazil nut production fluctuated between 23,000 tons and 42,000 tons per year, strongly linked to natural variation of annual fruit production and individual tree productivity [12]. Despite the high intensity of fruit collection, Brazil nut collection does not jeopardize the natural regeneration and maintenance of it populations [13]. Its total production has been increasing year after year, jumping from R$ 9.6 million in 1998 to R$ 105 million in 2017 [14].

The large number of people involved in collecting Brazil nuts and their economic importance underscore the need to unveil the impacts of this activity on the livelihoods of Amazonian extractivists. This chapter aims to present socioeconomic and cultural characteristics of Brazil nut collectors from three different regional contexts in the Brazilian Amazon, depicting their production systems and identifying individual perceptions about the effects of this activity on their livelihoods and on the conservation of forest ecosystems.

2. Extractivism of Brazil nut as a livelihood for Amazonian peoples

2.1 Methodology

The study was carried out in the Brazilian Amazon, in the municipalities of Almeirim, Manicoré and Cotriguaçu, in the states of Pará (PA), Amazonas (AM), and Mato Grosso (MT), respectively (**Figure 1**). The choice was due to the possibility of access to Brazil nut collectors, made possible through partnerships with local institutions. Data collection was carried out through in-depth interviews [15] with Brazil nut collectors, either men, women or young people directly involved in the activity.

In order to identify possible differences of livelihood outcomes, data collection was stratified in relation to their market-oriented organization levels: (i) collectors selling directly to middlemen, called informal; (ii) collectors organized in local associations; and (iii) collectors organized in local and regional cooperatives. In the municipality of Almeirim-PA, community associations have played an important role in organizing the collective commercialization of nuts, at the same time, many collectors still sell directly to middlemen. In Manicoré-AM, the collectors interviewed are part of a cooperative called Cooperativa Verde de Manicoré (COVEMA). In Cotriguaçu, the study included

Figure 1.
Location of the target cities of the surveys.

producers organized in the Associação dos Coletores de Castanha-do-brasil do PA Juruena (ACCPAJ), as well as indigenous people of the Rikbaktsa ethnic group from the Terra Indígena do Escondido, who sell nuts informally, although starting a formal association. In total, 119 collectors (n) were interviewed, of which 50 (42%) trade with middlemen, 39 (32.8%) trade through associations, and 30 (25.2%) through cooperatives.

Interviews were composed of two parts. The first part was designed to characterize the socioeconomic context of producers and their production and marketing systems. The second part aimed at identifying collectors' individual perceptions of the effects of extractive activities on their livelihood capitals, for which indicators related to five capitals (human, social, physical, financial, and natural) were developed based on the Sustainable Livelihood Framework [16, 17].

In addition to Brazil nut collectors, key informants were interviewed, such as members of the board of directors of community associations, community leaders, and/or former members of the community. These interviews were designed to raise general aspects of the communities in relation to their infrastructure, access to public services, population, main economic activities, collective institutions engaged in community organization, and a brief historical view.

Contextual data and overall information collected from key informants were analyzed using descriptive statistics and exploratory data analysis with graphical representations of the results and cross-referencing techniques. Interviews on collectors' individual perceptions of the effects of extractive activity on their livelihood capitals were analyzed using the "sustainable livelihoods approach" [16–18]. Each indicator was assessed through open-ended questions, from which the answers were subsequently categorized. The focus was on the impacts of each production stage until the commercialization. The answers translate the interviewees' perceptions of the existence, nature, and intensity of impacts on livelihood capital, being represented by ordinal scores as very negative (0), negative (0.25), neutral (0.5), positive (0.75), and very positive (1).

For each interview, a value was determined for each capital based on the simple arithmetic average of the indicators (questions) referring to that capital. From the value of each interview, the value of capital was determined, separating each of the situations studied. The averages referring to each indicator and each capital were compared using non-parametric statistics, as the Kruskal-Wallis followed by Wilcoxon Two-Sample Test, using R computation environment [19].

3. Results

3.1 Characterization of Brazil nut collectors

Brazil nut collectors can be considered as "people dependent on forests" [20], as they derive benefits from the forests where they live and work. Agriculture, extractivism, and fishing are the main livelihoods related to subsistence and income (**Figure 2A**). In addition to Brazil nuts, they also harvest more than 30 non-timber forest products (NTFPs), with emphasis on açaí (*Euterpe* spp.), piquiá (*Caryocar villosum*), and uxi (*Endopleura uchi*) (**Figure 2B**). In addition to serving for family use and consumption, these NTFPs are also a source of income for 62% of the interviewed collectors.

Brazil nut collection is carried out based on traditional knowledge, which is usually transmitted orally, from generation to generation. Most respondents (91%) reported that their parents already worked with nut extraction and other NTFPs, from whom they learned to carry out the activity. While on the one hand, Brazil nut collectors have good access to traditional knowledge, on the other hand, poor access to formal education is evident. Only 15.5% of respondents had completed elementary school (**Table 1**), and this percentage reaches 51% of the adult population in Brazil [21]. Another shortcoming of collectors was their low access to the internet, being only 26.4%, while the national average is 64.7% [22]. The lack of access to basic services also contrasts with the great access to government income programs.

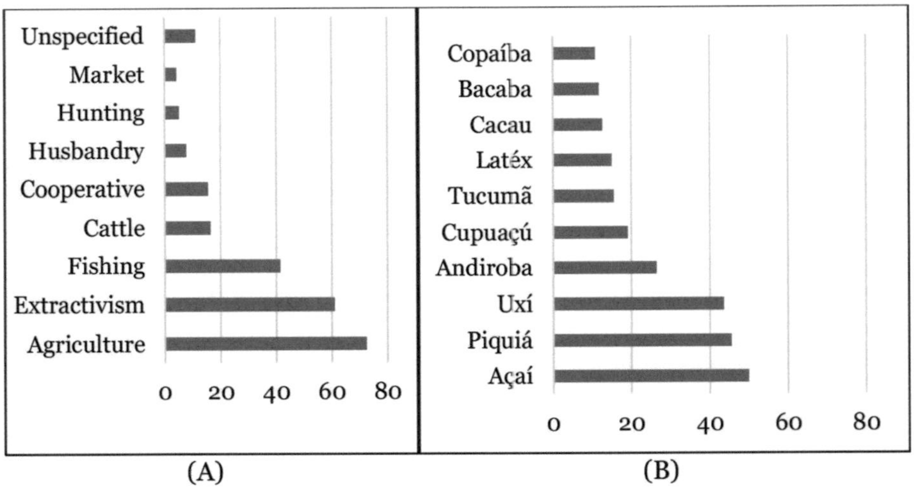

Figure 2.
(A) Main activities carried out by the interviewees in addition to collecting Brazil nuts (a) and main NTFPs used and/or sold (b). Source: Prepared by the authors, 2022. (B) Copaifera spp.; Oenocarpus baacaba; Theobroma cacao; Rubber (Hevea brasiliense); Astrocaryum spp.; Theobroma grandiflorum; Carapa guianensis; Endopleura uchi; Caryocar villosum; Euterpe oleracea and Euterpe precatoria.

Level of formal education	%	Attributes of collectors	%
Did not study	10.3	Use computer	11.7
Incomplete elementary school	74.1	Access to internet	26.4
Complete elementary school	4.3	Access to government programs	59.2
Complete high school	9.5	Benefit from government income transfer (*Bolsa Família*)	49.2
Higher	1.7	Benefit from Ecosystem Service Payment scheme (*Bolsa Floresta*)[1]	17.5
		Fisheries protection insurance	4.2

[1]Specific program for residents of Protected Areas in the State of Amazonas, such as a REDD+ scheme. Source: Prepared by the authors, 2022.

Table 1.
Personal characteristics of nut collectors expressed as a percentage of respondents.

Approximately 59% of collectors, more than double the national average of 22% [23], have their income complemented by social programs, such as the Bolsa Família program (**Table 1**).

Average interviewees' age was 43 years old (±13.24), which may indicate low involvement of young people (under 30 years old) with this activity. Likewise, only 7.4% of the interviewees were women, which could point to low participation of women in the collection of Brazil nuts.

3.2 Productive processes and commercialization of Brazil nuts

Brazil nut production is basically characterized by the collection of fruits, popularly known as "*ouriços*" (**Figure 3**), straight from the forest floor. Fruits have a rounded shape, about 20 cm in diameter, and with a very resistant shell, inside which there are approximately 15 seeds. The seeds, in turn, are composed of a hard woody coat, and the almonds are the edible and the main commercialized part.

Dispersion of *B. excelsa* fruits occurs in the rainy season, from December and July, most concentrated in February, March, and April throughout the Amazon basin. To carry out the collection, extractivists need to enter the forest, either by land or river, which are usually located far away from their homes, taking from a few hours to more than a day to arrive to reach *B. excelsa* groves. Aiming at greater efficiency and effectiveness collection, most of the interviewed collectors (76%) make temporary camps, staying for up to 14 days in the forest, some of the interviewees remain in the camps for up to 3 months. While still in the forest the seeds are extracted from the fruits and stored until they can be taken to the communities. An interesting partnership between one association of Brazil nut producers and farmers who own large tracts of forests was reported. In Cotriguaçú-MT, some of these farmers, who often carry out legal logging, make their forests available for extractivists during the rainy season, (especially when Brazil nuts fruits are available), and in some cases, they even provide accommodation or support structures to extractivists.

B. excelsa collectors require a high level of traditional knowledge regarding the local forests, their historically established limits, their ecological and geographic characteristics, and use of traditional techniques and tools. As an example, we can cite: (i) the use of the *paneiro*, a handmade basket from *titica* vine (*Heteropsis flexuosa*) used to

Figure 3.
(a) Brazil nut tree; (b) breakage of the fruits; (c) Fruits and almonds. Source: (a) authors, (b) Embrapa Portal and (c) OPA health blog.

carry the fruits and seeds, as well as the *câmbito*, a tool made of a wooden stick used to grab the fruits from the ground without having to bend down to catch them; (ii) geographical knowledge of the *B. excelsa* groves location, which includes recognition of trails and access paths, clusters and their main productive trees; (iii) prior determination of the production potential of each year through the systematic observation of flowering; and (iv) opening the fruits by cutting them with a machete, ax or sickle, which requires high technical and manual skills. In addition to several other indirect knowledge indispensable to carrying out the work in the forest, such as hunting and gathering forest products for subsistence during the harvesting period; moving and flow of production through rivers, knowledge about vessels, among others.

Management practices of *B. excelsa* groves identified among collectors' responses were: (i) cleaning, which consists of mowing the trails traveled by people and eliminating competition from the surroundings of individuals of *B. excelsa* and (ii) cutting of vines established in productive individuals, which aims to increase the individual production of Brazil nut trees (**Table 2**). Such practices can promote improvements in work, reducing waste and increasing productivity. No practices were identified to establish collection limits.

The climatic conditions of the forest (heat and humidity) and contact with the soil favor aflatoxin contamination. In order to reduce the risk of this contamination, some procedures called "good practices" can be adopted in the collection and storage of seeds, which basically consist of separating good quality seeds, washing, drying, and storing them properly. It was found that collectors established in organized groups (associations and cooperatives) are more prone to adopting good practices than informal collectors (**Table 2**).

The commercialization of Brazil nuts, at least in the regions studied, is still influenced by the old *aviação* system and the historical changes that this system has

Activities		Manicoré Coop. (%)	Almerim Assoc. (%)	Cotriguaçu Assoc. (%)	Informal (%)
Stand Management practices	Cleaning	52	81	nc[1]	60
	Liana thinning	17	72	nc	74
	Inventory	4	3	nc	3
Best processing practices	Fully application	60	72	0	22
	Partial application	40	22	0	10
	No	0	6	100	68
Primary buyer	Middleman	47	63	57	100
	Industry	0	56	57	0
	Cooperative	93	0	0	0
	Special commercial partnership	40	19	14	0
Receive technical assistance		71	88	18	23
Know Prog. of government		73	69	nc	23

[1]*Data not collected.*
Source: Prepared by the authors based on data collected in the field, 2022.

Table 2.
Activities and characteristics related to Brazil nut productive process according to the percentage of citations by collectors in each location and marketing organization.

undergone, where middlemen play an important role. Middlemen are present, even when associations and cooperatives work (**Table 2**). Eventually, when it proves to be more advantageous, collectors who participate in an association or cooperative may choose to sell to the middleman.

In Almeirim-PA, there is only one out of four communities studied, that interviewees declared they sell straight to the industry. This sale is due to the action of the community association whose main objective is to organize the sale of the *B. excelsa*. Through this association, collectors access the Banco do Brasil credit line for Sustainable Rural Development and other federal public policies such as the National Forestry Program (PRONAF). The Association stood out for accessing credit for the collection of Brazil nuts for 9 years without default, a rare occurrence for community associations. Access to funding lines was identified in just one other community.

In Manicoré - AM, the collectors interviewed are part of the Cooperativa Verde de Manicoré (COVEMA) founded on June 5, 2006. The cooperative has a processing plant where Brazil nuts are processed, until the final product, which is a dried Brazil nut vacuum packaged. The Cooperative established partnerships with government and private institutions in order to guarantee the purchase and quality. Despite the great socioeconomic importance of the Cooperative, it was identified that 47% of the cooperative members interviewed also sell their production to middlemen.

In Cotriguaçu - MT, the study included producers organized in the Associação dos Coletores de Castanha-do-brasil do PA Juruena (ACCPAJ), as well as indigenous people who sell nuts informally, both groups sell their production to middlemen, or directly to the small and large industries that carry out the final processing.

3.3 Commercialization and livelihoods

The survey results show that collecting Brazil nuts is an activity that contributes positively to the five livelihood capitals of collectors (**Figure 4**).

Two indicators of physical capital, acquisition of household goods and acquisition of work materials were made possible by the income from marketing Brazil nuts, had the highest scores in the survey. And these were followed by the natural capital indicator: forest conservation. Strengthening and expansion of the communities' external relations; income regularity; and acquisition of new knowledge, were the main highlights in social, financial, and human capital, respectively (**Table 3**).

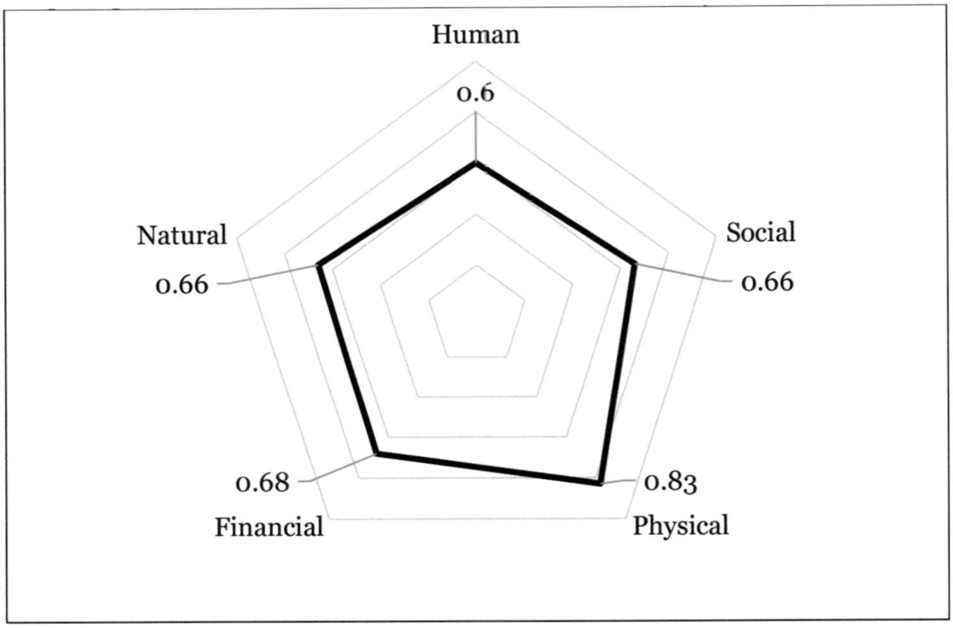

Figure 4.
Average score of each livelihood capital of collectors obtained through the perception of the collectors themselves in the three study areas. Source: Prepared by the authors, 2022.

(Capital) Indicator	General	Informal		Associations		cooperative	
	Ind.	Ind.	Cap.	Ind.	Cap.	Ind.	Cap.
(H1) Safety at work	0.41	0.32b -	0.53b -	0.60 to	0.68 to	0.30b -	0.58b -
(H2) Food safety	0.59	0.53b -		0.65 to		0.60 ab	
(H3) Traditional knowledge	0.69	0.66 to		0.74 to		0.67 to	
(H4) New knowledge	0.7	0.60b -		0.74 to		0.75 to	
(S1) Opportunity for young people	0.57	0.62 to	0.60b -	0.57 to	0.75 to	0.53 to	0.62b -
(S2) Participation of women	0.64	0.53b -		0.79 to		0.61 ab	
(S3) Social participation	0.68	0.66b -		0.79 to		0.58b -	
(S4) External relations	0.74	0.60c -		0.86 to		0.76b -	

(Capital) Indicator	General	Informal		Associations		cooperative	
	Ind.	Ind.	Cap.	Ind.	Cap.	Ind.	Cap.
(F1) Community infrastructure	0.76	0.65b -	0.80b -	0.78 ^{to}	0.89 ^{to}	0.84 ^{to}	0.81 ^{b}
(F2) Household goods	0.92	0.94 ^{ab}		0.95 ^{to}		0.87b -	
(F3) Working materials	0.82	0.80b -		0.94 ^{to}		0.71c -	
(Fin1) Price	0.68	0.49b -	0.58c -	0.77 ^{to}	0.77 ^{to}	0.78 ^{to}	0.69b -
(Fin2) Income regularity	0.73	0.79 ^{to}		0.82 ^{to}		0.59b -	
(Fin3) Put options	0.62	0.45b -		0.72 ^{to}		0.68 ^{to}	
(N1) Resource stock	0.61	0.59 ^{ab}	0.70 ^{to}	0.66 ^{to}	0.66 ^{to}	0.57 ^{b}	0.62 ^{to}
(N2) Access to the resource	0.63	0.76 ^{to}		0.63 ^{ab}		0.49 ^{b}	
(N3) Forest conservation	0.75	0.75 ^{to}		0.70 ^{to}		0.80 ^{to}	

H, S, F, Fin, and N refers to the capitals: human, social, physicist, financial, and natural. Values of the same letter did not show significant differences in t. by Mann–Whitney. Source: Prepared by the authors, 2022.

Table 3.
Average scores and comparisons of livelihood capitals and indicators according to the organization for commercialization: informal, association, and cooperative.

These results also allowed a comparative analysis between the different publics studied (**Table 3**). The intermediate level of organization, represented by associations, showed significantly better results than the cooperative level of organization and informal sales (direct to middlemen), for financial, human, social, and physical capital. In natural capital, the results were similar for the three groups. In a general analysis, of the 17 studied indicators, the associates surpassed the cooperative ones in 7 (41%) and the informal ones in 10 indicators (59%).

4. Discussion

4.1 Brazil nut collectors: people who live with the forest

Brazil nut collectors interviewed in this research can be considered as Traditional Communities or Peoples, defined as:

> *... culturally differentiated groups that recognize themselves as such, that have their own forms of social organization, that occupy and use territories and natural resources as a condition for their cultural, social, religious, ancestral and economic reproduction, using knowledge, innovations and practices generated and transmitted by tradition. [24]*

The livelihoods of Brazil nut collectors are strongly linked to the environment in which they live. Subsistence agriculture, forest extraction and the use of rivers are among the main work activities developed. Each collector follows a specific strategy characterized by a complex combination of different production and income-generating activities. Most of the activities are developed in a very close relationship with nature and require the use of traditional knowledge, which directly influences the outcomes and processes of the activity.

Brazil nut plays a historical role of paramount importance for the livelihoods of the Amazonian peoples. Hundreds of years ago, Amerindian peoples were directly responsible for the creation of dense Brazil nut groves [2, 25]. At that time, the Amazon forest was not under the threat of deforestation, as has been happening since the 1960s. In this sense, collecting Brazil nuts is a counterpoint to deforestation, directly contributing to forest conservation, as pointed out in the natural capital.

The indicator on forest conservation points out that the presence of Brazil nut trees encourages the protection of forests, preventing the forest to be cut down for other land uses or incentivizing its protection. In addition, 20% of the collectors interviewed stated that they were planting *B. excelsa* trees. These facts place the collectors in a central role in the conservation of the forest acting directly in the axis of the conservation of biodiversity and the mitigation of climatic changes.

The Brazilian Amazon has special importance in the discussion on environmental policies due to its cultural diversity, biodiversity, and its role in the global climate [26]. The danger of global warming urges to reduce greenhouse gas emissions and/or increase atmospheric carbon fluxes to biomass or soil. Due to its great extension, the carbon stock and its capacity to retain and release carbon, the Amazon forest is a fundamental component in this context. Greenhouse gas emissions caused by deforestation and forest degradation account for 20% of total emissions [27].

The *Bolsa Floresta* Program (**Table 1**), a program run by Fundação Amazônia Sustentável, can be considered a payment scheme carried out for the reduction of greenhouse gas emissions from deforestation and forest degradation including forest conservation, sustainable forest management, and maintenance of forest stock (REDD+). *Bolsa Floresta* beneficiaries are the only ones who receive this type of benefit among the groups studied, although reducing emissions via REDD+ constitutes an important strategy for mitigating climate change, particularly in developing countries with large forest cover [28].

In this sense, the activity of Brazil nut collectors is part of the international debate on REDD in terms of social issues, making the path to ensure that those who really need and act directly to reduce gas emissions are actually the beneficiaries [27, 29, 30]. Despite the uncertainties and methodological difficulties that would involve an assessment of the contribution of forests maintained by Brazil nut collectors, payment for this environmental service would be a challenge for governments, research institutions, donors, and conservationist organizations.

In addition to the conservationist role that Brazil nut collectors play, nut extraction ensures the necessary income for subsistence and investments in improving their quality of life. The income regularity, which harvests provide, is essential to reduce financial vulnerability in which they live, allowing investments in physical capital such as household goods, work materials, and community infrastructure. It was with the drop in international demand for Brazilian rubber, in the 1920s, that Brazil nuts began to gain economic significance [31]. And, from 1986, it became part of the official data of the Brazilian Institute of Geography and Statistics, having reached a volume of 34,644 thousand tons in 2016, generating an income of 110.1 million reais [14].

Brazil nut is a product that resists the climatic conditions of the Amazon, and can be transported and stored without the need for high investments in infrastructure and equipment, however, good collection and storage practices are recommended. Due to its natural durability, it has established itself as a product for the national and international market, even with very few incentives for public action, unlike other agricultural activities that rely on incentives, from technical assistance and technological development to financing. Only recently, programs such as the Minimum Price Guarantee Policy for Sociobiodiversity Products (PGPM-Bio) and the Food Acquisition Program (PAA), provided possibilities for the commercialization of Brazil nuts to find complementary or alternative paths [32].

Cattle raising is increasingly becoming a common livelihood in several contexts across Amazonian agricultural frontiers and it sets the forest at risk, leading to forest conversion, as has been happening with rubber tappers in Acre or with settlers in southern Pará [33]. Cattle ranching is one of the factors that can lead to an increase in deforestation rates, which has been increasing in the Brazilian Amazon in 2019 and 2022. The decline in extractive activities can increase forest vulnerability. Factors such as technological progress, emergence of economic alternatives, population growth, reduction of stocks, wage levels in the economy, changes in relative prices, put extractivism in check [11]. The longevity of the nut collection and commercialization systems contrasts with the extractivism cycle pointed out by Homma [33], marked by initiation, rise, stagnation and decline.

Aiming at conserving the forest, it is important to value the production of nuts within natural forests and the work carried out by collectors. Nut production brings improvements to local livelihoods and contributes to forest conservation. This research corroborates the results found by Guariguata et al. [7] where socio-ecological systems do not require major changes to maintain yield, actions may be indicated to preserve, diversify and intensify the production of Brazil nut groves, making them increasingly integrated into human modified landscapes.

4.2 The role of associativism and cooperativism

The organization of production and commercialization processes, when carried out through associations, proved to be more beneficial for collectors, when compared to the Cooperative system and commercialization through middlemen. COVEMA, the cooperative to which collectors in the Manicoré region are linked, has a more complex organizational structure, with a larger number of collectors and representing a larger geographic area, being responsible for the administration and management of a nut processing plant that sells Brazil nut at vacuum packages. To establish a viable business structure, an initial phase of 10–20 years would be required, followed by a consolidation phase of similar duration [34]. In this sense, the Cooperative, created in 2006, would still have a time of at least 20 years for its consolidation.

In addition to the intrinsic complexity of managing a cooperative, the historical development context of social groups in the region does not favor the work developed by complex collaboration networks. The Amerindian peoples suffered a relatively recent process of social and cultural disruption. The colonization of the Americas followed the expansion patterns of other empires in history, characterized by wars, enslavement, deportation and genocide [35]. In the Brazilian Amazon, this process occurred more sharply from the 18th century onwards, in the search for forest products [36], when the extermination of indigenous peoples in the Brazilian Amazon took place.

The Rubber cycle had a social, economic and cultural impact on the Amazon. Approximately 500,000 Brazilians were transported from the northeast region to the north of the country to work in the rubber plantations. The rubber tappers, descendants of northeastern migrants, were incorporated into an organized system to control the scarce workforce, where the rubber tappers were forced to sell and buy in the rubber tapper's shed and were tied to the boss who supplied them with the goods [10]. Changes to break with this system began to intensify from the 1970s and 1980s, with movements led by rubber tappers such as the *Aliança dos Povos da Floresta*. In Manicoré, where COVEMA was established, the work of the National Council of Rubber Tappers (CNS) was fundamental to boost community organization in the municipality [37].

The collection and commercialization of Brazil nuts follows the traces of this same system. Most of the collectors interviewed (92%) had their parents involved in extracting resins/gums (75.2%) and/or Brazil nuts (76.2%). However, the results found in financial capital show that collectors who are organized in associations or cooperatives managed to expand their sales options (**Tables 2** and **3**), reducing dependence on middlemen, accessing differentiated markets and obtaining better prices. Of the organized collectors, 56% declared that they still sell to middlemen. In this case, as they have commercialization options, they sell to whoever offers the best prices. Collectors who sell directly to middlemen have few or no sales options, adopting the value stipulated by the middlemen, which has historically been an unfavorable system for collectors [37, 38].

Among the levels of organization studied, associativism is the one that most resembles the social organization of Amerindians, based on the reciprocity of family networks, increasing proximity and the level of trust between individuals [34]. Higher organizational levels are usually more fragile [35]. However, associations have no legal attribution to market and earn profit. The creation of cooperatives or another type of legal representation that enables commercialization must permeate the discussion process of formalized groups.

The organization of producers brought improvements in several studied indicators, showing its importance in overcoming historical difficulties. One aspect still to be considered is the low presence of young people, showing a tendency for future generations to be removed from collection, therefore, strategies that will promote the participation of a younger public can come to collaborate with continuity in the transmission of knowledge and the strengthening of the activity. For this, it is necessary to offer better opportunities for education and work, as well as other socioeconomic conditions that are more favorable to the maintenance of young people in their communities of origin.

5. Conclusions

Brazil nut collectors have extensive traditional, ecological and technical knowledge about the activity they perform, in addition to making use of a wide range of forestry products. The importance of the forest in their livelihood means that, in addition to being people who depend on the forest, collectors play the role of guardians of the forest. The forest provides livelihoods for people, who contribute to its conservation. However, the difficulty of accessing formal education, technological development and internet, can lead to the removal of these actors from living with the forest, increasing the risks of its destruction. Recognition of this gap in service provision is essential to propose ways of including this group in public policies for the development of the Amazon.

The associations, representing an intermediate level of organization for marketing, between direct sales to the middleman and the cooperative, brought the most benefits to the collectors' livelihoods. The improvements found in indicators such as the choice of buyers, price negotiation, expansion of external relations and infrastructure, show the importance of social organization for the production and commercialization of Brazil nuts, overcoming historical difficulties.

This research showed that Brazil nut collection is an activity that improves the livelihood of local populations and contributes to forest conservation, with associations and cooperatives playing an important role in this process. Increasing business management capacity, promoting plantations, strengthening local and regional markets and seeking income alternatives for times of insufficient harvest, may be some of the actions that should be promoted.

Acknowledgements

To Professor Edson Vidal of the University of São Paulo (USP) by supporting the coordination of the Project. To PhD. Renata Bergamo Camarez, who coordinated data collection and analysis of Almeirim-PA. To M.Sc. Fabrício Tinto, who coordinated data collection in Manicoré-AM.

Notes/thanks/other declarations

Special thanks to all collectors and other participants who were willing to give interviews and supported data collection, and to all the traditional peoples and communities that have been keeping the forests standing and planting Brazil nut trees for hundreds of years. This book chapter is a translation of the original article "B. excelsa: espécie chave para meios de vida sustentáveis e conservação florestal".

Author details

Philippe Waldhoff[1]* and Saulo Eduardo Xavier Franco de Souza[2]

1 Federal Institute of Education, Science and Technology of Amazonas (IFAM), Manaus, AM, Brazil

2 Center for International Forestry Research (CIFOR) and World Agroforestry (ICRAF), Belém, PA, Brazil

*Address all correspondence to: philippe.tim@gmail.com

References

[1] Thomas E, Caicedo CA, McMichael CH, Corvera R, Loo J. Uncovering spatial patterns in the natural and human history of Brazil nut (*Bertholletia excelsa*) across the Amazon Basin. Journal of Biogeography. 2015;**42**:1367-1382

[2] Scoles R. Do rio Madeira ao rio Trombetas: novas evidências ecológicas e históricas da origem antrópica dos castanhais amazônicos. Novos Cadernos NAEA, Belém. 2011;**14**(2):265-282

[3] Mori SA. The Brazil nut industry: Past, present and future. In: Plotkin M, Farmocare L, editors. Sustainable Harvest and Marketing of Rain Forest Products. Washington: Island Press; 1992. pp. 241-251

[4] Zuidema PA, Boot RGA. Demography of the Brazil nut tree (*Bertholletia excelsa*) in the Bolivian Amazon: Impact of seed extraction on recruitment and population dynamics. Journal of Tropical Ecology, Cambridge. 2002;**18**:1-31

[5] Ribeiro MBN, Jerozolimski A, Robert P, Magnussom WE. Brazil nut stock and harvesting at different spatial scales in southeastern Amazonia. Forest Ecology and Management, Melbourne. 2014;**319**:67-74

[6] Clay JW. Brazil nuts. The use of a keystone species for conservation and development. In: Freese CH, editor. Harvesting Wild Species: Implications for Biodiversity Conservation. Baltimore: The John Hopkins University Press; 1997. pp. 246-282

[7] Guariguata MR, Cronkleton P, Duchelle AE, Zuidema PA. Revisiting the 'cornerstone of Amazonian conservation': A socioecological assessment of Brazil nut exploitation. Biodiversity and Conservation. 2017;**1**(21):2007-2027

[8] Ortiz EG. Brazil nuts (*Bertholletia excelsa*). In: Shanley PAR, Pierce SA, Laird AGUILLEN, editors. Tapping the Green Market: Certification and Management of Non-Timber Forest Products. London, UK: Earthscan Publications Ltd.; 2002. pp. 61-74

[9] Wadt LHO, Kainer KA, Staudhammer CL, Serrano ROP. Sustainable forest use in Brazilian extractive reserves: Natural regeneration of Brazil nut in exploited populations. Biological Conservation. 2008;**141**:332-346

[10] Allegretti M. A construção social de políticas públicas. Chico Mendes e o movimento dos seringueiros. Desenvolvimento e Meio Ambiente, Curitiba. 2008;**18**:39-59

[11] Homma AKO. Extrativismo vegetal ou plantio: qual a opção para Amazônia? In: Homma AKO, editor. Extrativismo vegetal na Amazônia: história, ecologia, economia e domesticação. Brasília, DF: Embrapa; 2014. p. 468

[12] Kainer KA, Wadt LHO, Staudhammer CL. Explaining variation in Brazil nut fruit production. Forest Ecology and Management. 2007;**250**(3):244-255

[13] Wadt LHO, Faustino CL, Staudhammer CL, Kainer KA, Evangelista JS. Primary and secondary dispersal of *Bertholletia excelsa*: Implications for sustainable harvests. Forest Ecology and Management. 2018;**415/416**:98-105

[14] IBGE. Produção da extração vegetal e da silvicultura. Rio de Janeiro: Instituto Brasileiro de Geografia e Estatística; 2018 Available from: https://www.ibge.gov.br/estatisticas-novoportal/economicas/agricultura-e-pecuaria/9105-producao-da-extracao-vegetal-e-da-silvicultura.html?=&t=resultados [Accessed: June 14, 2020]

[15] Bernard HR. Research Methods in Anthropology: Qualitative and Quantitative Approaches. 4th ed. New York: Altamira Press; 2006. p. 821 Available from: http://www.antropocaos.com.ar/Russel-Research-Method-in-Anthropology.pdf [Accessed: March 22, 2014]

[16] DFID. Sustainable Livelihoods Guidance Sheets. London: Department For International Development; 1999. p. 22 Available from: www.eldis.org/vfile/upload/1/document/0901/section2.pdf [Accessed: October 12, 2012]

[17] Kusters K, Belcher B, Ruíz-Pérez M, Achdiawan R. A method to assess the outcomes of forest product trade on livelihoods and environment. CIFOR, Bogor. 2005;**32**:1-26 Available from: http://www.cifor.org/ntfpcd/pdf/owp7.pdf [Accessed: January 10, 2012]

[18] Bebbington A. Capitals and capabilities: A framework for analyzing peasant viability, rural livelihoods and poverty. World Development. 1999;**27**(12):2021-2044

[19] R CORE TEAM. R: A Language and Environment for Statistical Computing. Vienna: R Foundation for Statistical Computing; 2020 Available from: http://www.r-project.org/index.html [Accessed: June 28, 2020]

[20] Newton P, Miller DC, Byenka MAA, Agraval A. Who are de forest-dependent people: A taxonomy to aid livehood and lande use decision-making in forested regions. Land Use Policy, Guilford. 2017;**57**:388-395

[21] IBGE. PNAD Contínua 2016: 51% da população com 25 anos ou mais do Brasil possuíam no máximo o ensino fundamental completo. Rio de Janeiro: Instituto Brasileiro de Geografia e Estatística; 2018 Available from: https://agenciadenoticias.ibge.gov.br/agencia-noticias/2013-agencia-de-noticias/releases/18992-pnad-continua-2016-51-da-populacao-com-25-anos-ou-most-of-brazil-have-only-complete-fundamental-education.html [Accessed: June 20, 2018]

[22] Gomes HS. Brasil tem 116 milhões de pessoas conectadas à internet, diz IBGE. G1, Rio de Janeiro, 21 fev. Rio de Janeiro. 2018 Available from: https://g1.globo.com/economia/tecnologia/noticia/brasil-tem-116-milhoes-de-pessoas-conectadas-a-internet-diz-ibge.ghtml [Accessed: May 8, 2019]

[23] BRAZIL 2018. Ministério do Desenvolvimento Social. Available from: www.brasil.gov.br/noticias/cidadania-e-inclusao/2018/05/cerca-de-46-6-milhoes-de-brasileiros-sao-beneficiados-pelo-bolsa-family [Accessed: May 8, 2019]

[24] BRAZIL. Decreto Federal N° 6.040, de 07 de fevereiro de 2007: Institui a Política Nacional de Desenvolvimento Sustentável dos Povos e Comunidades Tradicionais. Brasília, DF: Presidency of the Republic; 2007 Available from: http://www.planalto.gov.br/ccivil_03/_ato2007-2010/2007/decreto/d6040.htm [Accessed: April 11, 2012]

[25] Scoles R, Gribel R. Human influence on the regeneration of the Brazil nut tree (*Bertholetia excelsa* Bonpl., Lecythidaceae) at Capanã Grande Lake, Manicoré, Brazil. Human Ecology. 2015;**43**(6):843-854

[26] Fearnside PM. The involving context of Brazil's environmental policies in Amazonia. New NAEA Notebooks, Belém. 2013;**16**(2):9-25

[27] Angelsen A. Moving Ahead with REDD: Issues, Options and Implications. Bogor: CIFOR; 2009. p. 172 Available from: http://www.cifor.org/publications/pdf_files/Books/BAngelsen 0801.pdf [Accessed: September 30, 2014]

[28] Arima EY, Barreto P, Araujo E, Filho BS. Public policies can reduce tropical deforestation: Lessons and challenges from Brazil. Land Use Policy. 2014;**41**:465-473

[29] Karsenty A, Vogel A, Castell F. "Carbon rights", REDD+ and payments for environmental services. Environmental Science & Policy, Exeter. 2014;**35**:20-29

[30] Mahanty S, Such H, Tacconi L. Access and benefits in payments for environmental services and implications for REDD+: Lessons from seven PES schemes. Land Use Policy, Guildford. 2013;**31**:38-47

[31] Barbosa M, Moret AS. Produção e comercialização da castanha-do-brasil: economia e disponibilidade financeira (subsistência das famílias residentes em Reservas Extrativistas). Management Scare Ambient, Florianópolis. 2015;**4**(2):413-428

[32] BRAZIL, 2020. Entenda como funciona o PAA. Available from: https://www.gov.br/pt-br/noticias/agricultura-e-pecuaria/2020/01/entenda-como-funciona-o-programa-de-aquisicao-de-alimentos [Accessed: January 25, 2021]

[33] Homma AKO, Walker RT, Carvalho RA, Couto AJ, Ferreira SAP. Políticas agrícolas e econômicas para a conservação de recursos naturais: o caso de castanhais em lotes de colonos no sul do Pará. In: Homma AKO, editor. Extrativismo vegetal na Amazônia: história, ecologia, economia e domesticação. Brasília, DF: Embrapa; 2014. pp. 177-191

[34] Jong W, Cornejo C, Pacheco P, Pokorny B, Stoian D, Sabogal C, et al. Opportunities and challenges for community forestry: Lessons from Tropical America. In: Mery G, Katila P, Galloway G, Alfaro RI, Kanninen M, Lobovikov M, Varjo J, editors. Forest and Society: Responding Global Drivers of Change. Vienna: IUFRO; 2010. pp. 299-314 (World Series, 25)

[35] Harari YN. Sapiens – A Brief History of Mankind. Translated by Marcoantonio J. 32nd ed. Porto Alegre: L&PM; 2018. p. 464

[36] Roller HF. Expedições coloniais de coleta e a busca por oportunidades no sertão amazônico, c. 1750-1801. Revista de História de São Paulo, São Paulo. 2013;**168**:201-243

[37] Amaral R, Aleixo J, Franco M. Organização da produção na Amazônia: a experiência de comercialização coletiva da castanha em Manicoré, AM. IEB: Brasília, DF; 2012. p. 86

[38] CHAVES. MPSR De "cativo" a "liberto": o processo de constituição sócio-histórica do seringueiro do Amazonas. Manaus: Valer; 2011. p. 164

Section 5

Culture and Education

Chapter 17

The Amazon Forest in Classes of "Guamá Bilíngue" Project

Douglas Rodrigues Pinheiro da Costa and Rita de Cássia Paiva

Abstract

This chapter describes at first the program Guamá Bilíngue (name referred to the river along the university bank which is part of the Amazon Basis) that, from the Universidade Federal do Pará (UFPA in Portuguese acronym meaning the Federal University of Para State) and its purpose of social transformation through the foreign languages teaching/learning process. Then, is to divulge the proposal to produce a textbook for future English classes in the program, especially because it is where students can discuss about the Brazilian Amazon Forest to generate critical consciousness about taking care of it. So, it will be presented some steps to that creation considering the theoretical framework and the objectives that students must reach in this program. Some authors are mentioned as a researching base but are not present in this paper.

Keywords: Guamá Bilíngue, foreign languages textbook, Brazilian amazon forest, students in vulnerability, amazon conservation studies

1. Introduction

We are the fruit of a society whose unique technological moment is shaping us. It is really a very hard situation and as well as a very hard issue to talk and deal with all the (almost endless) conditions on the schooling in the riverbanks all over the world and assure those populations a quality education or even any education at all.

It is not—and it is never going to be—a beautiful sight. Brown children among wastes do not go to billboards, do not they? Even if they are cuties offered by the Amazon rainforest. However, the forest peoples are not only those who have lived in the forest. Students in the peripherical areas of the main cities are as uncared by the government as the autochthonal ones. And there comes the environmental/social/cultural education that can provide a new perspective to adolescents in public schools as a way of reaching the formal work market or even taking them out of the shantytown dirty streets.

Awaking a critical view in the enormous population of poor students is an impossible task if it is taken as a whole. But when a free language course is offered to such public, as a lure, to bring them to a new paradigm of thinking and learning, important themes can be discussed. So, a foreign language textbook is built especially for this public—adolescents in secondary public schools—aiming to provide them a game-changing.

In other words, the proposal is not a social experiment or just teaching languages. The discourse implicit in the whole project, whether in the texts, in the parallel research, or in the training to take over a classroom, is very simple: education, as a right guaranteed in the Brazilian constitution, offered in the classroom is, to a large extent, involved with the discourse of that elite that seeks to maintain the status quo of the minimum quality of public education and that it is up to Guamá Bilíngue[1] to raise its voice and offer didactic materials and activities that can provide the necessary inputs for students to find their academic path in the midst of what the contemporary school is offering in the way of education.

And there is an important advantage: environmental care in the Amazon region is as needed as the forest itself; then, teaching English while teaching respect and conservation may be, or shall be, an opportunity for adolescents, volunteers, and society.

2. Guamá Bilíngue project

Guamá Bilíngue is an extension program of the Universidade Federal do Pará (UFPA in the Portuguese acronym) and exists since 2010 with one main objective: to transform the lives of teenagers from the most peripheral areas of Belém, the capital of Pará state, in Brazilian Amazon Forest region. Those teenagers must be in public high school yet, but for any reason do not hold any perspective on the future. In this program, such students have the chance to study a foreign language and increase their world vision and the way they interpret their environment. In other words, the overall aim of the project is to reduce violence and social prejudice through education. At the same time, the underlying idea is to try to rescue young people from 15 to 18 years old who are in secondary education in the public network and help them to stay away from marginality using strategies in the theoretical framework of the Pedagogy of the Oppressed (Freire, 2012); the mental models of Lawson and Lawson (1993); Meaningful Learning (Ausubel, 1982); premises of Vygotsky (1998); and cognitive linguistics. For it, the classes have been designed to involve language properly and jointly with what is known as transversal themes or transversal subjects according to the PCN (Parâmetros Curriculares Nacionais in Portuguese) that align the course of Brazilian education. In addition, it collaborates with both the undergraduate students of language and literature as well as the adolescents who attend the classes, as it is proposed as a place of formation for future generations with a critical conscience and willingness to transform themselves and their work environment based on their new perception of the world.

In the first part, the chapter will present the same theories that are related to the elaboration of a textbook and its importance for a learning process, some concepts about the transversal themes, and what allows to be performed discussions in the classroom about subjects that are important to increase critical thinking. Then, it will be presented a suggestion for the elaboration of a unit for the textbook talking about the Brazilian Amazon Forest, the pedagogical steps that the unit will have and its justifications. Also, it aims alert people about the importance to start and initiate the English classes in the project and especially the relevance to involve the subject of Amazon Forest to reach the goal of fomenting the responsibility in students that sometimes do not think about their own environment in which they are inserted and conserves their own cultural background, their values, and responsibilities.

[1] *Guamá* is the name of one of the main rivers in the Amazon Basin and *Bilíngue* means bilingual.

As the realization of the entire book would not be possible in this chapter, it is going to present the steps for creating one unit about nature will have in the textbook considering the purpose that teachers have to achieve in these students and to explain how the text suggestion and the grammar part will foment the oral discussion during the classes when the English classes start, because until now the program just offers Spanish classes with a special textbook for this language and German whose beginning of the classes are settled to April/2023.

2.1 Theoretical framework

At the same time, the program offers some benefits to students with social vulnerability, undergraduate students with degrees in English Language, and Literature could practice their teaching skills with this program which becomes a place to put into practice all was learned during their career and reflecting about their professional actions in the classroom. Guamá Bilíngue program has been a teaching laboratory for 13 years. Therefore, the program, sustained by the university volunteers, has the power to motivate as many students that are learning the language as students from the degree in English Language and Literature. Paiva supports that "el proyecto propone, en el mismo afán, incentivar la motivación en los alumnos que se benefician del proyecto, el alumno de la carrera de letras/español, tal como el alumno joven en riesgo social" [1].

For that, the classes count with material that was elaborated by the volunteers, divided by lessons containing basically two parts: a grammatical one, which presents the structure of the language, vocabulary and exercises, and a reading part, which can be found a text that talks about any theme or subject which takes students to deliberate about the society, nature, technology, or literature. It is important to mention that the main goal of the program turns around the transformation and motivation of students to make them create their own critical perspectives and objectives for their lives. Teaching language becomes just an instrument, a tool to reach this goal. Clearly, students will learn the language during this process, but not without stimulating their critical thinking.

Each unit of the material available to the program contains a theme to be discussed where there is a text to be read and some questions for students to make a text comprehension. To keep the structure of the material, it will be presented a plan to elaborate a unit where the main theme to be discussed involves the environment. As known, Belém is the biggest city placed in the Amazon Forest, so it becomes much important to stimulate critical thinking about caring for the place and its benefits and also to encourage the responsibility of citizens of the forest themselves. That is why the objective of the program is not only to teach language, but also citizenship. It is a commitment to society, education, and nature. Silva says that "O ensino da língua inglesa deve ir além das quatro habilidades linguísticas, contribuindo para a formação de cidadãos críticos e transformadores da realidade social" ([2], p. 195).

Volunteer teachers of Guamá Bilingue have the mission to stimulate this critical thinking in the students as the first goal. The language and the subject discussions become a very important tools to reach it. Through it, students will have time to learn how to identify a problem in order to find a probable solution or even how to see possibilities that they are not used to see yet. "Critical thinking can develop if it is practiced, and there needs to be enough time in the learning process. One way to develop thinking skills in the learning process is by training the students to search and find the problem" ([3], p. 2).

Discussing subjects such as those, students have the chance to open their minds to points of view that they have not thought about yet. Concerning to conservation of nature, for example, they would be able to think about their own reality, precisely about the environment they are inserted (the Amazon region), and reflect about the problems and difficulties that this environment can present. For example, volunteer teachers would be able to make them think about the problems the Amazon region faces, know about dilemmas they do not know yet, know the law related to this environment, sustainable actions that could be put into practice seeking reforestation always facing the reality. Reality such as presented by Fearnside: "Sustainable Forest management has become a requirement of Brazilian legislation and an objective at least nominally espoused by all. But it faces fundamental contradictions between restraining harvest rates to levels that will allow the forest to regenerate and maximizing financial returns to loggers" ([4], p. 766).

In this case, it becomes important the process of making the material that works as a textbook to make teaching an increasing process of critical thinking. The textbook guides the teaching practice, the learning process and contains the subjects that students must study and have contact with. "O LD[2] é um guia para professores e alunos, orientando a prática pedagógica e também que é uma ferramenta que deve ser utilizada de maneira crítica pelos alunos e professores" ([2], p. 195).

These themes that are contained in each unit can be considered the main part of the class offered by the program because in this material we have an opportunity to create a space in the classroom where students can express themselves and have a chance to think of their knowledge about the world and even about their self-knowledge. Sometimes, this debate during the language class is the only moment that students must expose their opinion and practice the critical thinking about some subject. For it, the transversal themes get into action as a tool to make this discussion happen in class. "Os temas transversais dizem respeito à compreensão dos diferentes objetos de conhecimento, possibilitando a referência a sistemas construídos na realidade dos alunos" ([2], p. 198). In other words, transversal themes are subjects that explore the comprehension of students about various objects of knowledge.

When these themes are worked on during the classes, the teacher is responsible to manage the discussion and guide the opinion exposing in order to detect any sign where pedagogical intervention is necessary. It means that the teacher has the challenge to open the mind of students who needs to be encouraged to make himself or herself an active agent of society. Furthermore, the transversal themes are requirements within PCN that determines the methodological ways in the education of Brazil:

> *[Trabalhar os temas transversais, segundo os PCN, não significa que os professores devam parar sua programação para trabalhar os temas, mas sim de que explicitem as relações entre ambos e as incluam como conteúdo de sua área, articulando a finalidade do estudo escolar com as questões sociais, possibilitando aos alunos o uso dos conhecimentos escolares em sua vida extraescolar. ([2], p. 199)]*

Lopes mentions Hutchinson e Waters (1987) when these authors say consider some topics about the elaboration of pedagogical material. These authors mention the importance to analysis the language description and the Learning Theories which becomes much relevant to the Guamá Bilíngue program. It is about thinking which are

[2] Textbook is LD that means Livro Didático in Portuguese

the necessities of students and it can help the teacher to define the objectives of the class. As mentioned previously, this objective is to transform positively students' life. Teach the language is left in the background.

> *[A análise de necessidades, quando bem realizada, trará informações valiosas a respeito do aluno (como suas lacunas e expectativas), do meio onde o curso será ministrado, da tarefa a ser realizada na situação-alvo e das competências necessárias para bem desempenhá-la. Toda essa informação, no entanto, precisa ser organizada e direcionada ao aluno de forma que o objetivo, a priori determinado, seja alcançado. ([5], p. 22)]*

Reading can be not only a process of comprehension of random information. It must be meaningful to the student. A chance to learn some subject or lesson from that text. In the program Guamá Bilíngue, the reading process works as a window to see the world outside and a moment to reflect on this subject beyond learning the language. Braga "entende a leitura em língua estrangeira como uma atividade de solução de problemas, pode também favorecer a aquisição de língua via leitura." ([6], p. 6)

The program Guamá Bilíngue already has a textbook for the students of Spanish, the only language that has been taught until this year, 2022. To the next year the goal is to initiate English classes, and for that, it is necessary that the program also has a textbook containing texts in English and its respective grammatical and vocabulary topics. In unit 10 of the Spanish textbook, a text about the Patagonia is found. In it, it is possible to discuss about the nature, global warming, and foment the responsibility of the students to take care of the environment. This chapter seeks to present a proposal for the elaboration of a unit where students of English can reach this same objective about the natural theme.

2.2 Methodology

Originally the Spanish material should have 13 units, however, due to several issues that are not going to be related here, the book was finished with 10 units. The Guamá Bilíngue project is used as a language course as a laboratory for the future teachers who emerged from the UFPA language and literature degree. So, starting from the premise that students on under graduating Spanish Language and Literature courses will be future facilitators and representatives of the teaching of Spanish in everyday life in elementary and high schools, it is expected that their academic guidance should go beyond the teaching materials used up to secondary school level. How, then, can we train ESL teachers in Brazilian Language Courses who are linguistically competent and capable of going beyond the classroom, searching for solutions to the daily praxis of teaching Spanish?

In the Meaningful Learning theory, subsidies can be found to contribute to the design of a methodology which, based on the use of analogies, in order to allow students of Language Studies/Spanish to have a differentiated treatment when learning a foreign language which should be used as a working and professional improvement tool. In the context of this work, the focus is on the Romance language family, Iberian, focusing on Spanish and Portuguese. This proposal does not deal fundamentally with "Linguistics", nor does it intend to make or propose a new appreciation of this science. It only intends to suggest a new approach of teaching methodology for future Spanish teachers which will give them the tools to work in their profession with critical analysis and the ability to understand/teach Spanish at some of its levels: phonetics/phonology, morphology, syntax, and semantics. This proposal assesses professional training with a view to its application to the teaching of the Spanish language in degree courses in Languages.

Authors such Maffesoli (apud Bango; Stubbs; Gangé, 2002) refers to the specificity of analogical thinking as being "typified forms without any concern with their existence, in order to give greater prominence to what constitutes the essentials of the social plot". Therefore, it should be considered that, through the exercise of the comparison when it comes to the training of Spanish teachers, meaningful learning implies the organization of linguistic material in the student's cognitive structure. One of the most important professors and researchers on the language teaching/learning process Almeida Filho [7] explained several factors that affect the whole process in/out the classroom, so, **Figure 1** is adapted from his ideas:

In **Figure 1**, we can see how teaching material becomes a base for this process of teaching and learning process where factors like the quality of this material, the affective filter of the teacher and the student, the teachers' approach of teaching, and many other can intervene in the result of the students learning process.

When the idea was to build English material, it was necessary to "re-process" all the themes constructed on linguistic similarity. It was possible working the same "simple" methodology that uses structural analogs as a tool for language teaching, which allows an innovative approach to the teaching of English based on the Spanish textbook to students due to the philological proximity—among others—of the languages: English/Portuguese/Spanish. So, in this chapter it is demonstrated how was the process of creating only one unit of a textbook talking about the Amazon as a guideline for English classes, taking as basis the unit 10, from the Spanish textbook, that talks about Patagonia. In it, it is possible to find a text explaining the Patagonia of Chile followed by some questions of comprehension and at the final of the text comprehension part a question directed for the students where they will be able to think what their role in the world is to not contribute to global warming.

One of our main concerns about this was finding the grammatical part containing the Future Tense of Spanish, called "Futuro Simple" with its conjugation and exercises. The proposal of putting the future in this unit that talks about nature helps students to form sentences talking about how the nature in our future will be. Then it is found another text explaining the concept of global warming, its causes, its consequences, and possible solution to avoid it.

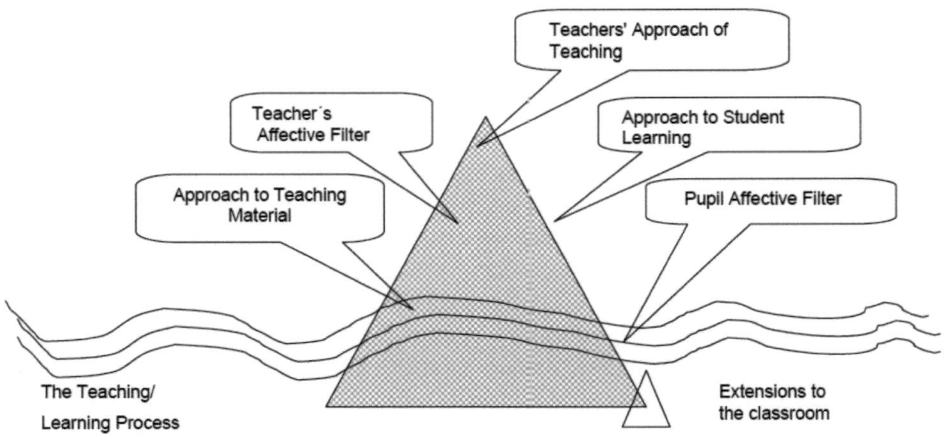

Figure 1.
Main factors intervening in the process of teaching and learning other languages.

The following figures are the 10th unit of the Spanish book using Chilean Patagonia as a biome to be taken care:

Figure 2 shows the front page of a unit from the textbook of Spanish classes of Guamá Bilíngue. The title and the pictures are the showiest when students come to this page, what helps them to know previously the subject and recognize the context. In the same page, a text about the nature of Patagonia where students can get vocabulary and start a discussion about that. This figure shows how is structured the Spanish material of the program and how could be structured the English material to meet the same proposal of Spanish.

The questions about the text have already addressed the climate change issue:

Figure 3 shows several questions in Spanish related to the text about Patagonia with the goal to test the comprehension of the information presented on the previous page. In this photo, it is possible to notice that at the bottom of the page, there is a question that is directed to the student, which is not possible to answer by finding responses in the text, but making a reflection on our own habits, due to the question: "¿Cuál es mi papel para no contribuir al calentamiento global?", which means "What is my role in not contributing to global warming?"

Patagonia Unidad 10

Montañas, islas y bosques vírgenes, una zona indómita y salvaje con todo su mágico poderío original es lo que ofrece este territorio de naturaleza pura. Pocos lugares en el mundo poseen el magnetismo de la Patagonia chilena. Territorio de pueblos aborígenes, aventuras y secretos que aún esperan ser revelados en la zona más austral de Sudamérica.

La Patagonia chilena es una subregión de la Patagonia que se extiende entre los paralelos 41,5º y 56º latitud Sur y comprende parte de las regiones de Los Lagos, Aysén y Magallanes. Es una de las superficies más extensas del país, sin embargo la menos poblada: 0.5 habitantes por km2. La naturaleza allí permanece prácticamente virgen.

En una salvaje y accidentada geografía, enormes masas de hielo y bosques prístinos se yerguen majestuosos. Aquí el ser humano no domina la naturaleza, convive y se complementa con ella. Los indígenas que habitaron la zona hace siglos cubrían sus cuerpos con aceite de lobo de mar y utilizaban pieles para protegerse de las lluvias y el frío intenso. Tal vez por esto el calor humano es tan poderoso y evidente en los pequeños pueblos y ciudades de Patagonia.

La historia del punto más extremo del continente siempre ha estado ligada a navegantes y aventureros, a descubridores y míticos habitantes originarios, como los patagones. El navegante portugués Hernando de Magallanes llegó a la zona en 1520 y la bautizó con el nombre Tierra del Fuego, tras observar las fogatas que hacían los Onas o Selknam en las riveras del estrecho. Se trata de extensos paisajes indómitos, islas y canales que guardan trazos de identidad legendaria.

A 312 kilómetros de Punta Arenas, se encuentra el parque nacional Torres del Paine, con sus imponentes cuernos que se levantan hacia el cielo. Algo más de 240 mil hectáreas de áreas silvestres y protegidas que reciben año a año a numerosos visitantes de todo el mundo.

La Patagonia es el territorio más cercano al continente antártico y al polo sur, lo que constituye una gran ventaja. Barcos y aviones cruzan los mares australes para llegar a territorio antártico y realizar visitas a bases científicas y breves trekkings observando la fauna autóctona. La zona ofrece atractivos insospechados que transforman este viaje en una experiencia para toda la vida.

Sin embargo, es una de las zonas más afectadas por el calentamiento global. Hasta hace 5 años, los barcos que iban al glaciar de Perito Moreno le sacaban hielo para las bebidas. Actualmente ya no se le pueden acercar por el riesgo de un derrumbe.

Fuente: http://www.thisischile.cl/Article.aspx?id=290&sec=277&eje=&t= 41

Figure 2.
Main text page 41 in the Spanish material.

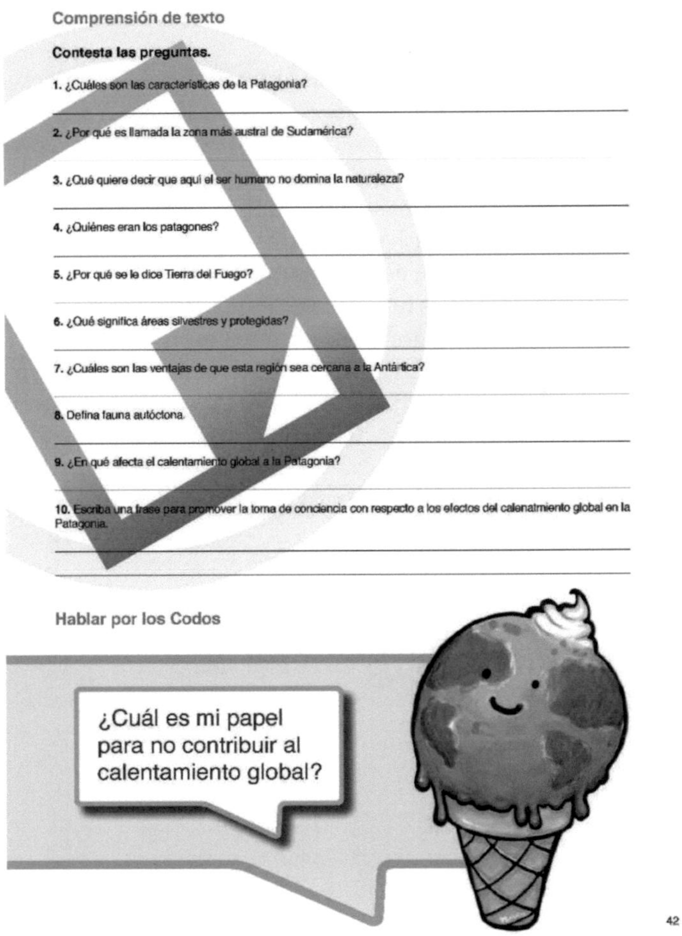

Comprensión de texto

Contesta las preguntas.

1. ¿Cuáles son las características de la Patagonia?
2. ¿Por qué es llamada la zona más austral de Sudamérica?
3. ¿Qué quiere decir que aquí el ser humano no domina la naturaleza?
4. ¿Quiénes eran los patagones?
5. ¿Por qué se le dice Tierra del Fuego?
6. ¿Qué significa áreas silvestres y protegidas?
7. ¿Cuáles son las ventajas de que esta región sea cercana a la Antártica?
8. Defina fauna autóctona.
9. ¿En qué afecta el calentamiento global a la Patagonia?
10. Escriba una frase para promover la toma de conciencia con respecto a los efectos del calenatmiento global en la Patagonia.

Hablar por los Codos

42

Figure 3.
Text comprehension questions, page 42 in the Spanish material.

Due to this perspective, for the English textbook, it was decided to consider the following suggestion: a unit directed to the environment discussion that will present the title "*The Brazilian Amazon Forest*", followed by a text taken from a new from the website Global Witness where a 24-year-old activist, called Txai Suruí, from the Brazilian Amazon, speaks about the reality of the forest. In the text, subjects such as indigenous representatives, worried about the Amazon forest state as well as the increase of deforestation in this area leading to climate change. It means that the students of the program will be able to discuss about these problems searching for possible solutions with help of the teacher, and furthermore, thinking of how we are taking care of our forest and all its particularities.

2.3 Discussing topics

After presenting and explaining some vocabularies and language structures, the unit will contain beside questions to test the text comprehension of students, these questions will be a way of brainstorming for the oral discussion part. The questions of this part of the unit and the objectives of each one will be presented below.

2.3.1 How many indigenous women were present at COP26 (the conference presented in the text)?

This question has the objective to make students think about the lack of opportunities that women have in spots that historically were thought to be for men, especially for indigenous people, and how important was to the Brazilian Amazonians to have Txai Suruí in the event representing the indigenous community.

2.3.2 According to Txai Suruí, what are the consequences lived by the indigenous people in the Amazon Forest?

Here, the objective is to make students think about these consequences that affect the indigenous community and how they suffer with it. It is also important to make them put themselves in indigenous people's shoes and imagine how they would deal with each situation that deforestation causes.

2.3.3 How much has deforestation increased in percentual?

This question will make the student not only search for statistical data but also consider what causes are to deforestation and who is responsible for this situation. Probably will be kind of hard to get the exact reply due to the complex knowledge necessary about politics and the economy in relation to the environment. It is worth reminding that the students from the program do not have access to education of quality due to their social vulnerability.

2.3.4 What is the profession of Txai Suruí presented in the text?

With this question, students will be able to recognize how important is to provide education to indigenous people and it is an opportunity for the teacher to foment in them the search for this education opening their minds to think about the academic context.

Clearly, during the process of elaboration of all of the units, it is possible to add more questions according to the needs of them. Other suggestion for this unit is a section where students will find some questions directed to them. Questions like: "How are you taking care of Amazon Forest?", "Are you against or in favor to deforestation?", "Do you agree that saving the Amazon Forest is important to the rest of the world?" The core of the textbook is given to Guamá Bilíngue students the total picture of the current situation of the Brazilian Amazon Forest focusing on lifestyle, predatory extraction, deforestation versus conservation, equality in education, especially indigenous and "quilombos original peoples" as much as it is possible in 10 units always keeping in mind that the analogical understanding process has to be used from the vocabulary choice to the grammar topics and language linguistic register.

After the text reading and comprehension, the grammar presented in the material contribute to the learning process of students. The suggestion is to follow the example of the Spanish textbook and insert the Future Tense of English in the unit. Although, the section part will not start right with the grammar explanation, some questions like: "Will the Amazon Forest disappear?", "Will indigenous people reach their goals with their activism?" "Will we suffer the consequences of deforestation too?" "Will global warming finally end some day?". The proposal of these questions is to introduce the future in a contextualized way and then to explain how to use the auxiliary verb "will" in the affirmative, negative, and interrogative forms with its respective exercises. With this, students can think about the future of nature and the actions that must be taken to guarantee a good future for us.

Here it is presented the grammar topic in the Spanish textbook which is Simple Future as it has been mentioned above:

In **Figure 4**, it is possible to notice how the grammar part is presented. It is worth reminding that grammar is not the main focus of the program but is a tool in the learning process for Guamá Bilíngue students.

One final item to be shown is that the final page in the Spanish textbook is already about global warming. Here it will be found a topic that seems not to belong to the ecological/sustainable tactic of the textbook. STD issues are a constant menace to Guamá Bilíngue students as much as to the forest peoples so that topic is somehow correlated with the even living in the relations with forest peoples.

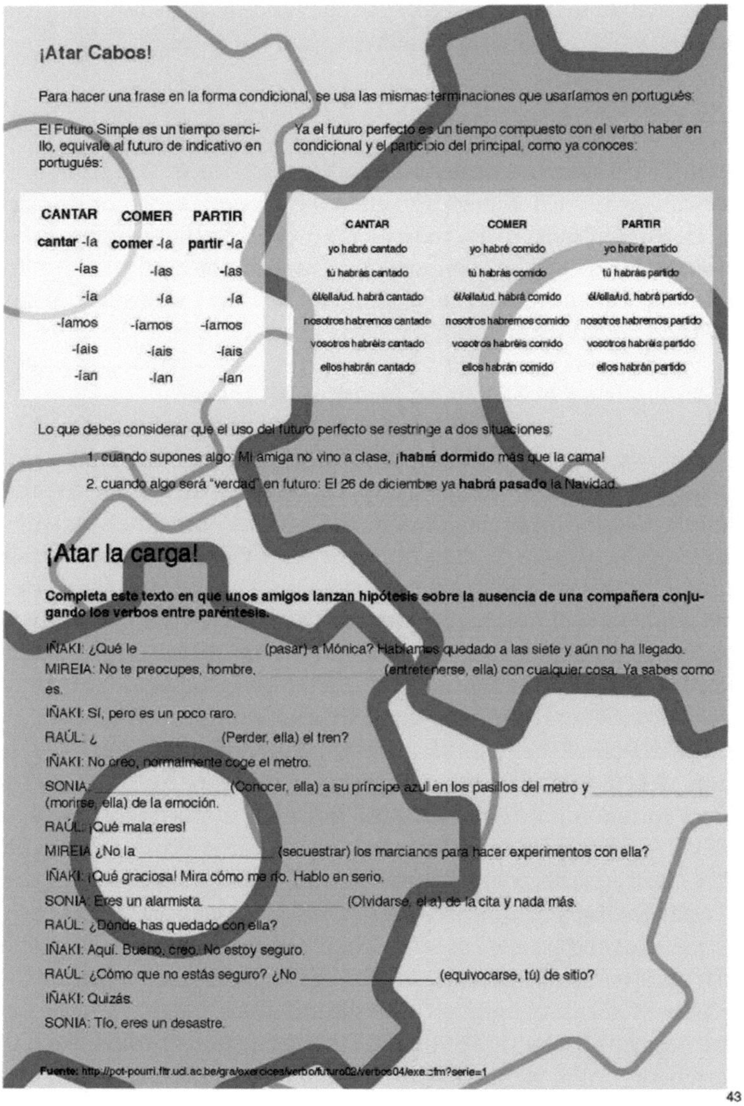

¡Atar Cabos!

Para hacer una frase en la forma condicional, se usa las mismas terminaciones que usaríamos en portugués:

El Futuro Simple es un tiempo sencillo, equivale al futuro de indicativo en portugués:

CANTAR	COMER	PARTIR
cantar -ía	**comer** -ía	**partir** -ía
-ías	-ías	-ías
-ía	-ía	-ía
-íamos	-íamos	-íamos
-íais	-íais	-íais
-ían	-ían	-ían

Ya el futuro perfecto es un tiempo compuesto con el verbo haber en condicional y el participio del principal, como ya conoces:

CANTAR	COMER	PARTIR
yo habré cantado	yo habré comido	yo habré partido
tú habrás cantado	tú habrás comido	tú habrás partido
él/ella/ud. habrá cantado	él/ella/ud. habrá comido	él/ella/ud. habrá partido
nosotros habremos cantado	nosotros habremos comido	nosotros habremos partido
vosotros habréis cantado	vosotros habréis comido	vosotros habréis partido
ellos habrán cantado	ellos habrán comido	ellos habrán partido

Lo que debes considerar que el uso del futuro perfecto se restringe a dos situaciones:

1. cuando supones algo: Mi amiga no vino a clase, ¡**habrá dormido** más que la cama!
2. cuando algo será "verdad" en futuro: El 26 de diciembre ya **habrá pasado** la Navidad.

¡Atar la carga!

Completa este texto en que unos amigos lanzan hipótesis sobre la ausencia de una compañera conjugando los verbos entre paréntesis.

IÑAKI: ¿Qué le ________________ (pasar) a Mónica? Habíamos quedado a las siete y aún no ha llegado.
MIREIA: No te preocupes, hombre, ________________ (entretenerse, ella) con cualquier cosa. Ya sabes como es.
IÑAKI: Sí, pero es un poco raro.
RAÚL: ¿ ________________ (Perder, ella) el tren?
IÑAKI: No creo, normalmente coge el metro.
SONIA: ________________ (Conocer, ella) a su príncipe azul en los pasillos del metro y ________________ (morirse, ella) de la emoción.
RAÚL: ¡Qué mala eres!
MIREIA: ¿No la ________________ (secuestrar) los marcianos para hacer experimentos con ella?
IÑAKI: ¡Qué graciosa! Mira cómo me río. Hablo en serio.
SONIA: Eres un alarmista. ________________ (Olvidarse, ella) de la cita y nada más.
RAÚL: ¿Dónde has quedado con ella?
IÑAKI: Aquí. Bueno, creo. No estoy seguro.
RAÚL: ¿Cómo que no estás seguro? ¿No ________________ (equivocarse, tú) de sitio?
IÑAKI: Quizás.
SONIA: Tío, eres un desastre.

Fuente: http://pot-pourri.fltr.ucl.ac.be/gra/exercices/verbo/futuro02/verbos04/exe.cfm?serie=1

43

Figure 4.
Present simple explanation and questions, page 43 in the Spanish material.

Figure 5 shows another part of the unit that contains another text about global warming. Here is a big important part of the class in the program. It is where students will know the causes and consequences of the destruction and deforestation of nature through the reading activity.

After the reading part, the teacher will start a discussion in order to increase the critical thinking of these students mentioning each cause of global warming presented in the text and trying to make them identify who causes it or how. Then, the discussion goes to the consequences, where students are able to identify how these consequences are already affecting us directly or indirectly, and maybe invite some

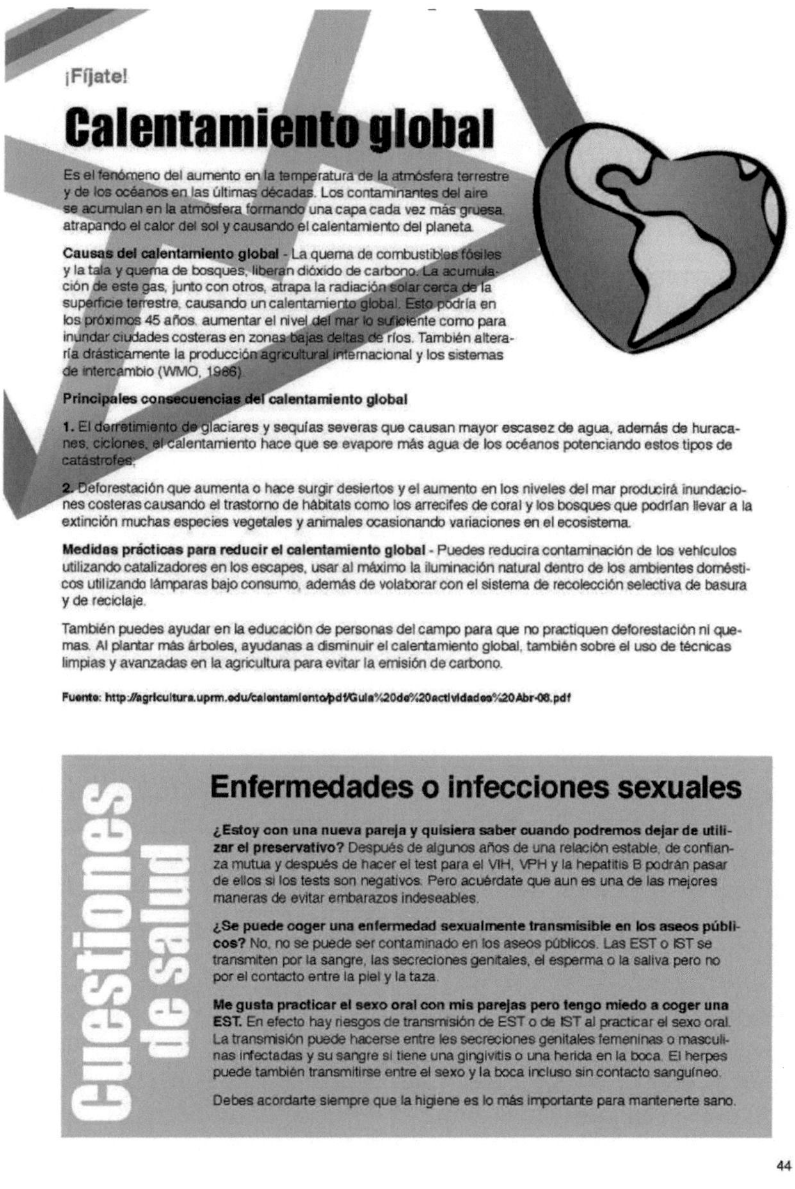

¡Fíjate!

Calentamiento global

Es el fenómeno del aumento en la temperatura de la atmósfera terrestre y de los océanos en las últimas décadas. Los contaminantes del aire se acumulan en la atmósfera formando una capa cada vez más gruesa, atrapando el calor del sol y causando el calentamiento del planeta.

Causas del calentamiento global - La quema de combustibles fósiles y la tala y quema de bosques, liberan dióxido de carbono. La acumulación de este gas, junto con otros, atrapa la radiación solar cerca de la superficie terrestre, causando un calentamiento global. Esto podría en los próximos 45 años, aumentar el nivel del mar lo suficiente como para inundar ciudades costeras en zonas bajas deltas de ríos. También alteraría drásticamente la producción agricultural internacional y los sistemas de intercambio (WMO, 1986).

Principales consecuencias del calentamiento global

1. El derretimiento de glaciares y sequías severas que causan mayor escasez de agua, además de huracanes, ciclones, el calentamiento hace que se evapore más agua de los océanos potenciando estos tipos de catástrofes;

2. Deforestación que aumenta o hace surgir desiertos y el aumento en los niveles del mar producirá inundaciones costeras causando el trastorno de hábitats como los arrecifes de coral y los bosques que podrían llevar a la extinción muchas especies vegetales y animales ocasionando variaciones en el ecosistema.

Medidas prácticas para reducir el calentamiento global - Puedes reducira contaminación de los vehículos utilizando catalizadores en los escapes, usar al máximo la iluminación natural dentro de los ambientes domésticos utilizando lámparas bajo consumo, además de volaborar con el sistema de recolección selectiva de basura y de reciclaje.

También puedes ayudar en la educación de personas del campo para que no practiquen deforestación ni quemas. Al plantar más árboles, ayudanas a disminuir el calentamiento global, también sobre el uso de técnicas limpias y avanzadas en la agricultura para evitar la emisión de carbono.

Fuente: http://agricultura.uprm.edu/calentamiento/pdf/Guia%20de%20actividades%20Abr-08.pdf

Cuestiones de salud

Enfermedades o infecciones sexuales

¿Estoy con una nueva pareja y quisiera saber cuando podremos dejar de utilizar el preservativo? Después de algunos años de una relación estable, de confianza mutua y después de hacer el test para el VIH, VPH y la hepatitis B podrán pasar de ellos si los tests son negativos. Pero acuérdate que aun es una de las mejores maneras de evitar embarazos indeseables.

¿Se puede coger una enfermedad sexualmente transmisible en los aseos públicos? No, no se puede ser contaminado en los aseos públicos. Las EST o IST se transmiten por la sangre, las secreciones genitales, el esperma o la saliva pero no por el contacto entre la piel y la taza.

Me gusta practicar el sexo oral con mis parejas pero tengo miedo a coger una EST. En efecto hay riesgos de transmisión de EST o de IST al practicar el sexo oral. La transmisión puede hacerse entre les secreciones genitales femeninas o masculinas infectadas y su sangre si tiene una gingivitis o una herida en la boca. El herpes puede también transmitirse entre el sexo y la boca incluso sin contacto sanguíneo.

Debes acordarte siempre que la higiene es lo más importante para mantenerte sano.

44

Figure 5.
Global warming and STD, page 44 in the Spanish material.

students to talk about their own experience, because sometimes some of them could have already lived some experience such as facing a flood in their neighborhood, or some kind of contamination of food.

This makes the discussion in class more personal about what involves the affective filter. Then, teachers encourage students to find solutions or even to start practicing actions that they are not used to practicing like saving water, energy, etc.

3. Conclusions

In a nutshell, it is possible to realize how the suggested unit for the English textbook to the Guamá Bilíngue will take students to a mood of reflection about the environment and the Amazon Forest, where they are interleaved. The textbook must be elaborated following the proposal of the program and the goals that teachers and students must reach. For this reason, is important to consider the needs that students have, the objectives they must reach and the learning process of the language that works as a tool for the achievement of the main proposal of the program: to encourage teenagers that are about to conclude the high school and have little or no perspective to their future.

All issues regarding conservation must be focused on the development of Guamá Bilíngue material for English learners. While we are still trying to find the best way to teach sustainability to these vulnerable students, we still keep growing in our way of seeing sustainability in Amazon Forest based on education from the elementary school until teachers under graduation course.

The material or textbook is a guide much relevant in any learning process and it must be elaborated thinking about the goal that the teacher wants to reach in their students. In this program, the textbook differs slightly from other textbooks of other languages courses where the main objective is to learn the language properly, whether to travel or to get communication with others. In Guamá Bilíngue, the textbook is configured to serve as a guide to mind opening process of students.

It is worth to mention that is important to select themes to contain in units of the textbook that serve correctly to this objective of social transformation. And the English becomes an issue of background to construct opinion and foment responsibility in relation to nature, the current situation of the Amazon Forest, and the problems that its indigenous communities are suffering. Something that has to be done about it by starting a guided discussion in the classroom in an attempt to make those teenagers create responsibility in themselves that was mentioned previously. In conclusion, when the English classes start finally in the program, by the middle of 2023, its benefice is expected to have a textbook available with units talking about the Amazon Forest and then continuing this social work that has been developed since 2010.

Acknowledgements

We would like to thank the Universidade Federal do Pará – UFPA –personified by the Magnificent Rector Emmanuel Zagury Tourinho for supporting us until the 13th anniversary in2022, as well as to thank the whole study group "Guamá Bilíngue", and all the students, whose accomplishments helped us to outline the elements we needed to terreplein a critical pathway to many future generations. Also, the authors wish to acknowledge the support provided by PROPESP/UFPA (PAPQ).

Author details

Douglas Rodrigues Pinheiro da Costa* and Rita de Cássia Paiva
Federal University of Pará – UFPA, Belém, Brazil

*Address all correspondence to: douglas.rpc@otmail.com

References

[1] de Paiva RC, Evelyn SPB, de Lima Jacqueline Augusta L. Proyecto de extensión: enseñanza de ele y la motivación como herramientas en la formación ciudadana del alumno en la amazonia; Vol. 2 n. 2, 2017. Blucher: São Paulo. pp. 745-754

[2] Silva S, Amanda S. Interdisciplinaridade e transversalidade em livros didáticos de língua inglesa. Revista Eletrônica do Instituto de Humanidades. Rio de Janeiro. 2018. Available from: http://publicacoes.unigranrio.edu.br/index.php/reihm/article/view/5035;**20**(46):194-209 Accessed: 2022-09-20

[3] Cintamulya I. Analysis of Students' Critical Thinking Skills with Reflective and Impulsive Cognitive Styles on Conservation and Environmental Knowledge Learning. Vol. 20. Jalan Manunggal: University of PGRI Ronggolawe Tuban; 2018. p. 1

[4] Fearnside P. Conservation policy in Brazilian Amazonia: Understanding the dilemmas. World Development. Great Britian. 2003;**31**(5):757-759

[5] Lopes R. A elaboração de material didático: instrumento de (re)constituição do professor de inglês. Dissertação (Mestrado em Linguística Aplicada e estudos da linguagem) – Pontifícia Universidade Católica de São Paulo. São Paulo. 2009. p. 229

[6] Braga DB. Ensino de Língua Inglesa via Leitura: Uma Reflexão Sobre a Elaboração de Material Didático Para Autoinstrução. Vol. 30. Campinas, SP: Trabalhos em Linguística Aplicada; 1997. Available from: https://periodicos.sbu.unicamp.br/ojs/index.php/tla/article/view/8639267 Accessed: 2022-09-20

[7] Filho A, Paes JC. Dimensões comunicativas no ensino de línguas. Campinas: Pontes; 1993